# Fundamental of Transport Phenome
# and Metallurgical Process Modeling

Sujay Kumar Dutta

# Fundamental of Transport Phenomena and Metallurgical Process Modeling

 Springer

Sujay Kumar Dutta
Metallurgical and Materials Engineering
Maharaja Sayajirao University of Baroda
Vadodara, India

ISBN 978-981-19-2158-2          ISBN 978-981-19-2156-8  (eBook)
https://doi.org/10.1007/978-981-19-2156-8

This Springer imprint is published by the registered company Springer Nature Singapore Pte Ltd.
The registered company address is: 152 Beach Road, #21-01/04 Gateway East, Singapore 189721,
Singapore

# Preface

The subject of *transport phenomena* is a part of *process metallurgy*, which deals with momentum, mass, and heat transfers. Heat, mass, and momentum transport processes are an integral part of all metallurgical operations. A good understanding of the principles involved can lead to many useful predictions about and improvements in industrial processing, operations, and practices.

Needless to say, that although few books are available, none discuss the subject matter of *transport phenomena* in easy-to-understand format. This textbook covers almost all the important basic concepts, derivations and numerical for undergraduate and graduate engineering students in simple and easy way. Even plant's engineers or operators, and research scientists can brush up their understanding on mass and heat balances and can be used as a source of reference.

This textbook is divided into six chapters: Chap. 1 of the book covers *Fluid Dynamics.* Chapter 2 covers *Mass and Momentum Balances* including derivation of Continuity and Navier–Stokes equations and their application. All types of *Heat Transfers* are discussed in Chap. 3. Chapter 4 covers *Mass Transfer. Basic Concept of Models and Applications* are covered in Chaps. 5 and 6.

Professor Sujay Kumar Dutta taught process metallurgy at M. S. University of Baroda, India, for 36 years. He expresses his gratitude to Prof. S. C. Koria (*Indian Institute of Technology*, Kanpur, India), Prof. B. N. Ballal (*Indian Institute of Technology Bombay*, Mumbai, India), and Prof. A. K. Lahari (*Indian Institute of Science,* Bangalore, India) for introducing him to the *transport phenomena* and its applications.

Despite taking all the possible care, there may be some errors or mistakes left out unnoticed. If so, please feel free to interact with me. I poured my long experience in it and collected the information from several sources. I am indebted to one and all, from whose valuable knowledge I have been benefited. Finally, I thank my wife (Gopa) for her cooperation during the preparation of the book.

Vadodara, India  
November 2021

Sujay Kumar Dutta

# Contents

# About the Author

**Sujay Kumar Dutta** is a former Professor and Head of the Department of Metallurgical & Materials Engineering at Maharaja Sayajirao University of Baroda, India. He obtained his Bachelor of Engineering (Metallurgy) from Calcutta University, India in 1975 and Master of Engineering (Industrial Metallurgy) from MS University of Baroda, India in 1980. He was awarded Ph.D. degree in 1992 by the Indian Institute of Technology, Kanpur, India. He has one-year practical industrial experience. He joined MS University of Baroda as a Lecturer in 1981 and subsequently promoted to Professor. He became Head (Department of Metallurgical & Materials Engineering) and Director (ME, Welding Course) from 2012 to 2016. His major areas of research interests include iron ore-coal composite pellets, micro-alloy steel development, DRI melting, and utilization of iron and steel plants' waste materials. Prof. Dutta has received several awards, including Essar Gold Medal (2006), Fellowship (2014) and Distinguished Educator Award (2015), all from The Indian Institute of Metals (IIM), Kolkata, in recognition of his distinguished service to the field of Metallurgical Education and to the Indian Institute of Metals. He was also awarded SAIL Award 2019 by The Institution of Engineers (India), Kolkata. The American Biographical Institute Board of International Research had nominated him for Man of the Year 2012, for distinctive accomplishments in the Metallurgical field.

Dr. Dutta has authored of six books {including: Basic Concepts of Iron and Steel Making (Springer Nature, Singapore, 2020), Extraction of Nuclear and Nonferrous Metals (Springer Nature, Singapore, 2018), Metallurgical Thermodynamics, Kinetics & Numerical (S Chand, New Delhi, 2012) Extractive Metallurgy: Processes and Applications (PHI, Delhi, 2018)}; and two chapters of Encyclopedia of Iron, Steel, and Their Alloys, (CRC Press, New York, 2016). He has published more than 130 research papers in national/international journals and conference proceedings. He served as member of editorial board and reviewer for several national/international journals.

# Symbols

| | |
|---|---|
| a | Acceleration |
| A | Area |
| $A_{cs}$ | Cross-sectional area |
| $a_v$ | Specific surface area |
| $\lambda$ | Correction factor |
| $\rho$ | Density |
| D | Diameter |
| $d_e$ | Equivalent diameter |
| D | Diffusion coefficient |
| $C_D$ | Discharge coefficient |
| $D_{AB}$ | Mass diffusivity for binary mixture of A and B |
| $E_f$ | Efficiency |
| $\epsilon, \varepsilon$ | Emissivity |
| E | Surface emissive power |
| $E_b$ | Emissive power of blackbody |
| $\epsilon_\lambda$ | Monochromatic emissivity |
| $\epsilon_n$ | Normal total emissivity |
| E | Energy |
| KE | Kinetic energy |
| PE | Potential energy |
| IE | Internal energy |
| BE | Bulk energy |
| $k_e$ | Equilibrium constant for reaction |
| F | Force |
| $F_g$ | Force due to gravity |
| $F_k$ | Frictional force |
| $F_n$ | Buoyant force |
| $f_{fr}$ | Friction factor or drag coefficient |
| $\rho$ | Fraction of incident radiation reflected (*reflectivity*) |
| $\tau$ | Fraction of incident radiation transmitted (*transmissivity*) |
| $\alpha$ | Fraction of incident radiation absorbed (*absorptivity*) |

| | |
|---|---|
| $\nu$ | Frequency of quantum |
| R | Gas constant |
| g | Gravitational force |
| Q | Heat absorbed |
| $q'_x$ | Heat flux |
| $q^\wedge$ | Heat absorbed per kg of fluid |
| $q^\cdot$ | Internal heat generation per unit volume and per unit time |
| $h_r$ | Radiation heat transfer coefficient |
| $h^-_L$ | Average heat transfer coefficient over a length L |
| $h^-$ | Total heat transfer coefficient for entire surface |
| U | Overall heat transfer coefficient |
| $R_c$ | Thermal resistance due to convection |
| t | Time |
| u | Velocity |
| $\frac{du}{dy}$ | Velocity gradient |
| $\delta$ | Velocity boundary layer |
| $u^-_z$ | Average velocity |
| $u_t$ | Terminal velocity |
| $u_E$ | Elutriation velocity |
| $\mu$ | Viscosity |
| H | Convection heat transfer coefficient |
| H, z | Height |
| L | Length |
| B | Magnetic flux intensity |
| M, m | Mass |
| $m_T$ | Total mass |
| $k_m$ | Mass transfer coefficient |
| $C_i$ | Molar concentration of component i |
| $M_i$ | Molecular weight of the component i |
| $M_g$ | Molecular weight of gas |
| $V_i$ | Molecular volume of component i |
| $X_i$ | Mole fraction of the component i |
| $J'_{ix}$ | Mass flux of component i in x direction |
| $J'_i$ | Relative or diffusive mass flux of component i |
| $J'^*_i$ | Molar flux of component i |
| $n''_i$ | Mass flux of component i |
| $n''$ | Total mass flux for mixture |
| $N''$ | Total molar flux |
| $m'$ | Rate of mass flow or mass flux |
| P | Pressure |
| $\Delta p$ | Pressure difference |
| $\mathcal{P}$ | Modified pressure |
| R | Radius |
| $r_h$ | Mean hydraulic radius |
| Re | Reynolds number |

| | |
|---|---|
| $Re_p$ | Reynolds number for particle |
| $Re_E$ | Reynold number used in Ergun equation |
| $Re_{mf}$ | Reynolds number for minimum fluidization |
| $\tau_{yx}$ | Shearing force per unit area |
| $C_p$ | Specific heat of the material |
| $I_{b\lambda}$ | Spectral intensity of the emitted radiation |
| $C_o$ | Speed of light in vacuum ($2.998 \times 10^8$ m/s) |
| $\sigma$ | Surface tension |
| $T$ | Temperature |
| $T_b$ | Bulk temperature |
| $T_{mp}$ | Melting temperature of material |
| $k$ | Thermal conductivity |
| $Ķ$ | Thermal conductance |
| $Ķ_c$ | Thermal conductance for convection |
| $Ķ_r$ | Thermal conductance for radiation |
| $\alpha$ | Thermal diffusivity of the material |
| $R_{con}$ | Conductive thermal resistance |
| $\varepsilon$ | Void fraction |
| $Q$ | Volumatic flow rate |
| $\lambda$ | Wavelength |
| $W$ | Work |
| $w_s$ | Work done |
| $\hat{w}_s$ | Work done per kg of fluid |
| $\hbar$ | Planck's constant ($6.625 \times 10^{-34}$ J.s) |
| $Ķ$ | Boltzmann constant ($1.3805 \times 10^{-23}$ J/K) |
| $\sigma$ | Stefan–Boltzmann constant ($5.669 \times 10^{-8}$ W. m$^{-2}$.K$^{-4}$) |

# Chapter 1
# Fluid Dynamics

Momentum transfer deals with flow of fluid and hence it is termed as fluid flow or fluid dynamics. Fluid flow influences mass transfer, heat transfer, mixing and homogenization. Metallurgists are concerned with the motion of fluids in many processes. Motion of fluids governs the effectiveness of mixing and homogenization of both composition and temperature in the reactors. Such mixing is of great importance in a variety of metal extraction and refining processes. Newton's law of viscosity, flow characterization, velocity profile, velocity boundary layer, pressure variation in a static fluid, energy balance, applications of overall energy balance, and overall mass balance are discussed in detail in this chapter.

## 1.1 Basic Concept

The subject of *transport phenomena* is a part of *process metallurgy*, which deals with momentum, mass and heat transfer; this means that transport phenomena is closely related to the following topics [1]:

(i)     Fluid dynamics,
(ii)    Mass transfer, and
(iii)   Heat transfer.

(i)     *Fluid dynamics* involves the transport of momentum. The dictionary meaning of *momentum* is quantity of motion for the moving body, it is a product of mass (m) and velocity (u), i.e. *mu*.
(ii)    *Mass transfer* is concerned with the transfer of materials, i.e. transport of mass of different chemical species.
(iii)   *Heat transfer* deals with the transport of heat energy.

S. K. Dutta, *Fundamental of Transport Phenomena and Metallurgical Process Modeling*,
https://doi.org/10.1007/978-981-19-2156-8_1

Heat, mass, and momentum transport processes are an integral part of all metallurgical operations. A good understanding of the principles involved can lead to many useful predictions about and improvements in industrial processing, operations, and practices.

All these three transport phenomena should be studied together for the following reasons:

- They frequently occur simultaneously in industrial problems,
- The basic equations that describe the three transport phenomena are closely related (similar equations can be solved by the same analogy, i.e. process of reasoning),
- The mathematical tools needed for describing these phenomena are very similar (students can learn how to use mathematical tools to solve the problem which will be advantageous for students by studying transport phenomena),
- The molecular mechanisms are very closely related. Since all materials are made up of molecules; motion of molecules and interactions are responsible for viscosity, thermal conductivity, and diffusion of materials.

Using the principles of thermodynamics, to know how far away the current state of a system is from the state of equilibrium. But thermodynamics does not talk about rates of the processes. Rates can be calculated by the driving forces and by the mechanisms of changes. Rates of processes are needed to understand: (i) how mass, work as well as energy are exchanged through the boundary of the system with time, (ii) the rate at which the state of the system goes toward equilibrium.

Transport phenomena along with reaction kinetics deal with these aspects of processes. Through the boundary of the system, mass can be transferred by bulk motion. Similarly, energy in the form of work, heat, or other forms can be exchanged through the system boundary. These aspects of transport of mass, momentum, and heat are dealt with together in transport phenomena.

Transport phenomena is an engineering subject and is essential for understanding, designing, and operating most of the engineering processes. Process metallurgy can be demonstrated through an example of a packed bed reactor (Fig. 1.1). Hot gas (e.g. carbon monoxide, CO) is fed from the bottom of the reactor and can rise through the packed bed of solid particles (e.g. iron ore), heating the bed as well as react with solid particles. Gas moving through the voids exerts a force on the particle, which in turn exerts an opposite force on the gas. To keep gas moving at a desired rate, a net force must be exerted on the gas in the form of pressure difference ($\Delta p$) across the bed. To estimate the required pressure difference, the forces acting on an individual particle should be understood, i.e. momentum transfer.

Hot gas moves up through the bed, the difference in temperature between the gas and the solid particle leads to thermal gradient, i.e. heat transfer from the hot gas to the colder solids particles takes place. Side by side mass transfer (i.e. oxygen) also takes place from the solid particle to the gas phase, as per reaction:

$$Fe_2O_3(s) + 3CO(g) = 2Fe(s) + 3CO_2(g).$$

**Fig. 1.1** Packed bed reactor

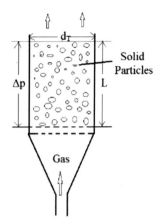

Considering materials properties of the gas and the solid in one hand, and process conditions, i.e. fluid flow, particle size and shape, and so on in other hand. Gas continuously cools as moves up through the bed and the solid particles are heated up. Temperature of the solid particles and the gas varies with time and position. The heat transfer is based on the principle of *conservation of energy*.

Engineers are employed in many industries, where raw materials are converted into the products by chemical reactions and physical changes [2]. Most industrial operations consist of a sequence of physical and chemical transformations, and this sequence is known as *process*. A process can be defined as that which brings about a change in the state of the system. The state of a system is defined using measurable properties such as density, pressure, temperature, and composition.

The principle, on which transport processes, is based on the *law of conservation*, i.e. balance between momentum, mass, and heat [3]. This means that any change in any of them in a system or its part will be equal to the amount exchanged by it with its surroundings.

Every physical quantity has some dimension by virtue of either its definition or its relationship with other quantities. As, for example, velocity is defined as distance traveled per unit time, dimensions: length (L) per unit time (t), i.e. $Lt^{-1}$.

There are four primary dimensions namely mass (M), length (L), time (t), and temperature (T). The dimensions of all other physical variables of any phenomenon can be obtained in terms of these four basic dimensions. These four dimensions/quantities are therefore called *basic dimensions/quantities*. Dimensions of these basic quantities are represented by the symbols M, L, t, and T, respectively.

Determination of dimensions of any physical quantity from its relationship with other quantities is based on the principle of dimensional homogeneity. As, for example, according to Newton's law: force (F) is related to mass (m) and acceleration (a) by the relation:

$$\text{Force (F)} = \text{mass (m)} \times \text{acceleration (a)} = ma \qquad (1.1)$$

According to the principle of dimensional homogeneity, both right- and left-hand sides of Eq. (1.1) must have the same dimensions. As acceleration has the dimension, $Lt^{-2}$ and so force (N) must have dimensions of $MLt^{-2}$ (according to Eq. (1.1)).

Since $F = Ma$                    (1.1).

So,

$$1N = \left[ \frac{1 \text{ kg} \cdot 1 \text{ m}}{s^2} \right] = MLt^{-2} \tag{1.2}$$

Similarly, work or energy = force x displacement = $MLt^{-2} \times L = ML^2 t^{-2}$.

Some of the frequently used variables along with their symbols and dimensions are given in Appendix I.

## 1.2  Fluid Dynamics

Momentum transfer refers to viscous momentum transfer, which comes under the science of *fluid mechanics*, i.e. laws of fluid motion for both viscous as well as non-viscous fluids [4]. The study of the fluid flow constitutes the most significant part of the analytical study of transport phenomena in metallurgical systems, particularly in reactors [3]. This is because such flow affects, to a great extent, the overall kinetics of the reactions through its effects on the heat and the mass transport steps. The main aim of the study of the flow of fluids in metallurgical processes is either to obtain the velocity or pressure profiles inside metallurgical reactors or liquid metal flow systems as a function of time or to find the pressure drop across a reactor for a given mass or volume flow rate.

Metallurgists are concerned with the motion of fluids in many processes. Air and other gases flow through pipelines, furnaces, and combustion chambers. Liquid metals and slags are in motion in the furnaces during tapping (furnace to ladle) and teeming (ladle to mold). Oxygen and other gases are blown into liquid metals, through melts and stir molten metal and slag in furnaces. Motion of fluids governs the effectiveness of mixing and homogenization of both composition and temperature in the reactors. Such mixing is of great importance in a variety of metal extraction and refining processes [4]. Some examples are as follows:

- Processes differ from one another due to the nature of fluid motion in the reaction vessel and consequent mixing pattern.
- Fluid flow also governs the extent of dispersion of phases (e.g. formation of emulsion and foam in steelmaking).
- Motion of fluid determines the rate of mass and heat transfer across phases.

Therefore, a knowledge of fluid flow is important in extractive metallurgy.

### 1.2.1  Newton's Law of Viscosity

Figure 1.2a shows large parallel plates (area A) separated by a distance Y in a fluid either a gas or a liquid. In the space between them is a fluid either a gas or a liquid. Initially system is at rest, but at time t = 0, the lower plate is set in motion in the positive x direction at a constant velocity, u.

As time proceeds, the fluid gains momentum, and ultimately the linear steady-state velocity profile is established (as shown in Fig. 1.2d). To maintain the motion of the lower plate [1], a constant force, F is required.

The magnitude of the force is directly proportional to the area of contact (A), the velocity (u), and inversely proportional to the distance (Y) between the plates.

The above statement is mathematically formulated as

$$F \alpha \frac{Au}{Y} \text{ or, } F = \mu \frac{Au}{Y} \tag{1.3}$$

where $\mu$ is the proportionality constant, defined as *viscosity* which is the property of the fluid.

Equation (1.3) can be rewritten as

$$\frac{F}{A} = \mu \frac{u}{Y} \tag{1.4}$$

Now, $\left(\frac{F}{A}\right)$ can be replaced by $\tau_{yx}$, which is shear force (or stress) in the x direction on a unit area perpendicular to the y direction. Again, $\left(\frac{u}{Y}\right)$ can be replaced by $\left(-\frac{\delta u_x}{\delta y}\right)$, then Eq. (1.4) becomes

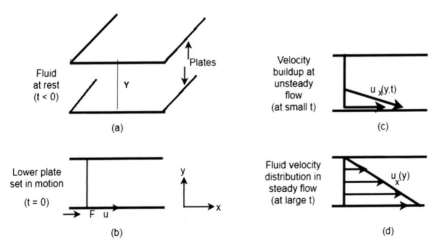

**Fig. 1.2**  Fluid in between large parallel plates

$$\tau_{yx} = -\mu\frac{\delta u_x}{\delta y} \tag{1.5}$$

where $\tau_{yx}$ is the shear force per unit area, $\frac{\delta u_x}{\delta y}$ is the velocity gradient of x component in the y direction, and $\mu$ is the proportionality constant, also known as viscosity of the fluid. The negative sign in Eq. (1.5) signifies that the shear force is such that it acts in the direction opposite to that of flow on the layer having higher velocity.

This Eq. (1.5) is known as *Newton's law of viscosity*, which quantifies the relationship between shear stress and shear deformation.

### 1.2.2  Flow Characterization

Fluid flow can be categorized as the follows [4]:

1. Laminar or turbulent,
2. Newtonian or non-Newtonian,
3. Viscous or non-viscous,
4. Incompressible or compressible,
5. Steady or unsteady,
6. Flow in a channel or flow around a submerged object.

#### 1.2.2.1  Laminar or Turbulent Flow

A classical experiment was performed by O. Reynolds in 1883. Reynolds used transparent pipes through which water was allowed to flow with different velocities and a stream of colored dye was injected parallel to the path of water flow as shown in Fig. 1.3.

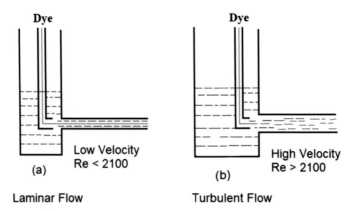

**Fig. 1.3** Experiment by Reynolds for fluid flow

(a)  At low velocities, it was observed that the dye moved in parallel straight lines with the path of water flow. This type of flow is called the *laminar* or *streamline flow*. The flow is called laminar because the adjacent layers of fluid (laminae) slide past one another in an orderly fashion [3]. In this case, the fluid flows, on a macroscopic scale, due to the movement of fine laminations relative to adjacent fluid layers in a parallel manner and fluid particles in each layer move in regular paths known as *streamlines* (Fig. 1.3a).

(b)  When the velocity of water increased, after a particular stage the flow became erratic, and the entire mass of water became colored with the dye (Fig. 1.3b). This is known as *turbulent flow*. Here the fluid particles move in irregular manner breaking down the streamlines and get intimately mixed.

A fluid flowing in a small tube or at low velocity does by the mechanism of laminar flow, also called viscous, or streamline flow. The layers of fluid slide over each other with no macroscopic mixing, and the velocity of this steady flow is constant at any point [2]. At higher velocities flow become turbulent, there are mixing by eddy motion between the layers, and even in overall steady flow the velocity at any point is about some mean value.

Laminar flow is obtained at low velocities and turbulent flow is obtained at higher velocities. The laminar flow is characterized by distinct streamlines with no cross mixing, whereas the turbulent flow is accompanied by extensive mixing [4]. Suppose velocity of fluid is measured at a particular location as function of time. Considering the flow is steady, i.e. velocity does not vary with time that means velocity is constant. Figure 1.4 shows the pattern of velocity vases time curves for both laminar as well as turbulent flows. Although the time averaged velocity is constant for turbulent flow, but the instantaneous velocity exhibits random fluctuations.

Imagine the fluid element M (Fig. 1.5), under turbulent flow, its instantaneous velocity fluctuates at random. Similar random fluctuations are exhibited by its neighbors. Since, in general, fluctuations of neighboring fluid elements are not in harmony with M, the latter is all the time receiving impacts from its neighbors and vice versa. This occasionally would throw M out of its location to region N. In exchange, fluid from location N may be imagined occupying location M. Such a process of exchange is visualized as an *eddy*. Such eddy like exchanges go on at random in all directions leading to extensive mixing in turbulent flow. That is why, turbulent flow is preferred in engineering processes [4].

**Fig. 1.4** Pattern of velocity versus time

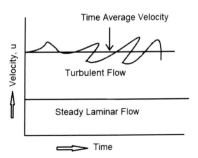

**Fig. 1.5** Eddy formation
under turbulent flow

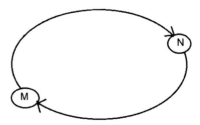

   Reynolds studied the flow of fluids of different physical properties, namely, the density and viscosity, at different velocities through horizontal pipes of different diameters [3]. Based on these studies, he concluded that whether a fluid flow was laminar or turbulent would depend on the magnitude of the parameter, $\left(\frac{\rho u d}{\mu}\right)$

where

   $\rho$ = density of fluid,
   u = average velocity of fluid,
   d = diameter of the pipe through which the fluid is flowing, and
   $\mu$ = viscosity of the fluid.

   This parameter $\left(\frac{\rho u d}{\mu}\right)$ is known as *Reynolds number* (Re), which is a dimensionless number.
   Therefore,

$$Re = \frac{\rho u d}{\mu} \tag{1.6}$$

e.g. unit: $\rho = ML^{-3}$, $u = Lt^{-1}$, $d = L$, $\mu = ML^{-1}t^{-1}$.
   Therefore, $Re = \frac{ML^{-3}Lt^{-1}L}{ML^{-1}t^{-1}} = M^{1-1}L^{-3+1+1+1}t^{-1+1} = M^0L^0t^0 =$ dimensionless number.
   Laminar flow of the fluid is maintained only at low velocities. Beyond some critical velocity, which is characteristic of a system, the flow becomes unstable and with further increase, become as turbulent flow. The dimensionless number that is used to characterize the transitions from laminar to unstable and further to turbulent flow is the Reynolds number [5]. Reynolds number characterizes the nature of the fluid flow. The flow through pipes, one can maintain laminar flow under normal circumstances for Reynolds number up to approximately 2100. At Reynolds number (Re) values above 5000–10,000, the flow becomes fully turbulent. In the intermediate region, the flow is said to be in transition: flow keeps alternating between laminar and turbulent states. In general, when Re < 2100 → flow is laminar and Re > 2100 → flow is turbulent; for smooth pipe. For other channels such as packed beds, different values of Reynolds number are applicable.

**Fig. 1.6** Velocity profile of a fluid passing on flat plate

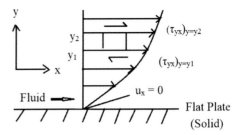

### 1.2.2.2  Newtonian or Non-Newtonian Flow

Unlike a solid, a fluid cannot sustain a shear stress. In other words, the fluid is completely deformable. If a shear stress is applied, then the fluid would undergo shear deformation continuously. Newton's law of viscosity is quantified the relationship between shear stress and shear deformation [4]. It is stated as

$$\tau_{yx} = -\mu \frac{\delta u_x}{\delta y} \tag{1.5}$$

where ($\tau_{yx}$) denotes shear stress acting on y plane (i.e. a plane normal to y axis) along x direction. $\mu$ is viscosity. Velocity in the x direction is zero at the solid surface since the fluid layer just at the surface cannot slip past the solid. Velocity increases as move along the y direction (Fig. 1.6).

This Eq. (1.5), which state that the shear force per unit area is proportional to the negative of the velocity gradient, is called *Newton's law of viscosity.*

It has been found that the resistance to flow of all gases and liquids, with molecular weight of <5000, is obeyed by Newton's law of viscosity (Eq. (1.5)) and such fluids are referred to as *Newtonian fluids.* The viscosity ($\mu$) is constant for *Newtonian fluids.*

Polymeric liquids, suspensions, pastes, slurries, and other complex fluids are not obeyed by Newton's law of viscosity (Eq. (1.5)) and they are referred to as *non-Newtonian fluids.* Many liquids, such as tar and polymer, which have high viscosity, do not exhibit a constant value of viscosity ($\mu$) and are known as *non-Newtonian fluids.*

### 1.2.2.3  Viscous or Non-viscous

Consider a fluid is in laminar flow. The adjacent layers of the fluid are moving with different velocities. The layer nearer to the solid surface has the lowest or zero velocity, whereas, the layer moves away from it, the velocity increase progressively. This difference in velocity causes an exertion of shearing force by one layer on the other adjacent to it.

In the field of metallurgy, peoples are concerned mainly with Newtonian fluids of all gases, liquid metal, and slag systems which follow Newton's law of viscosity.

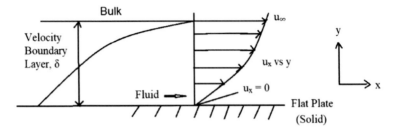

**Fig. 1.7** Typical velocity profile and boundary layer of a fluid passing on flat plate

Viscosity may be defined as the property of a fluid which determines its resistance to shear stresses. It is a measure of the internal fluid friction which causes resistance to flow. Viscosity of fluid is due to cohesion and interaction between particles.

An *ideal fluid* has no viscosity and surface tension. There is not a single fluid which can be classified as a perfectly ideal fluid. However, the fluids with very little viscosity are sometimes considered as ideal fluids.

In Eq. (1.5), $\frac{du}{dy}$ is the rate of shear strain or rate of shear deformation or velocity gradient.

Therefore,

$$\mu = \frac{\tau}{(du/dy)} \tag{1.7}$$

Thus, progressively viscosity ($\mu$) may be defined as the shear stress required to produce unit rate of shear strain.

If one of these is negligibly small, then $\tau$ can be ignored and the flow may be treated as *non-viscous* or *ideal flow*; otherwise it would be considered as *viscous flow*.

Figure 1.7 depicts a typical velocity profile adjacent to a solid surface, velocity of the fluid just at the surface is zero and it increases rapidly to a constant value ($u_\infty$) within a small distance. The region where the velocity is varying with distance is known as velocity boundary layer ($\delta$). Outside the velocity boundary layer is the bulk of the fluid [4]. As a general guideline, therefore, the motion of fluid within the boundary layer may be taken as viscous that in the bulk as non-viscous.

### 1.2.2.4   Incompressible or Compressible

When a fluid flows, considerable pressure differences may exist across the systems. If the fluid is a gas, such pressure difference would lead to variation of density and under such a situation, the flow is called a *compressible flow* [4]. If motion of gases at small pressure difference, i.e. without change of density, that fluids are known as *incompressible fluids*. A fluid is said to be incompressible, if the density of the same chunk of fluid in a flowing system does not change with time [3] in spite of change in external pressure.

The densities of fluids change with pressure, i.e. all real fluids are compressible. However, for many engineering flows, the change in density due to change in pressure during flow is small enough, that to be neglected [5]. Such flows are classified as *incompressible flows.*

### 1.2.2.5   Steady or Unsteady

A system is at *steady state,* when all the properties at any point (some of these such as velocity) in the system remain constant with respect to time [5]. If this condition is not satisfied, the system is said to be *unsteady* or in a *transient state.*

The steady state is defined as the state in which the density of the fluid does not change with time. In fact, the general definition of steady state includes the time independence of all intensive properties of the system. The concept of steady state is important in the case of continuous type of reactors, e.g. blast furnace for hot metal production, there is no accumulation of material, i.e. input is equal to output.

### 1.2.2.6   Flow in a Channel or Flow Around a Submerged Object

Analysis of fluid motion depends on geometry of flow:

(a)   Flow in channel: the simple examples of this when fluid is flowing through a straight circular pipe.
(b)   Flow around a submerged object: when fluid is flowing around a sphere.

These are discussed in detail at Sect. 2.9.

## *1.2.3   Velocity Profile*

When fluid is flowing at low pressure in a tube, the mean free path of a molecule of fluid is of the order of the tube diameter, if there is no slip at the tube wall. The velocity then increases from zero at the wall to a maximum at the center of the tube. The curve of velocity (u) versus distance from the wall (y) is referred as *velocity distribution or profile* (Fig. 1.8(a)).

At an enough large distance from the tube entrance, the velocity profile assumes a constant shape; the flow is said to be *developed* (Fig. 1.8b). The velocity distribution for developed flow depends on whether the flow is turbulent or laminar.

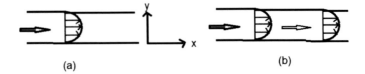

(a)                                                          (b)

**Fig. 1.8  a** Velocity profile and **b** developed velocity profile of a fluid flowing in a tube

## 1.2.4   Velocity Boundary Layer

Consider fluid flows over the flat plate as shown in Fig. 1.9. When fluid particles contact the solid surface, they assume zero velocity. These particles then act to retard the motion of particles in the adjoining fluid layer, which act to retard the motion of particles in the next layer, and so on until, at a distance $y = \delta$ from the surface, the effect becomes negligible [6]. This retardation of fluid motion is associated with shear stresses $\tau$ acting in planes that are parallel to the fluid velocity (Fig. 1.9). With increasing distance y from the surface, the x velocity component of the fluid, u, must then increase until it approaches the *free stream value*, $u_\infty$ (i.e. constant velocity at bulk). The subscript $\infty$ is used to designate conditions in the free stream outside the boundary layer at the bulk.

The quantity $\delta$ is termed the *velocity boundary layer*, where velocity profile refers to the manner in which u varies with y through the boundary layer. Accordingly, the fluid flow is characterized by two distinct regions, a thin fluid layer (i.e. the boundary layer) in which velocity gradients and shear stresses are large and a region outside the boundary layer (i.e. bulk) in which velocity gradients and shear stresses are negligible. With increasing distance from the leading edge, the effects of viscosity penetrate further into the free stream and the boundary layer grows ($\delta$ increases with x).

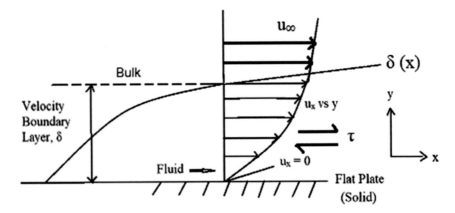

**Fig. 1.9**  Development of velocity boundary layer of a fluid passing on flat plate

## 1.2.5   *Pressure at a Point in Static Fluid*

The average pressure is calculated by dividing the normal force by the area (pushing against a plane area),
   i.e.

$$\text{i.e.} \quad Pressure = \frac{Force}{Area} = \left(\frac{F}{A}\right), \text{ Nm}^{-2} \tag{1.8}$$

The pressure at a point is the limit of the ratio of normal force to area, approaches zero;
   i.e.

$$\lim_{A \to 0}\left(\frac{F}{A}\right) = P_{\text{at a point}} \tag{1.9}$$

At a point a fluid, which is at rest, has the same pressure in all directions. This means that an element $\delta A$ of very small area, free to rotate about its center when submerged in a fluid at rest, will have a force of constant magnitude acting on either side of it, regardless of its orientation [7].

A small wedge-shaped free body of unit width is taken at the point $(x, y)$ in a fluid at rest (Fig. 1.10). Since there can be no shear forces, the only forces are the normal surface forces and gravity. So, forces acting on x and y directions are as follows:

$$x \text{ direction:} \quad p_x \, \delta y - p_s \, \delta s \sin\theta = 0 \tag{1.10}$$

$$y \text{ direction:} \quad p_y \, \delta x - p_s \, \delta s \cos\theta - \frac{\delta x \delta y}{2} \rho g = 0 \tag{1.11}$$

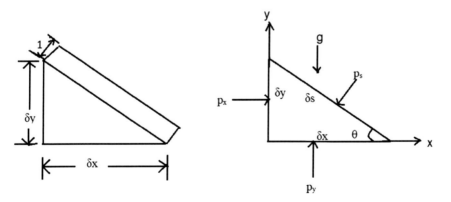

**Fig. 1.10**  Wedge-shaped free body in a fluid at rest

where $p_x$, $p_y$, $p_s$ are the average pressures on the three faces; $\delta x$, $\delta y$, $\delta s$ are the length of the three sides of the wedge, respectively. $\rho$ and g are the density and gravitational force.

Since, area of plane perpendicular to x direction $= \delta y . 1 = \delta y$ (since width of wedge $= 1$); area of plane perpendicular to y direction $= \delta x . 1 = \delta x$.

Again, $\sin\theta = \frac{\delta y}{\delta s}$, and $\cos\theta = \frac{\delta x}{\delta s}$, therefore $\delta y = \delta s \sin\theta$; and $\delta x = \delta s \cos\theta$.

Equation (1.10) becomes

$$p_x \, \delta y - p_s \, \delta y = 0, \ or \, p_x = p_s \tag{1.12}$$

Equation (1.11) becomes

$$p_y \, \delta x - p_s \, \delta x - \frac{\delta x \delta y}{2} \rho g = 0 \tag{1.13}$$

When the limit is taken, the free body is reduced to zero size by allowing the inclined face to approach (x, y) while maintaining the same angle ($\theta$). The last term of Eq. (1.13) will be very small and may be neglected.

So, Eq. (1.13) becomes

$$p_y = p_s \tag{1.14}$$

Therefore,

$$p_x = p_y = p_s \tag{1.15}$$

Since $\theta$ is an arbitrary angle, Eq. (1.15) proves that the pressure is the same in all directions at a point in a static fluid. This is the two-dimensional case; similarly, that can be proved for the three-dimensional case.

If the fluid is in motion, so that one layer moves relative to an adjacent layer, shear stresses are generated and the normal stresses, in general, are no longer the same in all directions at a point. The pressure is then defined as the average of any three mutually perpendicular normal compressive stresses at a point as follows:

$$p = \frac{p_x + p_y + p_s}{3} \tag{1.16}$$

In a frictionless fluid or ideal fluid, i.e. fluid with zero viscosity, no shear stresses occur for any motion of the fluid. So, at a point of pressure is the same in all directions.

## *1.2.6  Pressure Variation in Static Fluid*

In a fluid at rest, it is self-evident that any points within it that lie in the same horizontal plane, must be at equal pressure. If not, fluid motion would occur because of horizontal pressure gradients. However, vertical pressure gradients can be sustained within a contained fluid.

The forces acting on a control element of fluid at rest, Fig. 1.11, consist of surface forces and body forces. With gravity the only body force is acting on a control element of fluid.

Forces acting on x direction:

$$p_x \Delta y\, \Delta z - p_{x+\Delta x}\, \Delta y\, \Delta z = \rho\, \Delta x\, \Delta y\, \Delta z\, a_x \tag{1.17}$$

where $p_x$ is the pressure at the x direction. $\Delta x$, $\Delta y$, $\Delta z$ are the lengths of the control element in x, y, z directions. $a_x$ is the acceleration of fluid ($Lt^{-2}$).

[Dimensional analysis of Eq. (1.17): left-hand side: $ML^{-1}t^{-2} \cdot L^2 = MLt^{-2}$

and right-hand side:    $ML^{-3} \cdot L^3 \cdot Lt^{-2} = MLt^{-2}$

Therefore, both sides have same dimension, so Eq. (1.17) is dimensionally balanced.]

Similarly, forces acting on y and z directions:

$$p_y \Delta x\, \Delta z - p_{y+\Delta y} \Delta x\, \Delta z = \rho\, \Delta x\, \Delta y\, \Delta z\, a_y \tag{1.18}$$

$$p_z \Delta x\, \Delta y - p_{z+\Delta z} \Delta x\, \Delta y - \rho g\, \Delta x\, \Delta y\, \Delta z = \rho\, \Delta x\, \Delta y\, \Delta z\, a_z \tag{1.19}$$

Divided by ($\Delta x\, \Delta y\, \Delta z$) to Eqs. (1.17)–(1.19) become

**Fig. 1.11** Forces acting on a control element of fluid

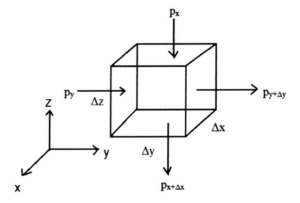

$$\left(p_x - p_{x+\Delta x}\right)\frac{1}{\Delta x} = \rho\, a_x$$

$$\left(p_y - p_{y+\Delta y}\right)\frac{1}{\Delta y} = \rho\, a_y$$

$$\left(p_z - p_{z+\Delta z}\right)\frac{1}{\Delta z} - \rho g = \rho\, a_z \qquad (1.20)$$

Since,

$$\frac{\delta p}{\delta x} = \lim_{\Delta x \to 0}\left[\left(p_{x+\Delta x} - p_x\right)\frac{1}{\Delta x}\right] \qquad (1.21)$$

Hence, Eq. (1.20) become

$$-\frac{\delta p}{\delta x} = \rho\, a_x, \quad -\frac{\delta p}{\delta y} = \rho\, a_y, \text{ and } -\frac{\delta p}{\delta z} - \rho g = \rho\, a_z \qquad (1.22)$$

In Eq. (1.22) is negative due to x, y, z directions.
Now if there is no acceleration, i.e. for static fluid, $a_i = 0$.
Therefore,

$$-\frac{\delta p}{\delta x} = 0, \quad -\frac{\delta p}{\delta y} = 0, \text{ and } -\frac{\delta p}{\delta z} - \rho g = 0 \qquad (1.23)$$

$$\frac{\delta p}{\delta z} = -\rho g \qquad (1.24)$$

Case I $\to$ If density is constant, i.e. for incompressible fluid:
By integration of Eq. (1.24):

$$p = p_0 - \rho g z \qquad (1.25)$$

That is, pressure is decreasing as fluid goes up.
For example at sea level, pressure of air is 760 mm of Hg; as goes up at Mount Everest (8848 m height), pressure of air is decreased to 270 mm of Hg.

(i)   For ideal gas:

$$pV = RT, \text{ so } V = \frac{RT}{p} \qquad (1.26)$$

Since Eq. (1.24):

$$\frac{\delta p}{\delta z} = -\rho g = -\left(\frac{M}{V}\right)g = -\left(\frac{pM}{RT}\right)g \qquad (1.27)$$

Therefore,

$$dz = -\frac{RT}{Mg}\left(\frac{dp}{p}\right) \tag{1.28}$$

By integration of Eq. (1.28) between limit: $\int_{z_0}^{z} dz = -\frac{RT}{Mg}\int_{p_0}^{p}\left(\frac{dp}{p}\right)$.

Or,

$$(z-z_0) = -\frac{RT}{Mg}\left[\ln\left(\frac{p}{p_0}\right)\right] \tag{1.29}$$

Therefore,

$$p = p_0\exp\left[-\left(\frac{z - z_0}{RT/Mg}\right)\right] \tag{1.30}$$

This Eq. (1.30) is for variation of pressure with height for an incompressible gas.

Case II → Density variation with pressure, i.e. compressible fluid:
When the fluid is a perfect gas at rest at constant temperature.
Then

$$\frac{p}{\rho} = \frac{p_0}{\rho_0} \text{ or } \rho = \left(\frac{p\rho_0}{p_0}\right) \tag{1.31}$$

From Eq. (1.24):
$\frac{\delta p}{\delta z} = -\rho g = -\left(\frac{p\rho_0}{p_0}\right)g$ (from Eq. (1.31)).
So

$$dz = -\left(\frac{p_0}{g\rho_0}\right)\left(\frac{dp}{p}\right) \tag{1.32}$$

By integration of Eq. (1.32) between limit: $\int_{z_0}^{z} dz = -\left(\frac{p_0}{g\rho_0}\right)\int_{p_0}^{p}\left(\frac{dp}{p}\right)$.

Or,

$$(z-z_0) = -\left(\frac{p_0}{g\rho_0}\right)\left[\ln\left(\frac{p}{p_0}\right)\right] \tag{1.33}$$

Therefore,

$$p = p_0\exp\left[-\left(\frac{z - z_0}{p_0/g\rho_0}\right)\right] \tag{1.34}$$

This Eq. (1.34) is for variation of pressure with height for a compressible isothermal gas.

**Fig. 1.12** Buoyancy force of submerged body in the fluid

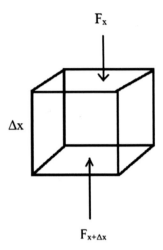

### 1.2.7 Buoyancy Forces

Consider a submerged body is in the fluid, as shown in Fig. 1.12. The fluid, in which the body is immersed, generates a net resultant force directed vertically upwards through the centroid.

Resolving forces:

$$F_B = F_{x+\Delta x} - F_x = A \cdot \frac{\delta p}{\delta x} \cdot \Delta x \quad \left[ \text{since } \frac{\delta_p}{\delta_p} = \rho g \right]$$

$$= A \cdot \rho g \cdot \Delta x \tag{1.35}$$

That is a vertical buoyancy force equal to the weight of displaced fluid acts on the submerged body.

## 1.3   Energy Balance

### 1.3.1   Chemical Analysis Laboratory

Case I → When door is open at the laboratory.

Suppose there is a chemical analysis laboratory, where a lot of chemical fumes are generated. In that laboratory, an exhaust fan will be installed to remove the chemical fumes. By energy balance, power of exhaust fan will be calculated for purchase purpose.

The region over which the balance is made is called the *control volume*. The control volume is chosen for energy balance as shown in Fig. 1.13. If there is no exhaust fan,

**Fig. 1.13** Metallurgical
laboratory with open door

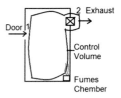

there is no flow of air, i.e. air in static condition in the laboratory. Presence of exhaust
fan, there is flowing of air, i.e. air is in motion or kinetic condition is prevailed in the
laboratory.

Now energy balance for control volume:

Kinetic energy at point 1 = Kinetic energy at point 2 + Work done by an exhaust
fan

$$\rightarrow \frac{m'u_1^2}{2} = \frac{m'u_2^2}{2} + w_s \tag{1.36}$$

where $m'$ is the rate of mass flow (kg/s), u is the average velocity (m/s) and $w_s$ is the
work done by the exhaust fan (J/s).

$$\rightarrow w_s = -m'\left(\frac{u_2^2 - u_1^2}{2}\right) \tag{1.37}$$

$$\rightarrow w_s^\wedge = -\left(\frac{u_2^2 - u_1^2}{2}\right) \tag{1.38}$$

where $w_s^\wedge$ is work done per kg of fluid upon the surroundings (J/kg).

Since velocity of air at point 2 (near the exhaust fan) is greater than velocity of
air at point 1 (near the door), i.e. $u_1 \ll u_2$. The velocity of air at point 1 is very small,
so considering $u_1$ is negligible, i.e. $u_1 \rightarrow 0$.

If Q is the volumetric flow rate of air, due to the exhaust fan ($m^3/s$).

Since,

$$Q = \text{velocity of air at point 2} \times \text{cross-sectional area}$$

$$= u_2\left(\frac{\pi d^2}{4}\right) \tag{1.39}$$

where d is the diameter of the exhaust fan.

Again

$$-w_s = m'\left(-w_s^\wedge\right) = \left\{(Q\,\rho_{air})\left(-w_s^\wedge\right)\right\} \tag{1.40}$$

Fan rating, i.e.

**Fig. 1.14** Metallurgical laboratory with close door

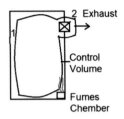

2 Exhaust

Control Volume

Fumes Chamber

$$\text{power of fan} = \frac{\text{Work done by fan}}{\text{Fraction of } \textit{efficiency of fan}} = \frac{-w_s}{E_f} \qquad (1.41)$$

where $E_f$ is the fraction of efficiency of the exhaust fan.

Case II → When door is closed at the laboratory.

Figure 1.14 shows the system with control volume. Since door is closed, that leads to a small vacuum in the laboratory room due to the leakage areas are small.

If the rate of exhaust gas remains the same.

(a)  For compressible gas:

$$pV = RT \text{ or } p = \frac{RT}{V} = \frac{RT}{M}\rho \qquad (1.42)$$

where V is the volume, R is the gas constant and T is the absolute temperature.

[since $V = \frac{M}{\rho}$, M and $\rho$ are the mass and density, respectively].

Now differentiate Eq. (1.42):

$$dp = \left(\frac{RT}{M}\right)d\rho \qquad (1.43)$$

For compressible fluid, density ($\rho$) varies with pressure (p).

Divided Eq. (1.43) by Eq. (1.42):

$$\frac{dp}{p} = \frac{d\rho}{\rho} \qquad (1.44)$$

(b)  For incompressible gas (i.e. $\rho$ is constant).

Now energy balance for control volume:

Kinetic energy at point 1 + Work done by the surrounding to the system.

= Kinetic energy at point 2 + Work done by the system to the surrounding + Fan work

$$\rightarrow \frac{m'u_1^2}{2} + \left(\frac{m'}{\rho}\right)p_1 = \frac{m'u_2^2}{2} + \left(\frac{m'}{\rho}\right)p_2 + w_s \qquad (1.45)$$

$$\rightarrow m'\left(\frac{u_2^2 - u_1^2}{2}\right) + m'\left(\frac{p_2 - p_1}{\rho}\right) = -w_s$$

$$\rightarrow \left(\frac{u_2^2 - u_1^2}{2}\right) + \left(\frac{p_2 - p_1}{\rho}\right) = -w_s^\wedge \qquad (1.46)$$

Since $u_1 \ll u_2$, the velocity of air at point 1 is very small; so considering $u_1$ is negligible,

i.e. $u_1 \rightarrow 0$.

Therefore, Eq. (1.46) becomes

$$\frac{u_2^2}{2} + \frac{\Delta p}{\rho} = -w_s^\wedge \qquad (1.47)$$

**Example 1.1** (a) To get about 30 m³/min of fresh air by installing an exhaust fan, what will be the power of exhaust fan? Also calculate the minimum amount of energy that the fan will deliver to the air. (b) If the pressure difference is 98.1 Nm$^{-2}$, what will be the power of exhaust fan? If the diameter of exhaust fan is 0.3 m and efficiency of fan is 30%. (Given density of air $= 1.16$ kg/m³ at 27 °C.)

**Solution**

(a)  The volumetric flow rate of air, $Q = 30$ m³/min $= 0.5$ m³/s.

From Eq. (1.39): $Q = u_2\left(\frac{\pi d^2}{4}\right)$.

Therefore, $u_2 = \left\{\frac{Q}{\left(\frac{\pi d^2}{4}\right)}\right\} = \frac{0.5 \times 4}{\pi (0.3)^2} = 7.08$ m/s.

From Eq. (1.38): $w_s^\wedge = -\left(\frac{u_2^2 - u_1^2}{2}\right)$.

or

$$-w_s^\wedge = \frac{u_2^2}{2} \quad \text{[By considering the value of } u_1 \text{ is negligible compared with } u_2.]$$

$$= \frac{7.08^2}{2} = 25.06 \text{ J/kg}$$

Hence, the minimum amount of energy that the fan will deliver to the air is **25 J/kg**. Again from Eq. (1.40):

$$-w_s = m'(-w_s^\wedge) = (Q\rho_{air})(-w_s^\wedge)$$
$$= \{(0.5 \times 1.16) \times 25.06\} = 14.53 \text{ w}$$

Now from Eq. (1.41): Fan rating, i.e. power of fan $= \frac{-w_s}{E_f} = \frac{14.53}{0.3} = 48.43$ w.

Hence, the power of exhaust fan will be **50 w**.

(b)   From Eq. (1.47): $-w_s^\wedge = 25 + \left(\frac{98.1}{116}\right) = 109.57\,\text{J/kg}$.

And from Eq. (1.40):

$$-w_s = \left\{(Q\rho_{air})\left(-w_s^\wedge\right)\right\}$$
$$= (0.5 \times 1.16 \times 109.57) = 63.55\ \text{w}$$

Therefore, the power of exhaust fan $= \left(\frac{63.55}{0.3}\right) = 211.83$ w.
Hence, the power of exhaust fan will be **215 w**.

### 1.3.2  Water Tank with Pump for Water Supply

Figure 1.15 shows the water tank with pump system. The levels of control volume 1 and 2 are same, so there is no pressure difference at points 1 and 2, i.e. $p_1 = p_2$.

Again $u_1 \ll u_2$. The velocity of water at point 1 is very small, so considering $u_1$ is negligible, i.e. $u_1 \rightarrow 0$.

Since there is no pressure difference at 1 and 2, i.e. $p_1 = p_2$.

From Eq. (1.46):

$$w_s^\wedge = -\left(\frac{u_2^2 - u_1^2}{2}\right) + \left(\frac{p_2 - p_1}{\rho}\right) \rightarrow -w_s^\wedge = \frac{u_2^2}{2} \qquad (1.48)$$

**Example 1.2** If the diameter of pipe is 60 mm and water is passed at the rate of 500 l/min. Find out the power of pump in hp. Given: the efficiency of water pump is 30% and density of water is 1000 kg/m³.

**Solution**

Volumetric flow rate of water, $Q = 500\,\text{L/min} = \frac{500 \times 0.001}{60} = 8.33 \times 10^{-3}\text{m}^3/\text{s}$ [since 1 L = 0.001m³].

From Eq. (1.39): $Q = u_2\left(\frac{\pi d^2}{4}\right)$.

Therefore, $u_2 = \left\{\frac{Q}{\left(\frac{\pi d^2}{4}\right)}\right\} = \left\{\frac{8.33 \times 10^{-3} \times 4}{\pi (0.06)^2}\right\} = \textbf{2.95 m/s}$.

Since there is no pressure difference at 1 and 2, i.e. $p_1 = p_2$.
Again $u_1 \ll u_2$. The velocity of water at point 1 is so small, i.e. $u_1 \rightarrow 0$.

**Fig. 1.15**  Water tank with pump

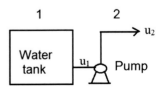

From Eq. (1.48): $-w_s^\wedge = \frac{u_2^2}{2} = \frac{2.95^2}{2} = 4.35$.
Again from Eq. (1.40):

$$-w_s = \{(Q\rho_{water})(-w_s^\wedge)\}$$
$$= \{(8.33 \times 10^{-3}) \times 1000 \times 4.35\} = 36.24 \text{ w}$$

Therefore, the power of pump $= (\frac{36.24}{0.3}) = 120.79$ w $= 0.12$ kw.
Hence, the power of pump will be 0.12 kw $= (0.12 \times 1.34) = \mathbf{0.16}$ hp [Since 1kw $= 1.34$ hp].

**Example 1.3** If the diameter of pipe is 60 mm and water supply to cool the mold at the rate of 500 L/min. Find out the power of pump.

**Solution**

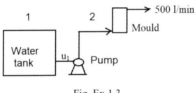

Fig. Ex.1.3

Figure Ex.1.3 shows the system with control volume. The levels of control volume 1 and 2 are not the same, so there is the pressure difference of 1atmosphere (i.e. 1.013 $\times 10^5$ N/m$^2$).

Since there is pressure difference of 1 atm, i.e. $p_2 - p_1 = \Delta p = 1.013 \times 10^5$ N/m$^2$.
Again $u_1 \ll u_2$. The velocity of water at point 1 is so small, i.e. $u_1 \rightarrow 0$.
Since, $u_2 = 2.95$ m/s [from Example 1.2].

From Eq. (1.46):
$$w_s^\wedge = -\left(\frac{u_2^2 - u_1^2}{2}\right) + \left(\frac{p_2 - p_1}{\rho}\right) = \left\{\frac{u_2^2}{2} + \frac{\Delta p}{\rho}\right\}$$
$$= \left\{\left(\frac{2.95^2}{2}\right) + \left(\frac{1.013 \times 10^5}{1000}\right)\right\} = 105.65$$

Again from Eq. (1.40): $-w_s = \{(Q\rho_{water})(-w_s^\wedge)\} =$
$\{(8.33 \times 10^{-3}) \times 1000 \times 105.65\} = 880.06$ w.

Therefore, the power of pump $= (\frac{880.06}{0.3}) = 2933.53$ w $= \mathbf{2.933}$ kw.
Hence, the power of pump will be 2.933 kw $= (2.933 \times 1.34) = \mathbf{3.9}$ hp.

### 1.3.3 Water Tank with Pump for Water Supply to the Tank at Roof

Water supply from ground tank to tank at roof is shown in Fig. 1.16. The levels

**Fig. 1.16** Water supply from
ground tank to roof tank

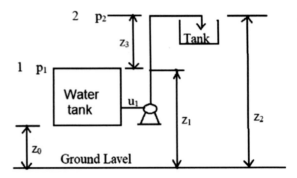

of control volumes 1 and 2 are not same. Apart from kinetic energy, here potential
energy is also taken into consideration.

Now energy balance for control volume:

Kinetic energy at point 1 + Work done by the surrounding to the system +
Potential energy point at 1 = Kinetic energy at point 2 + Work done by the system
to the surrounding + Potential energy point at 2 + Pump work

$$\frac{m'u_1^2}{2} + \left(\frac{m'}{\rho}\right)p_1 + m'g(z_1-z_0) = \frac{m'u_2^2}{2} + \left(\frac{m'}{\rho}\right)p_2 + m'g(z_2-z_0) + w_s \quad (1.49)$$

$$\rightarrow m'\left(\frac{u_2^2 - u_1^2}{2}\right) + m'\left(\frac{p_2 - p_1}{\rho}\right) + m'g(z_2-z_1) = -w_s$$

$$\rightarrow \left(\frac{u_2^2 - u_1^2}{2}\right) + \left(\frac{p_2 - p_1}{\rho}\right) + g(z_2-z_1) = -w_s^{\wedge} \quad (1.50)$$

Assumptions:

(1) Incompressible fluid, i.e. density is not varied with pressure or $\left(\frac{p_2-p_1}{\rho}\right) \rightarrow 0$.
(2) $u_1 \ll u_2$. The velocity of water at point 1 is very small, so considering $u_1$ is
     negligible, i.e. $u_1 \rightarrow 0$.

Hence, Eq. (1.50) becomes

$$-w_s^{\wedge} = \frac{u_2^2}{2} + g(z_2-z_1) \quad (1.51)$$

**Example 1.4** If the diameter of pipe is 60 mm and water supply to roof tank at the
rate of 500 L/min. Find out the power of pump. If heights of roof and roof + tank
are 7 m and 10 m, respectively.

**Solution**

Since, $u_2 = 2.95$ m/s [from Example 1.2].

From Eq. (1.51): $-w_s^\wedge = \frac{u_2^2}{2} + g(z_2-z_1) = \left\{\left(\frac{2.95^2}{2}\right) + (9.81 \times 3)\right\} = 33.78\,\text{J/kg}$.

Again from Eq. (1.40): $-w_s = \left\{(Q\rho_{water})(-w_s^\wedge)\right\} = \left\{(8.33 \times 10^{-3}) \times 1000 \times 33.78\right\} = \mathbf{281.39\,w}$.

Therefore, the power of pump $= \left(\frac{281.39}{0.3}\right) = 937.97\,\text{w} = \mathbf{0.94\,kw}$.

Hence, the power of pump will be $\mathbf{0.94}$ kw $= (0.94 \times 1.34) = \mathbf{1.3\,hp}$.

**Example 1.5** There is a water tank at the ground level. With the help of a pump, water can be lifted to a tank situated at the roof of a house at a rate 300 L/min by 60 mm diameter pipe. Height of the house is 10 m and tank height is 2 m. Neglect the pressure difference. Find out the rating of pump, if efficiency is 50%.

**Solution**

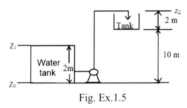

Fig. Ex.1.5

Figure Ex.1.5 shows the above problem.

From Eq. (1.51): $-w_s^\wedge = \frac{u_2^2}{2} + g(z_2-z_1)$.

Here $z_2 = 10 + 2 = 12$ m and $z_1 = 2$ m.

The volumetric flow rate, $Q = 300$ L/min $= 0.3\,\text{m}^3/\text{min} = 5 \times 10^{-3}\,\text{m}^3/\text{s}$.

Again, $u_2 = \frac{Q}{\frac{\pi d^2}{4}} = \left(\frac{4Q}{\pi d^2}\right) = \frac{4 \times 5 \times 10^{-3}}{3.14 \times (0.06)^2} = 1.77\,\text{m/s}$.

From Eq. (1.3):

$$-w_s^\wedge = \left(\frac{u_2^2}{2}\right) + g(z_2-z_1)$$

$$= \frac{1.77^2}{2} + 9.807\,(12-2) = 99.64\,\text{J/kg}$$

Therefore, $-w_s = Q.\rho.(-w_s^\wedge) = (5 \times 10^{-3}) \times 10^3 \times 99.64 = 498.18\,\text{W}$.

Hence, pump rating $= \left(\frac{498.18}{0.5}\right) = 996.36 = \mathbf{1000\,W}$.

## 1.3.4 General Equation for Overall Energy Balance

The system with control volume for fluid flow is shown in Fig. 1.17. For deriving the energy balance equations, the first law of thermodynamics can be applied, for a system with fluid flow, may be written as follows:

**Fig. 1.17** Control volume
for fluid flow

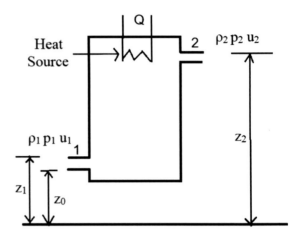

$[$(Amount of energy entering to the system with fluid per unit time)

$\quad -$(Amount of energy leaving  from the system with fluid per unit time)$]$

$= [$(Rate of change of energy of fluid in system)

$\quad +$ (Work   done by fluid per unit time on surrounding)$-$(Heat absorbed by fluid)$]$
$$(1.52)$$

A flowing fluid can carry energies in four forms: kinetic energy, potential energy, internal energy, and bulk energy. So

$$\text{total specific energy of the fluid} = KE + PE + IE + BE \qquad (1.53)$$

where KE is kinetic energy, PE is potential energy, IE is internal energy and BE is bulk energy.

Energy balance for steady state: Energy input $=$ Energy out put

$$\text{Energies entering to the system from point 1}$$
$$= \text{Energies leaving from the system from point 2} \qquad (1.54)$$

i.e. KE $+$ PE $+$ IE $+$ Work done by surrounding to the system $+$ Heat absorb by the system from the surrounding

$\quad =$ KE $+$ PE $+$ IE $+$ Work done by the system to surrounding $+$ Work done by shaft

$$\rightarrow \frac{m'u_1^2}{2} + m'g(z_1-z_0) + m'E_1 + \left(\frac{m'}{\rho_1}\right)p_1 + q$$
$$= \frac{m'u_2^2}{2} + m'g(z_2-z_0) + m'E_2 + \left(\frac{m'}{\rho_2}\right)p_2 + w_s \qquad (1.55)$$

where $q$ is the heat absorbed by the system from the surrounding, and $w$ is work done by the system on the surrounding.

$$\rightarrow m'\left(\frac{u_2^2 - u_1^2}{2}\right) + m'g(z_2-z_1) + m'(E_2 - E_1) + m'\left(\frac{p_2}{\rho_2} - \frac{p_1}{\rho_1}\right) + w_s - q = 0$$

$$(1.56)$$

$$\rightarrow \left(\frac{u_2^2 - u_1^2}{2}\right) + g(z_2-z_1) + (E_2 - E_1) + \left(\frac{p_2}{\rho_2} - \frac{p_1}{\rho_1}\right) + w_s^\wedge - q^\wedge = 0 \quad (1.57)$$

where $q^\wedge$ is heat absorbed per unit mass.

Equation (1.57) is the *general equation for overall energy balance.*

Case 1: If one-unit mass (i.e. 1 kg) of fluid comes into the system and goes out from the system

From thermodynamics:

$$\text{change of energy} = E_2 - E_1 = q - W \tag{1.58}$$

Or from 1st law of thermodynamics:

$$dE = \delta q - \delta W = \delta q - p dV + \delta\, w' \tag{1.59}$$

Internal energy change per unit mass: $dE = (E_2 - E_1) = q^\wedge - \int_1^2 p\, d\left(\frac{1}{\rho}\right) + w_{f_i}^\wedge$.
Therefore,

$$(E_2 - E_1) = q^\wedge - \left(\frac{p_2}{\rho_2} - \frac{p_1}{\rho_1}\right) + w_{f_i}^\wedge \tag{1.60}$$

where $w_{f_i}^\wedge$ is work done due to frictional effect.

[since $V = \frac{M}{\rho} = \frac{1}{\rho}$, if M is unit mass].

Now Eq. (1.57) becomes $\left(\frac{u_2^2 - u_1^2}{2}\right) + g(z_2-z_1) + \left\{q^\wedge - \left(\frac{p_2}{\rho_2} - \frac{p_1}{\rho_1}\right) + w_{f_i}^\wedge\right\} +$

$\left(\frac{p_2}{\rho_2} - \frac{p_1}{\rho_1}\right) + w_s^\wedge - q^\wedge = 0$

$$\rightarrow \left(\frac{u_2^2 - u_1^2}{2}\right) + g(z_2-z_1) + w_{f_i}^\wedge + w_s^\wedge = 0 \tag{1.61}$$

Case 2: Consider the following:

(i)    There is no friction, i.e. $w_{f_i}^\wedge = 0$,
(ii)   There is no shaft work, i.e. $w_s^\wedge = 0$,
(iii)  There is no change of internal energy, i.e. $dE = (E_2 - E_1) = 0$,

(iv)   There is no heat supply, i.e. $q^\wedge = 0$, and
(v)    Incompressible fluid, i.e. $\rho = \text{constant}$.

Hence, Eq. (1.57) becomes

$$\left(\frac{u_2^2 - u_1^2}{2}\right) + g(z_2 - z_1) + \left(\frac{p_2 - p_1}{\rho}\right) = 0 \qquad (1.62)$$

By further simplification:

$$\frac{u_1^2}{2g} + z_1 + \frac{p_1}{\rho g} = \frac{u_2^2}{2g} + z_2 + \frac{p_2}{\rho g} \qquad (1.63)$$

Equation (1.62) or Eq. (1.63) is known as *Bernoulli equation* for steady-state condition.

Bernoulli equation is important since it provides a working relationship between the velocity of the fluid, the change in potential energy, and the work done by the fluid.

**Example 1.6** A vacuum cleaner can create a vacuum of 3 kPa just inside the hose as shown in Figure Ex. 1.6. What is the maximum average velocity that would be expected in the hose?
(Given: Atm. pressure $= 101.3$ kPa and density of air $= 1.17$ kg/m$^3$).

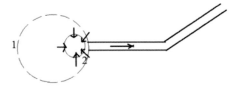

Fig. Ex.1.6

**Solution**

(a)   Energy balance:

K.E. at $1$ + Work done by the surrounding to the system = K.E. at $2$ + Work done by the system to the vacuum cleaner.

$$\rightarrow \frac{m'u_1^2}{2} + \left(\frac{m'}{\rho}\right)p_1 = \frac{m'u_2^2}{2} + \left(\frac{m'}{\rho}\right)p_2 \qquad (i)$$

where $m'$ is the rate of mass flow (kg/s), $u$ is velocity (m/s), $\rho$ is the density of air (kg/m$^3$) and p is the pressure (Pa).

$$\text{Or} \quad \left(\frac{u_2^2 - u_1^2}{2}\right) + \left(\frac{p_2 - p_1}{\rho}\right) = 0 \tag{ii}$$

Since $u_2 \gg u_1$, so $u_1 \rightarrow 0$.

Therefore, $\left(\frac{u_2^2}{2}\right) = -\left(\frac{p_2 - p_1}{\rho}\right) = -\left(\frac{-3000 - 101300}{1.17}\right) = \left(\frac{104300}{1.17}\right) = 89145.3$

Hence, $u_2 = \mathbf{422.24 \ m/s}$.

(b)   Directly from Bernoulli equation (Eq. (1.62)):

$$\left(\frac{u_2^2 - u_1^2}{2}\right) + g(z_2 - z_1) + \left(\frac{p_2 - p_1}{\rho}\right) = 0$$

Here $z_1 = z_2$, and getting equation as per Eq. (ii): $\left(\frac{u_2^2 - u_1^2}{2}\right) + \left(\frac{p_2 - p_1}{\rho}\right) = 0$.

Hence solve as above (a).

### 1.3.5  Applications of Overall Energy Balance

The rate of fluid flow through pipes can be measured with the help of instruments whose working are based on the overall energy balance, i.e. Bernoulli equation.

In these instruments, the fluid is either accelerated or retarded at the section used for measurement, and the change in kinetic energy is measured its motion by producing pressure difference.

Overall energy balance can be applied to the following flow meters:

1.   Orifice meter,
2.   Nozzle,
3.   Venturi meter,
4.   Pilot tube,
5.   Rotameter.

#### 1.3.5.1   Orifice Meter

The flow of fluids, through pipes, can be measured by passing the fluid through a constriction (i.e. obstacle), thereby increasing its kinetic energy; the actual flow rate can be determined by measuring the corresponding change in the pressure [8].

Orifice meter is shown in Fig. 1.18. The fluid approaching the orifice flows over a cross-sectional area ($A_1$, m$^2$) with a velocity ($u_1$, m/s), and at pressure ($p_1$). At the orifice, the fluid passes through a constriction of cross-sectional area ($A_2$) and is accelerated to a velocity ($u_2$).

The mass flow rate ($m'$, kg/s) maybe related to the pressure different ($p_1 - p_2$) by establishing an overall mass and energy balance at steady state over the planes 1 and 2 as follows:

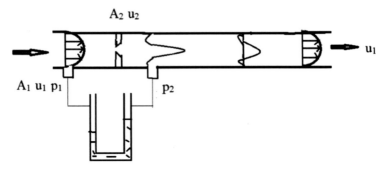

**Fig. 1.18** Orifice meter

$$u_1 \rho_1 A_1 = u_2 \rho_2 A_2 = m' = \text{constant} \qquad (1.64)$$

From Bernoulli equation (1.62):

For a horizontal pipe, $z_1 = z_2$ and for an incompressible fluid, $\rho_1 = \rho_2 = \rho = $ constant.

So Eq. (1.62) becomes

$$\frac{u_2^2}{2} - \frac{u_1^2}{2} + \left( \frac{p_2 - p_1}{\rho} \right) = 0 \qquad (1.65)$$

From Eq. (1.64):

$$u_1 = u_2 \left( \frac{A_2}{A_1} \right) \qquad (1.66)$$

Putting value of $u_1$ to the Eq. (1.65): $\frac{u_2^2}{2} - \left( \frac{1}{2} \right) \left( \frac{u_2 A_2}{A_1} \right)^2 + \left( \frac{p_2 - p_1}{\rho} \right) = 0$

$$\rightarrow \frac{u_2^2}{2} \left[ 1 - \left( \frac{A_2^2}{A_1^2} \right) \right] = - \left( \frac{p_2 - p_1}{\rho} \right) = \left( \frac{p_1 - p_2}{\rho} \right)$$

$$\rightarrow u_2^2 = \left[ \frac{2(p_1 - p_2)}{\rho \left\{ 1 - \left( \frac{A_2^2}{A_1^2} \right) \right\}} \right] \text{ or } u_2 = \left[ \frac{2(p_1 - p_2)}{\rho \left\{ 1 - \left( \frac{A_2^2}{A_1^2} \right) \right\}} \right]^{1/2} \qquad (1.67)$$

In practice, the minimum cross-sectional area of flow at the orifice ($A_2$) will be somewhat smaller than $A_0$, the actual area of the orifice, because of the formation of the *vena-contracta*. The fluid jet emerging from the hole is somewhat smaller than the hole itself. In highly turbulent flow this jet necks down to a minimum cross section at the vena-contracta. In order to relate the flow rate to the measurable quantity $A_0$, a correction factor ($C_D$) is introduced, which is commonly called the discharge coefficient.

Equation (1.67) becomes

$$u_2 = C_D \left[ \frac{2(p_1 - p_2)}{\rho\left\{1 - \left(\frac{A_0^2}{A_1^2}\right)\right\}} \right]^{1/2} \tag{1.68}$$

If the orifice diameter is much smaller than the pipe, i.e. $A_0^2 \ll A_1^2$, then $\left(\frac{A_0^2}{A_1^2}\right) \to 0$.

Equation (1.68) further simplifies to:

$$u_2 = C_D \left[ \frac{2(p_1 - p_2)}{\rho} \right]^{1/2} \tag{1.69}$$

The mass flow rate through the orifice can be calculated by multiplying $u_2$ (from Eq. (1.68)) by the orifice area and fluid density:

$$m' = u_2 \, A_0 \, \rho \tag{1.70}$$

Or,

$$m' = A_0 \, C_D \left[ 2 \left( p_1 - p_2 \right) \rho \right]^{1/2} \tag{1.71}$$

**Example 1.7** A 5 cm sharp edged orifice is installed in a 6 cm internal diameter water pipe to measure the flow rate of cooling water to an oxygen lance. The pressure drop across the orifice is 16 cm of Hg. Calculate the water flow rate. Given correction factor (CD) $= 0.8$, $g = 9.807$ m/s$^2$, $\rho_{H2O} = 1 \times 10^3$ kg/m$^3$ and $\rho Hg = 13.5 \times 10^3$ kg/m$^3$.

**Solution**

$A_0 = \frac{\pi d_0^2}{4}$ and $A_1 = \frac{\pi d_1^2}{4}$, therefore $\frac{A_0}{A_1} = \frac{d_0^2}{d_1^2}$ or $\left(\frac{A_0}{A_1}\right)^2 = \left(\frac{d_0^2}{d_1^2}\right)^2$.

Again, $p_1 - p_2 = 16$ cm of Hg $= 0.16 \times 13.5 \times 10^3 \times 9.807 = 21.18 \times 10^3$ kg/ms$^2$ $=$ N/m$^2$.

From Eq. (1.68): $u_2 = C_D \left[ \frac{2(p_1 - p_2)}{\rho\left\{1 - \left(\frac{A_0^2}{A_1^2}\right)\right\}} \right]^{1/2} = 0.8 \left[ \frac{2 \times 21.18 \times 10^3}{10^3 \left\{1 - \left(\frac{0.05^4}{0.06^4}\right)\right\}} \right]^{1/2} =$

**2.28 m/s.**

Again from Eq. (1.70):

water flow rate $= m' = u_2 \, A_0 \, \rho = u_2 \rho \left( \frac{\pi d_0^2}{4} \right)$

$$= 2.28 \times 10^3 \times \left( \frac{\pi 0.05^2}{4} \right) = \textbf{4.47 kg/s.}$$

### 1.3.5.2 Nozzle and Venturi Meter

Orifice meter, nozzle, and venturi meter are all measuring devices. The fluid is accelerated by causing it to flow through an obstacle. Thus, the kinetic energy is increased at the cost of the pressure energy. The flow rate is obtained by measuring the pressure difference between the points at 1, the inlet of the meter; and at point 2, a point of reduced pressure [3]. shows the venturi meter and nozzle.

Like orifice meter, same treatment can be given for nozzle and venturi meters for velocity and mass flow rate determinations Fig. 1.19.

**Example 1.8** The venturi meter shown reduced the pipe diameter from 10 to 5 cm as shown in Figure Ex.1.8. Calculate the flow rate and mass flux assuming ideal conditions.

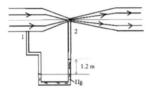

1.2 m

Hg

Fig. Ex.1.8

**Solution**

Material balance:

Overall mass in at $1$ = Overall mass out at 2.

Therefore, $A_1\rho_1 u_1 = A_2\rho_2 u_2$ (where A = area, $\rho$ = density, u = velocity)

$$\rightarrow \quad \frac{\pi d_1^2}{4}u_1 = \frac{\pi d_2^2}{4}u_2 \quad (\text{since } \rho_1 = \rho_2)$$

$$\rightarrow \quad \frac{u_1}{u_2} = \left(\frac{d_2}{d_1}\right)^2 \tag{i}$$

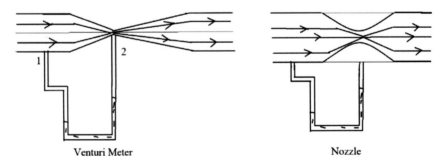

Venturi Meter                              Nozzle

**Fig. 1.19** Venturi meter and nozzle

From Bernoulli equation:

$$\left(\frac{u_2^2 - u_1^2}{2}\right) + g(z_2 - z_1) + \left(\frac{p_2 - p_1}{\rho}\right) = 0 \tag{ii}$$

For a horizontal pipe, $z_1 = z_2$ and for an incompressible fluid, $\rho_1 = \rho_2 = \rho = $ constant.

So Eq. (ii) becomes

$$\frac{u_2^2}{2} - \frac{u_1^2}{2} + \left(\frac{p_2 - p_1}{\rho}\right) = 0 \tag{iii}$$

From Eq. (i):

$$u_1 = u_2 \left(\frac{d_2}{d_1}\right)^2 \tag{iv}$$

Putting value of $u_1$ to the Eq. (iii): $\frac{u_2^2}{2} - \left(\frac{1}{2}\right)\left(\frac{u_2^2 d_2^4}{d_1^4}\right) + \left(\frac{p_2 - p_1}{\rho}\right) = 0$

$$\rightarrow \frac{u_2^2}{2}\left[1 - \left(\frac{d_2^4}{d_1^4}\right)\right] = -\left(\frac{p_2 - p_1}{\rho}\right) = \left(\frac{p_1 - p_2}{\rho}\right)$$

$$\rightarrow u_2^2 = \left[\frac{2(p_1 - p_2)}{\rho\left\{1 - \left(\frac{d_2^4}{d_1^4}\right)\right\}}\right] \text{ or } u_2 = \left[\frac{2(p_1 - p_2)}{\rho\left\{1 - \left(\frac{d_2^4}{d_1^4}\right)\right\}}\right]^{1/2} \tag{v}$$

So, $p_1 - p_2 = 1.2 \times (13.5 \times 10^3) \times 9.807 = 158.87 \times 10^3$ kg/m $\cdot$ s$^2$ = N/m$^2$

Where $\rho_{Hg} = 13.5 \times 10^3$ kg/m$^3$, g = 9.807 m/s$^2$, and $\rho_{H2O} = 10^3$ kg/m$^3$.

Therefore, From Eq. (v): $u_2 = \left[\frac{2(158.87 \times 10^3)}{10^3\left\{1 - \left(\frac{0.05^4}{0.1^4}\right)\right\}}\right]^{1/2} = \left[\frac{317.74}{0.9375}\right]^{1/2} = \textbf{18.41 m/s}$.

Hence mass flux $(m') = A_2 \rho u_2 = \frac{\pi d_2^2}{4}\rho u_2 = \left(\frac{3.14 \times 0.05^2}{4}\right) \times 10^3 \times 18.41 = $ **36.13 kg/s**.

**Example 1.9** Derive the energy equation for fluid flow of a gas following the relation $pV^n = C$, where n is not equal to one and C is constant.

**Solution**

From Bernoulli's Eq. (1.61): $\left(\frac{u_2^2 - u_1^2}{2}\right) + g(z_2 - z_1) + \left(\frac{p_2 - p_1}{\rho}\right) = 0$.

For compressible fluid, $\rho$ is constant; so

$$\left(\frac{p_2 - p_1}{\rho}\right) = \frac{\Delta p}{\rho} = \frac{\Delta p}{1/V} = V dp \tag{a}$$

[since $\rho = 1/V$ for unit mass of fluid].
Again, $pV^n = C$, therefore,

$$V = \left(\frac{C}{p}\right)^{1/n} \tag{b}$$

Putting V values in Eq. (a):

$$\left(\frac{p_2 - p_1}{\rho}\right) = V dp = \left(\frac{C}{p}\right)^{1/n} dp \tag{c}$$

Therefore by integrating,

$$\int_{p_1}^{p_2} \left(\frac{C}{p}\right)^{1/n} dp = C^{1/n} \int_{p_1}^{p_2} (p)^{-1/n} dp = C^{1/n} \left[\frac{p^{\{1-(\frac{1}{n})\}}}{\{1-(\frac{1}{n})\}}\right]_{p_1}^{p_2}$$

$$= \left[\left(\frac{n}{n-1}\right) C^{1/n} \left\{p_2^{\frac{n-1}{n}} - p_1^{\frac{n-1}{n}}\right\}\right] \tag{d}$$

Putting Eq. (d) to Eq. (1.61):

$$\left(\frac{u_2^2 - u_1^2}{2}\right) + g(z_2 - z_1) + \left[\left(\frac{n}{n-1}\right) C^{1/n} \left\{p_2^{\frac{n-1}{n}} - p_1^{\frac{n-1}{n}}\right\}\right] = 0 \tag{e}$$

Equation (e) is the energy equation for fluid flow of a gas.

### 1.3.5.3  Pilot Tube

Pilot tube measures the velocity of a fine filament of liquid, therefore, to measure velocity profile that can be used. In a pilot tube a small amount of fluid is brought to rest at an orifice situated at right angles to the direction of flow. The flow rate is obtained by measuring the difference between the impact and the static pressures [3].

The pilot tube (Fig. 1.20) consists of two concentric tubes, arranged parallel to the direction of flow. The impact pressure is measured on the open end of the inner tube. The end of the outer tube is sealed. By neglecting frictional losses, application of Bernoulli's equation:

$$\left(\frac{u_2^2 - u_1^2}{2}\right) + g(z_2 - z_1) + \left(\frac{p_2 - p_1}{\rho}\right) = 0 \tag{1.62}$$

For horizontal pipe, $z_1 = z_2$ and $u_2 = 0$ (since fluid is static at point 2).

**Fig. 1.20** Pilot tube

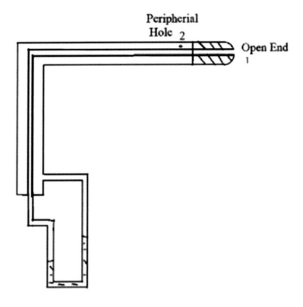

Equation (1.62) becomes $-\left(\frac{u_1^2}{2}\right) + \left(\frac{p_2-p_1}{\rho}\right) = 0$ or $\left(\frac{u_1^2}{2}\right) = \left(\frac{p_2-p_1}{\rho}\right)$.

Therefore,

$$u_1 = \sqrt{2\left(\frac{p_2 - p_1}{\rho}\right)} \qquad (1.72)$$

In general (for take care of frictional losses):

$$u_1 = C\sqrt{2\left(\frac{p_2 - p_1}{\rho}\right)} \qquad (1.73)$$

where C is constant.

#### 1.3.5.4 Rota Meter

Earlier meters, the area of obstacle or orifices is constant, and the pressure drop was expressed as a function of flow rate. However, a rota meter (Fig. 1.21) is a variable area meter. It consists of a vertical glass tube with a very small taper towards the lower end. The fluid passes upwards and flow rate is indicated by the position of the float.

Since pressure = force/area.

Therefore,

**Fig. 1.21** Rota meter

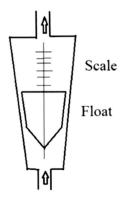

Scale

Float

$$(p_1 - p_2) = \left[ \frac{V_f (\rho_f - \rho) g}{A_f} \right] \tag{1.74}$$

where $V_f$ is the volume of float, $\rho_f$ and $\rho$ are the density of the float and fluid; $A_f$ is the maximum cross-sectional area of the float.

From Eq. (1.68):

$$u_2 = C_D \left[ \frac{2(p_1 - p_2)}{\rho \left\{ 1 - \left( \frac{A_2^2}{A_1^2} \right) \right\}} \right]^{1/2} = C_D \left[ \frac{2g V_f (\rho_f - \rho)}{\rho A_f \left\{ 1 - \left( \frac{A_2^2}{A_1^2} \right) \right\}} \right]^{1/2} \tag{1.75}$$

where $C_D$ is the discharge coefficient for the rota meter, depends on shape of the float and Reynold's number. $A_2$ is the annular area between the float and the tube. $A_1$ is the cross-sectional area of the tube on the down-stream side of the float.

The range of rota meter can be increased by using different materials (i.e. different densities) of floats.

### 1.3.6 Other Application

#### 1.3.6.1 Chimney Draft

Chimneys are provided to carry away the flue gases of the furnaces. Chimneys produce a draft, i.e. natural draft, which helps to draw in atmospheric air into the furnace to assist in the combustion process of the fuel (Fig. 1.22). The magnitude of this draft is proportional to the pressure drop caused by the chimney and can be calculated by applying the energy balance.

From Bernoulli's equation for compressible fluid:

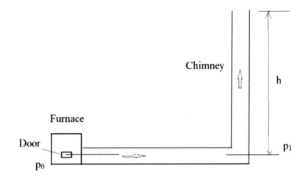

**Fig. 1.22** Furnace and chimney

$$\left(\frac{u_2^2 - u_1^2}{2}\right) + g(z_2 - z_1) + \left(\frac{p_2}{\rho_2} - \frac{p_1}{\rho_1}\right) = 0 \qquad (1.76)$$

If $u_1 = u_2$ and $z_2 - z_1 = h$ (height of chimney).

Again $p_1 = p_0$ (pressure of air at the door of the furnace) and $\rho_2 = \rho_1 = \rho_a$ (density of air at room temperature).

From Eq. (1.76): $gh + \left(\frac{p_2}{\rho_a} - \frac{p_0}{\rho_a}\right) = 0$ or $\frac{p_0}{\rho_a} = \frac{p_2}{\rho_a} + gh$.

Or

$$p_0 = p_2 + \rho_a gh \qquad (1.77)$$

Again, considering flue gases, if $u_1 = u_2$, $z_2 - z_1 = h$, and $\rho_2 = \rho_1 = \rho_g$ (density of hot flue gas at T °C).

Equation (1.76) becomes $gh + \left(\frac{p_2}{\rho_g} - \frac{p_1}{\rho_g}\right) = 0$ or $\frac{p_1}{\rho_g} = \frac{p_2}{\rho_g} + gh$.

Or

$$p_1 = p_2 + \rho_g\, gh \qquad (1.78)$$

From Eqs. (1.77) to (1.78):

$$(p_0 - p_1) = (\rho_a - \rho_g)\, gh \qquad (1.79)$$

where $(p_0 - p_1)$ is the pressure drop between the furnace and the bottom of the chimney, which generates the draft, which is also known as *buoyancy pressure*.

If $\rho_g^0$ is the density of the flue gas at room temperature, i.e. 25 °C.
It can be written as

$$\frac{\rho_{T_2}}{\rho_{T_1}} = \frac{T_1}{T_2} \tag{1.80}$$

where $T_1$ is the room temperature and $T_2$ is the hot gas temperature.
Equation (1.80) becomes $\frac{\rho_g}{\rho_g^0} = \frac{273+25}{273+T} = \frac{298}{273+T}$.
Or

$$\rho_g = \rho_g^0 \left( \frac{298}{273 + T} \right) \tag{1.81}$$

Substitute the value of $\rho_g$ and $\rho_a = \rho_g$ from Eq. (1.79):

$$(p_0 - p_1) = \Delta p = \left[ \rho_g - \rho_g^0 \left( \frac{298}{273 + T} \right) \right] gh \tag{1.82}$$

Equation (1.82) can be used to calculate the chimney draft, when all the other parameters in the equation are known.

- If the draft is small which is insufficient to draw in adequate quantity of combustion air. Hence, fan or blower have to be provided with the furnace to generate forced draft for complete combustion.

**Example 1.10** Find out the chimney draft of a 15 m height chimney. If density of air and the density of the flue gas at room temperature are 1.16 kg/m³ and 1.05 kg/m³, respectively. The temperature of hot flue gas is 327 °C.

**Solution**

Here $\rho_a = 1.16$ kg/m³, $\rho_g^0 = 1.05$ kg/m³, T = 327 °C, g = 9.81 m/s², h = 15 m.
From Eq. (1.82): $(p_0 - p_1) = \left[ 1.16 - 1.05 \left( \frac{298}{273+327} \right) \right] 9.81 \times 15 = \mathbf{94.18\,Pa}$.

### 1.3.7 Overall Mass Balance

Consider fluid flows through a container of complex shape (Fig. 1.23), and the whole container as the control volume.

**Fig. 1.23** Fluid flow
through a complex shape
container

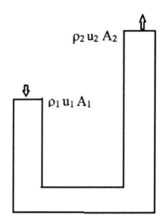

$\rho_2\, u_2\, A_2$

$\rho_1\, u_1\, A_1$

The mass balance is as follows [8]:

(Rate of total material input to the system) − (Rate of total material output from the system)
= (Accumulation of material within the system) (1.83)

If the material enters the system only through surface $A_1$ and the material leaves
only through surface $A_2$.

Thus, Eq. (1.83) can be written as

$$\rho_1 u_1\, A_1 - \rho_2\, u_2\, A_2 = \frac{dm_T}{dt} \quad (1.84)$$

where $u_1$ and $u_2$ are the mean or average velocities of the material perpendicular to
the surfaces $A_1$ and $A_2$, respectively; $m_T$ is the total mass within the system at any
given time.

If consider same material coming in and coming out, then $\rho_1 = \rho_2 = \rho$ (i.e. density
of material).

So Eq. (1.84) become

$$\rho\, u_1\, A_1 - \rho\, u_2\, A_2 = \frac{dm_T}{dt} \quad (1.85)$$

$$\text{Since the total mass flow} = m' = \rho\, u\, A \quad (1.86)$$

Therefore,

$$m'_1 = \rho\, u_1\, A_1 \text{ and } m'_2 = \rho\, u_2\, A_2 \quad (1.87)$$

Thus Eq. (1.85) may be written as

$$\frac{dm_T}{dt} = m_1' - m_2' = -\Delta m' \tag{1.88}$$

For steady-state condition, $\frac{dm_T}{dt} = 0$.
Therefore, the overall mass balance reduces to

$$\Delta m' = 0 \tag{1.89}$$

For transient situation, Eq. (1.88) is integrated as follows:

$$\int_0^t dm_T = -\int_0^t \Delta m' dt \tag{1.90}$$

Therefore,

$$m_{T,t} = m_{T,0} - \int_0^t \Delta m' dt \tag{1.91}$$

where $m_{T,0}$ is the total mass in the control volume at time, $t = 0$; and $m_{T,t}$ is the total mass in the control volume at time, $t$.

**Example 1.11** Molten steel is discharge from a ladle, through a 7.5 cm diameter nozzle located at the bottom of the ladle. Calculate the initial velocity and the time required for emptying the ladle.

The linear velocity in the nozzle may be expressed by $u_0 = (2gh)^{1/2}$, where h is the liquid steel height above the nozzle (Given: Diameter of ladle = 3.0 m, initial depth of liquid steel = 3.4 m, density of liquid steel = $7.6 \times 10^3$ kg/m³, g = 9.81 m/s²).

**Solution**

Since there is no inflow, i.e. $m_1' = 0$
Therefore, Eq. (1.87) may be written as

$$\frac{dm_T}{dt} = -m_2' \tag{a}$$

Again from Eq. (1.86):

$$m_2' = \rho u_0 A_2 = \rho A_2 (2gh)^{1/2} \quad \text{(since } u_0 = (2gh)^{1/2} \tag{b}$$

Therefore, $-\frac{dm_T}{dt} = \rho A_2 (2gh)^{1/2} = \rho A_2 \left(\frac{2gm_T}{A\rho}\right)^{1/2}$ [since $m_T = A\rho h$, so h = $(m_T / A\rho)$].
Therefore,

$$-\frac{dm_T}{m_T^{1/2}} = A_2 \left(\frac{2g\rho}{A}\right)^{1/2} dt \tag{c}$$

Now integrate Eq. (c):

$$-\int_{m_{T,0}}^{m_{T,t}} m_T^{-1/2} dm_T = A_2 \left(\frac{2g\rho}{A}\right)^{1/2} \int_0^t dt \tag{d}$$

$$\rightarrow \left(m_{T,t}^{1/2} - m_{T,0}^{1/2}\right) = -\left(\frac{A_2}{2}\right)\left(\frac{2g\rho}{A}\right)^{1/2} t \tag{e}$$

$\rightarrow$ Since at time, t: $m_{T,t} = 0$ (lade will be empty)

$\rightarrow$ So, $m_{T,0}^{1/2} = \left(\frac{A_2}{2}\right)\left(\frac{2g\rho}{A}\right)^{1/2} t$

$\rightarrow$ Therefore, $t = \left(\frac{2}{A_2}\right)\left(\frac{Am_{T,0}}{2g\rho}\right)^{1/2}$ \hfill (f)

Now $A_2 \frac{\pi d_2^2}{4} = \left[\frac{\pi (0.075)^2}{4}\right] = 4.42 \times 10^{-3} m^2$.

and $A = \frac{\pi (d)^2}{4} = \left[\frac{\pi (3)^2}{4}\right] = 7.07 \, m^2$.

Now $m_{T,0} = A\rho h = 7.07 \times (7.6 \times 10^3) \times 3.4 = 182.69 \times 10^3 \, kg = 182.69$ tonne.

Therefore, from Eq. (f): $t = \left(\frac{2}{4.42 \times 10^{-3}}\right) \times \left(\frac{7.07 \times 182.69 \times 10^3}{2 \times 9.81 \times 7.6 \times 10^3}\right)^{1/2} = 1.3317 \times 10^3 = 1331.75$ s.

Hence, time requires to empty the ladle = **22.2 min**.

Initial velocity of liquid steel, $u_0 = (2gh)^{1/2} = (2 \times 9.81 \times 3.4)^{1/2} = $ **7.53 m/s**.

**Example 1.12** The liquid steel is teemed from a ladle through a nozzle of 6 cm diameter at its bottom. If the volume of the liquid steel in the initial stage is 2.5 m³ and the linear velocity of discharge is 100 cm/s. Calculate the time required to empty the ladle.

**Solution**

According to the law of conservation of mass to inlet and outlet stream leads as follows:

[(Amount of liquid entering ladle per unit time)

− (Amount of liquid leaving ladle per unit time)]

= (Amount of liquid retained in ladle per unit time) \hfill (1)

Amount of liquid entering ladle per unit time = $\rho_1 \cdot u_1 \cdot A_1$ \hfill (2)

$$\text{Amount of liquid leaving ladle per unit time} = \rho_2 \cdot u_2 \cdot A_2 \tag{3}$$

where $\rho$, $u$, $A$ are the density, velocity of the liquid, and area of the ladle respectively.

$$\text{Amount of liquid retained in ladle per unit time} = \rho \frac{\delta V}{\delta t} \tag{4}$$

where $V$ is the volume of the ladle and $\frac{\delta V}{\delta t}$ is the rate of change of volume of liquid with time at constant density.

Now substituting Eqs. (2)–(4) into Eq. (1):

$$\rho_1.u_1.A_1 - \rho_2.u_2.A_2 = \rho \frac{\delta V}{\delta t} \tag{5}$$

Since there is no liquid is entering to the ladle, so $\rho_1 . u_1 . A_1 = 0$.
The density of liquid is constant, i.e. $\rho_2 = \rho$, and Eq. (5) becomes

$$-\rho \cdot u_2 \cdot A_2 = \rho \frac{\delta V}{\delta t} \quad \text{or} \quad dV = -u_2 \cdot A_2 \cdot dt \tag{6}$$

Integrating Eq. (6) between $t = 0$ and $t = t$, where $V = V$ and $V_t = 0$, respectively.

$$\int_{V}^{V_t} dV = -u_2.A_2 \int_{0}^{t} dt \tag{7}$$

Therefore, $V = u_2 \cdot A_2 \cdot t$ or

$$t = \frac{V}{u_2 A_2} \tag{8}$$

Here $V = 2.5 \text{ m}^3$, $u_2 = 100 \text{ cm/s} = 1.0 \text{ m/s}$, $A = \frac{\pi(d)^2}{4} = \left[\frac{\pi(0.06)^2}{4}\right] = 2.83 \times 10^{-3}\text{m}^2$.

From Eq. (8): $t = \left\{\frac{2.5}{1.0 \times 2.83 \times 10^{-3}}\right\} = 883.39 \text{ s} = \textbf{14.7 mins}$.

**Example 1.13** A water pump has one inlet and two outlets as shown in Figure Ex. 1.13, all are at the same level. What pump power is required if the pump is 85% efficient?

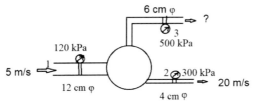

Fig. Ex.1.13

## Solution

Mass balance: Water in = Water out.

Therefore, $u_1 \, \rho \, A_1 = u_2 \, \rho \, A_2 + u_3 \, \rho \, A_3$ (where A = area, $\rho$ = density, u = velocity)

$$\rightarrow \quad u_1 \frac{\pi d_1^2}{4} = u_2 \frac{\pi d_2^2}{4} + u_3 \frac{\pi d_3^2}{4}$$

Putting values to the above equation: $5 \times \left\{ \frac{3.14 \times (0.12)^2}{4} \right\} = 20 \times \left\{ \frac{3.14 x \times (0.04)^2}{4} \right\} +$

$u_3 \times \left\{ \frac{3.14 \times (0.06)^2}{4} \right\}.$

Therefore, $u_3 = \frac{0.0565 - 0.0251}{2.826 x 10^{-3}} = \mathbf{11.11 \, m/s}.$

Energy balance:

K.E. at 1 + Work done by the surrounding to the system = K.E. at 2 + K.E. at 3 + Work done by the system to the surrounding at 2 and 3 + Pump work

$$\rightarrow \quad m_1' \left( \frac{u_1^2}{2} \right) + \left( \frac{m_1'}{\rho} \right) p_1 = m_2' \left( \frac{u_2^2}{2} \right) + m_3' \left( \frac{u_3^2}{2} \right) + \left( \frac{m_2'}{\rho} \right) p_2 + \left( \frac{m_3'}{\rho} \right) p_3 + w_s \tag{1}$$

$$\rightarrow \quad m_1' \left( \frac{u_1^2}{2} + \frac{p_1}{\rho} \right) = m_2' \left( \frac{u_2^2}{2} + \frac{p_2}{\rho} \right) + m_3' \left( \frac{u_3^2}{2} + \frac{p_3}{\rho} \right) + w_s \tag{2}$$

Again, $m' = u.A.\rho.$

So,

$$m_1' = u_1 \cdot A_1 \cdot \rho = u_1 \frac{\pi d_1^2}{4} \rho = \left[ 5 \times \left\{ \frac{3.14 \times (0.12)^2}{4} \right\} \times 1000 \right] = 56.52 \text{ kg/s}$$

$$m_2' = u_2 \cdot A_2 \cdot \rho = u_2 \frac{\pi d_2^2}{4} \rho = \left[ 20 \times \left\{ \frac{3.14 \times (0.4)^2}{4} \right\} \times 1000 \right] = 25.12 \text{ kg/s}$$

$$m_3' = u_3 \cdot A_3 \cdot \rho = u_3 \frac{\pi d_3^2}{4} \rho = \left[ 11.11 \times \left\{ \frac{3.14 \times (0.06)^2}{4} \right\} \times 1000 \right] = 31.40 \text{ kg/s}$$

From Eq. (2): $56.52 \times \frac{5^2}{2} + \frac{120 \times 10^3}{1000} = 25.12 \times \left(\frac{20^2}{2} + \frac{300 \times 10^3}{1000}\right) + 31.4 \times$
$\left(\frac{11.11^2}{2} + \frac{500 \times 10^3}{1000}\right)$.

Therefore, $w_s = 7488.9 - (12560.0 + 17637.88) = -22705.01$ W.

Hence, pump power $= (-w_s/0.85) = (22705.01/0.85) = \mathbf{26.71\ kW}$.

**Example 1.14** Determine the power output of the turbine as shown in Figure Ex.1.14 for a water flow rate of 0.6 m$^3$/s. (Given: $\rho_{Hg} = 13.5 \times 10^3$ kg/m$^3$, $\rho_{H2O} = 1000$ kg/m$^3$ and $g = 9.807$ m/s$^2$).

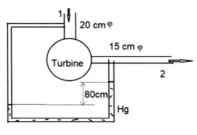

Fig. Ex.1.14

**Solution**

Flow rate $(Q) = u_1.A_1$,
   So

$$u_1 = \frac{Q}{A_1} = \frac{Q}{\frac{\pi d_1^2}{4}} = \frac{0.6}{\frac{3.14 \times 0.2^2}{4}} = 19.11 \text{ m/s} \tag{1}$$

K.E. at 1 + Work done by the surrounding to the system at 1 = K.E. at 2 + Work done by the system to the surrounding at 2 + Pump work.
   Or

$$m_1'\left(\frac{u_1^2}{2}\right) + \left(\frac{m_1'}{\rho}\right)p_1 = m_2'\left(\frac{u_2^2}{2}\right) + \left(\frac{m_2'}{\rho}\right)p_2 + w_s \tag{2}$$

$$\rightarrow \quad -w_s^\wedge = \frac{u_2^2 - u_1^2}{2} + \frac{p_2 - p_1}{\rho} \tag{3}$$

Since water in = water out or $u_1.A_1. \rho = u_2.A_2. \rho$

Therefore, $u_2 = u_1.\frac{A_1}{A_2} = u_1.\frac{d_1^2}{d_2^2} = 19.11 \times \left(\frac{0.2}{0.15}\right)^2 = 33.97$ m/s.

From Eq. (3): $-w_s^\wedge = \frac{33.97^2 - 19.11^2}{2} + \frac{0.8 \times 13.5 \times 10^3 \times 9.807}{1000} = 394.38 + 105.92 =$
500.3 J/kg.

Therefore, $-w_s = [m'(-w_s^\wedge)] = [(Q.\rho)(-w_s^\wedge)] = [0.6 \times 1000 \times 500.3] = \mathbf{300.18\ kW}$.

## 1.4 Exercises

**Problem 1.1**: (a) To get about 25 m$^3$/min of fresh air by installing an exhaust fan, what will be the power of exhaust fan? Also calculate the minimum amount of energy that the fan will deliver to the air. (b) If the pressure difference is 90 Nm$^{-2}$, what will be the power of exhaust fan? If diameter of exhaust fan is 0.5 m and efficiency of fan is 50%. (Given density of air $= 1.16$ kg/m$^3$ at 27 °C.)

[Ans: 2.2 w]

**Problem 1.2**: If the diameter of pipe is 50 mm and water supply to cool the mold at the rate of 1000 L/min. Find out the power of pump. Given: the efficiency of water pump is 50% and density of water is 1000 kg/m$^3$.

[Ans: 1.2 kw]

**Problem 1.3**: A water tank at the ground level, height of the tank is 2.5 m With a pump, water is lifted to another tank which is situated at the roof of the house at the rate of 350 L/min by 50 mm diameter pipe. Height of the house is 15 m and roof tank height is 2 m. Find out the rating of pump (if efficiency is 60%), and neglect the pressure difference.

[Ans: 1425 w]

**Problem 1.4**: There is a water tank at the ground level, tank height is 1m With the help of a pump, water can be lifted to a tank which is situated at the roof of the house at the rate of 300 l/min by 5 cm diameter pipe. Height of the house is 10 m and roof tank height is 1.5 m. Find out the rating of pump (if efficiency is 40%), and neglect the pressure difference.

[Ans: 1328 w]

**Problem 1.5**: A 9 cm sharp edged orifice is installed in a 10 cm internal diameter water pipe to measure the flow rate of cooling water to an oxygen lance. The pressure drop across the orifice is 20 cm of Hg. Calculate the water flow rate. Given correction factor $(C_D) = 0.8$, g $= 9.807$ m/s$^2$, $\rho_{H2O} = 1 \times 10^3$ kg/m$^3$ and $\rho_{Hg} = 13.5 \times 10^3$ kg/m$^3$.

[Ans: 68.37 kg/s]

**Problem 1.6**: Find out the chimney draft of a 10 m height chimney. If the density of air and the density of the flue gas at room temperature are 1.16 kg/m$^3$ and 0.95 kg/m$^3$, respectively. The temperature of hot flue gas is 357 °C.

[Ans: 69.71 Pa]

**Problem 1.7**: The liquid steel is teemed from a ladle through a nozzle of 5 cm diameter at its bottom. If the diameter and height of the ladle are 2 m and 1m, respectively. The linear velocity of discharge is 100 cm/s. Calculate the time required to empty the ladle.

[Ans: 26.7 mins]

## 1.5   Questions

Q1. What do you understand by transport phenomena?

Q2. What do you mean by fluid dynamics? What are the main aims of fluid dynamics?

Q3. What is the importance of transport phenomena in metallurgical fluid dynamics?

Q4. What are the characterizations of fluid motion?

Q5. What do you understand by Newtonian and Non-Newtonian fluids?

Q6. What do you understand by ideal and non-ideal fluids?

Q7. What are the basic quantities of dimensions?

Q8 What do you understand by dimensionless numbers?

Q9. What do you understand by velocity boundary layer?

Q10. What do you understand by buoyancy force?

Q11. Compared between the following:

(a) Laminar and turbulent, (b) Newtonian and Non-Newtonian, (c) Viscous and non-viscous, (d) Steady and unsteady, (e) Compressible and incompressible.

Q12. What do you understand by Reynolds number? Give the relationship and significant of it.

Q13. What is the function of Orifice meter?

Q14. Derive equation for Newton's law of viscosity.

Q15. Derive equation for pressure variation in a static isothermal gas.

Q16. Derive Bernoulli equation from the overall energy balance.

Q17. Determine the velocity of fluid at the orifice in Orifice meter.

## References

1. Bird RB, Stewart WE, Lightfoot EN (2006), Transport phenomena, 2nd edn. Wiley (Asia) Pvt Ltd, Singapore
2. Bennett CO, Myers JE (1983) Momentum, heat and mass transfer, 3rd edn. McGraw-Hill Book Company, Singapore
3. Mohanty AK (2012) Rate processes in metallurgy. PHI Learning Pvt Ltd., New Delhi
4. Ghosh A, Ray HS (1991) Principle of extractive metallurgy. Wiley Eastern, New Delhi
5. Nurni VN, Ballal BN (2014) Chap 4.1: Rate phenomena in process metallurgy. In: Treatise on process metallurgy, vol 1. Elsevier Ltd.
6. Incropera FP, Dewitt DP (2006) Fundamentals of heat and mass transfer, 5th edn. Wiley (Asia) Pvt Ltd., Singapore
7. Streeter VL, Wylie EB. Fluid mechanics, 1st SI Metric edn., McGraw Book Co, Auckland
8. Szekely J, Themalis NJ (1971) Rate phenomena in process metallurgy. Wiley Interscience, New York

# Chapter 2
# Mass and Momentum Balances

Mass and momentum balance are the basic of fluid flow. Differential mass balance (continuity equation), and differential momentum balance or equation of motion are derived. Application of continuity and Navier–Stokes equations are described. Mechanism of momentum transport, dimensional analysis, friction factor, steady flow around a submerged solid sphere, fluid flow in circular pipe, fluid flow through a packed bed of solids, and force balance for packed bed of solids are also discussed in detail in this chapter.

## 2.1 Introduction

These balances may be considered as basic in the formulation of various aspects of fluid flow. In general, the expressions for these balances which may be considered microscopic in nature are derived by considering a small volume element fixed in shape in the flowing fluid. This volume element is usually known as *control volume*. The change in any variable in this control volume can be either due to its interaction with the surroundings or due to some reaction inside the control volume or due to change in the value of this variable in the control volume [1].

In this section only non-reacting systems, i.e. those in which there is no change because of chemical reactions, will be considered for the sake of simplicity. The balances have been worked out by considering control volume in the *rectangular coordinate* system or *cartesian coordinate* (x, y, z). There are other coordinate systems that are also available, i.e. *cylindrical or polar coordinate* (r, θ, z) and *spherical coordinate* (r, θ, φ).

© The Author(s), under exclusive license to Springer Nature Singapore Pte Ltd. 2023     47
S. K. Dutta, *Fundamental of Transport Phenomena and Metallurgical Process Modeling*,
https://doi.org/10.1007/978-981-19-2156-8_2

## 2.2  Different Types of Derivatives

### 2.2.1  Simple Derivative $\left(\frac{df}{dt}\right)$

$$\frac{df}{dt} = \lim_{\Delta x \to 0}\left[\frac{f(x + \Delta x) - f(x)}{\Delta x}\right] \tag{2.1}$$

Apart from that several different time derivatives may be encountered in transport phenomena. This can be illustrated with an example, to observe the concentration of fish in the river. Because fishes swim around the river, the concentration of fish is a function of position (x, y, z) and time (t).

### 2.2.2  Partial Time Derivative $\left(\frac{\delta}{\delta t}\right)$

Suppose observer stands on a bridge and observing the concentration of fish just below the bridge as a function of time. Observer records the time rate of change of fish's concentration at a fixed location.

The result is $\left(\frac{\delta C}{\delta t}\right)_{x, y, z}$, the partial derivative of concentration (C) with respect to time (t) at constant x, y, and z.

### 2.2.3  Total Time Derivative $\left(\frac{d}{dt}\right)$

Now suppose observer jumps into a speed boat and moves around on the river, sometimes going up-stream, sometimes down-stream and sometimes across the current of river. All the time observer is observing the fish's concentration. At any instant, the time rate of change of the observed fish's concentration is given as

$$\frac{dC}{dt} = \left(\frac{\delta C}{\delta t}\right)_{x, y, z} + \frac{dx}{dt}\left(\frac{\delta C}{\delta x}\right)_{y, z, t} + \frac{dy}{dt}\left(\frac{\delta C}{\delta y}\right)_{z, x, t} + \frac{dz}{dt}\left(\frac{\delta C}{\delta z}\right)_{x, y, t} \tag{2.2}$$

where $\frac{dx}{dt}$, $\frac{dy}{dt}$ and $\frac{dz}{dt}$ are the velocity of the boat in x, y, and z direction.

### 2.2.4  Substantial Time Derivative $\left(\frac{D}{Dt}\right)$

Next observer sits on a small boat, and he is not feeling energetic for boating. He just floats along with the current of river and observing fish's concentration. Here, the velocity of the observer's boat is the same as the velocity of the river, which has components $u_x$, $u_y$ and $u_z$. At any instant, the time rate of change of fish's concentration can be given as

$$\frac{DC}{Dt} = \frac{\delta C}{\delta t} + u_x\left(\frac{\delta C}{\delta x}\right) + u_y\left(\frac{\delta C}{\delta y}\right) + u_z\left(\frac{\delta C}{\delta z}\right) = \frac{\delta C}{\delta t} + u.\nabla C \qquad (2.3)$$

Meaning of substantial derivative is the change of time when reporting as one moves along with the substance.

## 2.3 Differential Mass Balance (Continuity Equation)

The *continuity equation* is the mathematical expression for the law of conservation of mass for flowing fluids. This is universally applicable to all types of fluid flow, such as laminar or turbulent with only slight variation, as long as the fluid forms a continuum.

To derive the equation, considering a differential volume element as shown in Fig. 2.1, in which fluid enters from the faces touching the point P (x, y, z) and the fluid leaves from the faces touching the point Q (x + δx, y + δy, z + δz).

According to the law of conservation of mass, it can be written the following general relation for a non-reacting fluid [2]:

[(Rate of mass of fluid entering the volume element)—(Rate of mass of fluid leaving the volume element)]

$$= \text{(Rate of mass of fluid accumulation in the volume element).} \qquad (2.4)$$

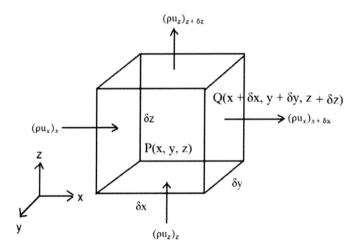

**Fig. 2.1** Differential volume element

Rate of material flows (kg/s) through any one of the faces of cube (i.e. volume element) is represented by the product [(density of fluid) x (velocity perpendicular to the face) × (cross-sectional area)].

Rate of mass of fluid entering the volume element

$$= (\rho u_x)_x \cdot \delta y \delta z + \left(\rho u_y\right)_y \cdot \delta x \delta z + \left(\rho u_z\right)_z \cdot \delta x \delta y \qquad (2.5)$$

where $\rho$ is the density of fluid, and $u_x$, $u_y$, $u_z$ are the components of the fluid velocity in x, y, and z directions, respectively.

Similarly, rate of mass of fluid leaving the volume element

$$= (\rho u_x)_{x+\delta x} \cdot \delta y \delta z + \left(\rho u_y\right)_{y+\delta y} \cdot \delta x \delta z + (\rho u_z)_{z+\delta z} \cdot \delta x \delta y \qquad (2.6)$$

Rate of mass of fluid accumulation in the volume element

$$= [(\text{volume of element}) \times (\text{rate of change of density}) = (\delta x \delta y \delta z)\left(\frac{\delta \rho}{\delta t}\right) \qquad (2.7)$$

Now substitute of Eqs. (2.5–2.7) in Eq. (2.4) and divided both sides by ($\delta x$ $\delta y$ $\delta z$): Therefore,

$$-\left[\left\{\frac{(\rho u_x)_{x+\delta x} - (\rho u_x)_x}{\delta x}\right\} + \left\{\frac{(\rho u_y)_{y+\delta y} - (\rho u_y)_y}{\delta y}\right\} + \left\{\frac{(\rho u_z)_{z+\delta z} - (\rho u_z)_z}{\delta z}\right\}\right] = \left(\frac{\delta \rho}{\delta t}\right) \quad (2.8)$$

Since the dimensions of the volume element are infinitely small, and by taking the limit as these dimensions approach zero, Eq. (2.8) can be expressed in the form of the partial differential equation:

$$\left(\frac{\delta \rho}{\delta t}\right) = -\left[\left\{\frac{\delta(\rho u_x)}{\delta x}\right\} + \left\{\frac{\delta(\rho u_y)}{\delta y}\right\} + \left\{\frac{\delta(\rho u_z)}{\delta z}\right\}\right] \qquad (2.9)$$

This Eq. (2.9) is the *equation of continuity (for cartesian coordinates)*, which describes time rate of change of fluid density at a fixed point resulting from the changes in the mass velocity vector, $\rho u$.

Equation (2.9) can be written more conveniently in vector-tensor notation:

$$\left(\frac{\delta \rho}{\delta t}\right) = -(\nabla \rho u) \qquad (2.10)$$

where $\nabla$ has the dimensions of reciprocal length.

The equation of continuity for turbulent flow is like that for laminar flow. This Eq. (2.9) is the general equation of continuity in three dimensions and is applicable to any type of flow and for any fluid whether compressible or incompressible.

The equation of continuity for *cylindrical coordinates* and *spherical coordinates* are shown in *Appendix II*.

## 2.3.1 Continuity Equation for Steady State

The steady state is defined as the state in which the density of the fluid does not change with time. In fact, the general definition of steady state includes the time independence of all intensive properties of the system.

So, according to steady state:

$$\left(\frac{\delta \rho}{\delta t}\right) = 0 \tag{2.11}$$

Equation (2.9) becomes

$$\left[\left\{\frac{\delta(\rho u_x)}{\delta x}\right\} + \left\{\frac{\delta(\rho u_y)}{\delta y}\right\} + \left\{\frac{\delta(\rho u_z)}{\delta z}\right\}\right] = 0 \tag{2.12}$$

Equation (2.12) is the continuity equation for steady state.

Or in vector-tensor notation Eq. (2.10) becomes

$$(\nabla \rho u) = 0 \tag{2.13}$$

(a) For steady flow $\left(\frac{\delta \rho}{\delta t} = 0\right)$ and for incompressible fluids (i.e. $\rho = $ constant), continuity Eq. (2.12) reduces to

$$\left\{\frac{\delta u_x}{\delta x} + \frac{\delta u_y}{\delta y} + \frac{\delta u_z}{\delta z}\right\} = 0 \tag{2.14}$$

(b) For two-dimensional flow ($u_z = 0$), Eq. (2.14) becomes

$$\frac{\delta u_x}{\delta x} + \frac{\delta u_y}{\delta y} = 0 \tag{2.15}$$

(c) For one-dimensional flow ($u_y = 0$ and $u_z = 0$), Eq. (2.14) becomes

$$\frac{\delta u_x}{\delta x} = 0 \tag{2.16}$$

### 2.3.2   Continuity Equation for Incompressible Fluid

A fluid is said to be incompressible, if the density of the same chunk of fluid in a flowing system does not change with time in spite of change in external pressure [1].

In general, the density of a flowing fluid can be assumed to be a function of time and position, i.e.

$$\rho = f(t, x, y, z) \tag{2.17}$$

In the differential form:

$$d\rho = \frac{\delta\rho}{\delta t}dt + \frac{\delta\rho}{\delta x}dx + \frac{\delta\rho}{\delta y}dy + \frac{\delta\rho}{\delta z}dz \tag{2.18}$$

Therefore,

$$\frac{d\rho}{dt} = \frac{\delta\rho}{\delta t} + \frac{\delta\rho}{\delta x}\frac{dx}{dt} + \frac{\delta\rho}{\delta y}\frac{dy}{dt} + \frac{\delta\rho}{\delta z}\frac{dz}{dt} \tag{2.19}$$

Now substitution of $\left(\frac{\delta\rho}{\delta t}\right)$ from Eq. (2.9):

$$
\begin{aligned}
\frac{d\rho}{dt} &= -\left[\left\{\frac{\delta(\rho u_x)}{\delta x}\right\} + \left\{\frac{\delta(\rho u_y)}{\delta y}\right\} + \left\{\frac{\delta(\rho u_z)}{\delta z}\right\}\right] + \left\{\frac{\delta\rho}{\delta x}\frac{dx}{dt} + \frac{\delta\rho}{\delta y}\frac{dy}{dt} + \frac{\delta\rho}{\delta z}\frac{dz}{dt}\right\} \\
&= -\left[\left(u_x\frac{\delta\rho}{\delta x} + \rho\frac{\delta u_x}{\delta x}\right) + \left(u_y\frac{\delta\rho}{\delta y} + \rho\frac{\delta u_y}{\delta y}\right) + \left(u_z\frac{\delta\rho}{\delta z} + \rho\frac{\delta u_z}{\delta z}\right)\right] \\
&\quad + \left\{u_{x0}\frac{\delta\rho}{\delta x} + u_{y0}\frac{\delta\rho}{\delta y} + u_{z0}\frac{\delta\rho}{\delta z}\right\}
\end{aligned}
$$

where $u_{x0} = \frac{dx}{dt}$, $u_{y0} = \frac{dy}{dt}$, and $u_{z0} = \frac{dz}{dt}$.

Therefore,

$$\frac{d\rho}{dt} = -\left[(u_x - u_{x0})\frac{\delta\rho}{\delta x} + (u_y - u_{y0})\frac{\delta\rho}{\delta y} + (u_z - u_{z0})\frac{\delta\rho}{\delta z}\right] - \rho\left[\frac{\delta u_x}{\delta x} + \frac{\delta u_y}{\delta y} + \frac{\delta u_z}{\delta z}\right] \tag{2.20}$$

Equation (2.20) is helpful in calculating the change in density with time in a volume element supposed to be moving with a velocity having $u_{x0}$, $u_{y0}$, and $u_{z0}$ as its components in the x, y, and z directions, respectively.

If the volume element has the same velocity components as the flowing fluid, i.e. $u_x = u_{x0}$, $u_y = u_{y0}$ and $u_z = u_{z0}$.

Then Eq. (2.20) becomes

$$\frac{d\rho}{dt} = -\rho\left[\frac{\delta u_x}{\delta x} + \frac{\delta u_y}{\delta y} + \frac{\delta u_z}{\delta z}\right] \tag{2.21}$$

Now $\frac{d\rho}{dt}$ has been replaced by $\frac{D\rho}{Dt}$ and is called the *substantial derivative of density*. Then Eq. (2.21) becomes

$$\frac{D\rho}{Dt} = -\rho\left[\frac{\delta u_x}{\delta x} + \frac{\delta u_y}{\delta y} + \frac{\delta u_z}{\delta z}\right] \tag{2.22}$$

[This derivative $\left(\frac{D\rho}{Dt}\right)$ is a special kind of time derivative or more logically the *derivative following the motion* [2]. It is related to the partial time derivative as follows:

$$\frac{D\rho}{Dt} = \frac{d\rho}{dt} + \left\{u_{x0}\frac{d\rho}{dx} + u_{y0}\frac{d\rho}{dy} + u_{z0}\frac{d\rho}{dz}\right\} \tag{2.23}$$

It should be remembered that $\frac{d\rho}{dt}$ is the derivative at a fixed point in space and $\frac{D\rho}{Dt}$ is a derivative computed by an observer floating down-stream along with the fluid.]

Equation (2.22) can be expressed in the following alternative form in vector-tensor notation:

$$\frac{D\rho}{Dt} = -\rho\nabla u \quad \text{or} \quad \frac{D\rho}{Dt} + \rho\nabla u = 0 \tag{2.24}$$

According to the definition of an incompressible fluid, which requires to concentrate on the same chunk of fluid:

$$\frac{D\rho}{Dt} = 0 \tag{2.25}$$

This relation on substitution in Eq. (2.22) leads to

$$\left[\frac{\delta u_x}{\delta x} + \frac{\delta u_y}{\delta y} + \frac{\delta u_z}{\delta z}\right] = 0 \tag{2.26}$$

Or in vector-tensor notation form, for an incompressible fluid:

$$\nabla u = 0 \tag{2.27}$$

**Example 2.1** The velocity distribution for two-dimensional flow of an incompressible fluid is given by: $u_x = 5 - x$ and $u_y = 4 + y$. Show that the requirements of the continuity equation is satisfied.

**Solution**

For two-dimensional flow of an incompressible fluid, the continuity Eq. (2.26) is modified as $\frac{\delta u_x}{\delta x} + \frac{\delta u_y}{\delta y} = 0$.

Given: $u_x = 5 - x$, therefore $\frac{\delta u_x}{\delta x} = -1$; and $u_y = 4 + y$, therefore $\frac{\delta u_y}{\delta y} = 1$.

Therefore, $\frac{\delta u_x}{\delta x} + \frac{\delta u_y}{\delta y} = -1 + 1 = 0$.

This satisfies the requirement of the continuity equation.


## 2.4  Differential Momentum Balance or Equation of Motion

The concept of a differential momentum balance on a control volume, in a flowing fluid, is based on Newton's second law of motion which states that the rate of change of momentum is equal to the force acting on the system [1].

Considering the control volume in cartesian (i.e. rectangular) coordinates shown in Fig. 2.2. According to Newton's second law of motion, for momentum balance on the volume element:

$$\{(\text{Rate of momentum input}) - (\text{Rate of momentum output})\}$$

$$= \begin{cases} (\text{Rate of change of momentum on the control volume}) \\ -(\text{Sum of net forces acting on the control volume}) \end{cases} \quad (2.28)$$

As both momentum (i.e. mass x velocity) and force (i.e. mass x acceleration) are vector, the above balance is to be considered separately for each of the three directions which will ultimately lead to three momentum balance equations. A relation for momentum balance for its component in the x direction will be derived for cartesian coordinates and the relations for momentum components in the y and z directions will be written considering the symmetry properties.

$$\text{The } x \text{ component of momentum for unit volume} = \rho u_x \quad (2.29)$$

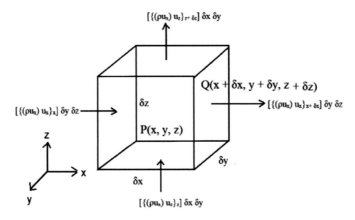

**Fig. 2.2** Control volume

where $\rho$ is the fluid density and $u_x$ is the velocity of fluid in the x direction.

This x component of momentum enters the control volume (as shown in Fig. 2.2) from all three faces (adjacent to point P) through which the fluid enters and the fluid leaves this control volume from the rest of the faces (adjacent to point Q). Hence, the amount of x component of momentum entering the control volume will be expressed by the relation:

$$\left( \begin{array}{l} \text{x component of momentum accompanying the} \\ \text{fluid entering the control volume per unit time} \end{array} \right)$$

$$= \left[ \{ (\rho u_x) u_x \}_x \right] \delta y \delta z + \left[ \{ (\rho u_x) u_y \}_y \right] \delta x \delta z + \left[ \{ (\rho u_x) u_z \}_z \right] \delta x \delta y \qquad (2.30)$$

(x component of momentum accompanying the fluid leaving the control volume per unit time)

$$= \left[ \{ (\rho u_x) u_x \}_{x+\delta x} \right] \delta y \delta z + \left[ \{ (\rho u_x) u_y \}_{y+\delta y} \right] \delta x \delta z + \left[ \{ (\rho u_x) u_z \}_{z+\delta z} \right] \delta x \delta y$$

$$(2.31)$$

$$(\text{Rate of change of x component of momentum in the control volume}) = \left\{ \frac{\delta (\rho u_x)}{\delta t} \right\} \delta x \delta y \delta z \quad (2.32)$$

The total forces acting in the x direction $(F_x)$ on the volume element can be classified broadly into two categories: (i) bulk forces and (ii) superficial forces.

(i)    Bulk forces: Bulk forces, which may be defined as those acting on the whole volume of the element, can be further subdivided into two categories: (a) gravitational forces and (b) those due to the pressure of the matter inside the volume element.

(ia)   For gravitational force in the x direction $(F_g)$ can be written as

$$F_g = \rho g_x \delta x \delta y \delta z \qquad (2.33)$$

where $g_x$ is the acceleration due to gravity in the x direction.

(ib)   The force due to pressure $(F_P)$ is equal to the difference in pressure on two faces of the element whose normal lies in the x direction, that can be written as

$$F_P = \{ (p_x) \delta y \delta z - (p_{x+\delta x}) \delta y \delta z \} = -\left( \frac{\delta p}{\delta x} \right) \delta x \delta y \delta z \qquad (2.34)$$

(ii)   Superficial forces: The superficial forces can also be classified into two categories: (a) normal and (b) tangential stresses.

(iia)  The net force $(F_{xx})$, acting due to normal component of stress on the yz plane, is expressed as

$$F_{xx} = \{ (\tau_{xx})_x \} \delta y \delta z - \{ (\tau_{xx})_{x+\delta x} \} \delta y \delta z = -\left( \frac{\delta \tau_{xx}}{\delta x} \right) \delta x \delta y \delta z \qquad (2.35)$$

where $\tau_{xx}$ is the normal stress per unit area in the x direction.

In symbol $F_{xx} \rightarrow$ the first subscript denotes the direction of the normal to the plane on which the stress is acting. The second subscript denotes the direction in which the stress acts.

(iib)   Similarly, the net force ($F_{yx}$), acting due to tangential component of stress on the xz plane is obtained as

$$F_{yx} = \left\{(\tau_{yx})_y\right\}\delta x \delta z - \left\{(\tau_{yx})_{y+\delta y}\right\}\delta x \delta z = -\left(\frac{\delta \tau_{yx}}{\delta y}\right)\delta x \delta y \delta z \qquad (2.36)$$

The net force ($F_{zx}$), acting due to tangential component of stress on the xy plane is obtained as

$$F_{zx} = \left\{(\tau_{zx})_z\right\}\delta x\, \delta y - \left\{(\tau_{zx})_{z+\delta z}\right\}\delta x\, \delta y = -\left(\frac{\delta \tau_{zx}}{\delta z}\right)\delta x\, \delta y\, \delta z \qquad (2.37)$$

Thus, the total forces acting in the x direction ($F_x$) on the volume element is the sum of forces:

$$F_x = F_g + F_P + F_{xx} + F_{yx} + F_{zx}$$
$$= \left\{(\rho g_x) - \left(\frac{\delta p}{\delta x}\right) - \left(\frac{\delta \tau_{xx}}{\delta x}\right) - \left(\frac{\delta \tau_{yx}}{\delta y}\right) - \left(\frac{\delta \tau_{zx}}{\delta z}\right)\right\}\delta x\, \delta y\, \delta z \qquad (2.38)$$

Now substitutes of Eqs. (2.30), (2.31), (2.32) and Eq. (2.38) in Eq. (2.28) leads to

$$\rightarrow \left(\left[\{(\rho u_x)u_x\}_x\right]\delta y\, \delta z + \left[\{(\rho u_x)u_y\}_y\right]\delta x\, \delta z + \left[\{(\rho u_x)u_z\}_z\right]\delta x\, \delta y\right)$$
$$- \left(\left[\{(\rho u_x)u_x\}_{x+\delta x}\right]\delta y\, \delta z + \left[\{(\rho u_x)\right.\right.$$
$$\left.\left.u_y\}_{y+\delta y}\right]\delta x\, \delta z + \left[\{(\rho u_x)u_z\}_{z+\delta z}\right]\delta x\, \delta y\right)$$
$$= \left[\left\{\frac{\delta(\rho u_x)}{\delta t}\right\}\delta x\delta y\delta z\right] - \left[\left\{(\rho g_x) - \left(\frac{\delta p}{\delta x}\right) - \left(\frac{\delta \tau_{xx}}{\delta x}\right) - \left(\frac{\delta \tau_{yx}}{\delta y}\right) - \right.\right.$$
$$\left.\left.\left(\frac{\delta \tau_{zx}}{\delta z}\right)\right\}\delta x\delta y\delta z\right]$$
$$\rightarrow \left(\left\{\frac{\delta(\rho u_x)}{\delta t}\right\}\delta x\delta y\delta z\right) = \left[\{(\rho u_x)\cdot u_x\}_x - \{(\rho u_x)\cdot u_x\}_{x+\delta x}\right]\cdot \delta y \cdot \delta z$$
$$+ \left[\{(\rho u_x)\cdot u_y\}_y - \{(\rho u_x)\cdot u_y\}_{y+\delta y}\right]\cdot \delta x \cdot \delta z \cdot$$
$$+ \left[\{(\rho u_x)u_z\}_z - \{(\rho u_x)u_z\}_{z+\delta z}\right]\delta x\delta y$$
$$+ \left\{(\rho g_x) - \left(\frac{\delta p}{\delta x}\right) - \left(\frac{\delta \tau_{xx}}{\delta x}\right) - \left(\frac{\delta \tau_{yx}}{\delta y}\right) - \left(\frac{\delta \tau_{zx}}{\delta z}\right)\right\}\delta x\delta y\delta z$$

$$\left\{\frac{\delta(\rho u_x)}{\delta t}\right\} = -\left[\left\{\frac{\delta(\rho u_x u_x)}{\delta x}\right\} + \left\{\frac{\delta(\rho u_x u_y)}{\delta y}\right\} + \left\{\frac{\delta(\rho u_x u_z)}{\delta z}\right\}\right]$$

$$+ \left[ (\rho g_x) - \left( \frac{\delta p}{\delta x} \right) \right] - \left[ \left( \frac{\delta \tau_{xx}}{\delta x} \right) + \left( \frac{\delta \tau_{yx}}{\delta y} \right) + \left( \frac{\delta \tau_{zx}}{\delta z} \right) \right] \quad (2.39)$$

On the basis of symmetry, the following equations can be derived for components of momentum balance in the y and z directions:

$$\left\{ \frac{\delta(\rho u_y)}{\delta t} \right\} = - \left[ \left\{ \frac{\delta(\rho u_y u_x)}{\delta x} \right\} + \left\{ \frac{\delta(\rho u_y u_y)}{\delta y} \right\} + \left\{ \frac{\delta(\rho u_y u_z)}{\delta z} \right\} \right]$$

$$+ \left[ (\rho g_y) - \left( \frac{\delta p}{\delta y} \right) \right] - \left[ \left( \frac{\delta \tau_{xy}}{\delta x} \right) + \left( \frac{\delta \tau_{yy}}{\delta y} \right) + \left( \frac{\delta \tau_{zy}}{\delta z} \right) \right] \quad (2.40)$$

$$\left\{ \frac{\delta(\rho u_z)}{\delta t} \right\} = - \left[ \left\{ \frac{\delta(\rho u_z u_x)}{\delta x} \right\} + \left\{ \frac{\delta(\rho u_z u_y)}{\delta y} \right\} + \left\{ \frac{\delta(\rho u_z u_z)}{\delta z} \right\} \right]$$

$$+ \left[ (\rho g_z) - \left( \frac{\delta p}{\delta z} \right) \right] - \left[ \left( \frac{\delta \tau_{xz}}{\delta x} \right) + \left( \frac{\delta \tau_{yz}}{\delta y} \right) + \left( \frac{\delta \tau_{zz}}{\delta z} \right) \right] \quad (2.41)$$

Equations (2.39), (2.40), (2.41) are the *equations of motion* (in terms of τ), which can be written in the following vector-tensor notation form:

$$\left\{ \frac{\delta(\rho u)}{\delta t} \right\} = -[\nabla(\rho u u)] - (\nabla p) - (\nabla \tau) + (\rho g) \quad (2.42)$$

where $\frac{\delta(\rho u)}{\delta t}$ = Rate of increase of momentum per unit volume,
  $\nabla(\rho u u)$ = Rate of momentum gain by convection per unit volume,
  $\nabla p$ = Pressure force on element per unit volume,
  $\nabla \tau$ = Rate of momentum gain by viscous transfer per unit volume, and.
  $\rho g$ = Gravitational force on element per unit volume.
  It should be noted that $[\nabla(\rho u u)]$ and $(\nabla \tau)$ are not simple divergences because of the tensorial nature of $(\rho u u)$ and τ. The physical interpretation is, however, analogous to that of $[\nabla(\rho u)]$ in Eq. (2.10) for equation of continuity; whereas $[\nabla(\rho u)]$ represents the rate of loss of mass (a scalar) per unit volume by fluid flow, the quantity $[\nabla(\rho u u)]$ represents the rate of loss of momentum (a vector) per unit volume of fluid flow [2].
  Equation (2.39) may be rearranged with the help of substantial derivative form:

$$\rho \left\{ \frac{Du_x}{Dt} \right\} = - \left( \frac{\delta p}{\delta x} \right) - \left[ \left( \frac{\delta \tau_{xx}}{\delta x} \right) + \left( \frac{\delta \tau_{yx}}{\delta y} \right) + \left( \frac{\delta \tau_{zx}}{\delta z} \right) \right] + \rho g_x \quad (2.43)$$

Similar rearrangements can be made for the y and z components.
  When all three components are added together vector-tensor notation, then that can be written as

$$\rho \left\{ \frac{Du}{Dt} \right\} = -\nabla p - (\nabla \tau) + \rho g \quad (2.44)$$

where $\rho\{\frac{Du}{Dt}\}$ = Mass per unit volume, times acceleration.

In this form *equation of motion* states that a small volume element moving with the fluid is accelerated because of the forces acting upon it. In other words, this is a statement of Newton's second law in the form: {(mass × acceleration) = sum of forces}. It is observed that the momentum balance is completely equivalent to Newton's second law of motion.

The above set of equations contains thirteen dependent variables, namely, pressure, p; three velocity vectors, $u_x$, $u_y$, and $u_z$; and nine stress variables, i.e. $\tau_{xx}$, $\tau_{yy}$, $\tau_{zz}$, $\tau_{yx}$, $\tau_{zx}$, $\tau_{xy}$, $\tau_{zy}$, $\tau_{xz}$, and $\tau_{yz}$. Thus, in order to arrive to a solution, there must have, in all, thirteen simultaneous equations. But there are only four equations, one continuity Eq. (2.9) and three equations of motion (2.39, 2.40, and 2.41).

In order to arrive at the remaining nine equations, now must consider Newton's second law of motion. Second principle is devoted to the rate and the direction towards which a process will proceed under a given set of condition. According to it, the *momentum flux* or the rate of transfer of momentum from one point to the other is proportional to the momentum concentration or the momentum per unit volume. For example, considering the x direction momentum flux along the y direction; according to Newton's second law of motion:

$$\left\{\frac{\delta(\rho u_x)}{\delta t}\right\} \infty \left\{\frac{\delta(\rho u_x)}{\delta y}\right\} \tag{2.45}$$

Or

$$\left\{\frac{\delta(\rho u_x)}{\delta t}\right\} = -\alpha\rho\left\{\frac{\delta(u_x)}{\delta y}\right\} \tag{2.46}$$

where $\alpha$ is the proportionality constant and $\rho$, density, is considered to be constant; and the minus sign signifies that the flux will be down the gradient.

According to Newton's second law of motion, the rate of change of momentum is equal to the force. As in the present case, the change is in the x direction, but the gradient of change is considered along the y direction, i.e. on the xz plane; so this force will be a shear force, i.e. $\tau_{yx}$.

Further the value of the term ($\alpha$ $\rho$) depends upon the type of fluid and forms the most important kinetic property of the fluid called the *viscosity* which is denoted by the symbol $\mu$. Hence, Eq. (2.46) can be rewritten as

$$\tau_{yx} = -\mu\left(\frac{\delta u_x}{\delta y}\right) \tag{2.47}$$

Equation (2.47) correlates the shearing force with the velocity gradient in a two-dimensional system. This equation can be extended to the three-dimensional cases as well, and the resultant equations obtained for various shearing forces will be as follows:

$$\tau_{xx} = -2\mu \left( \frac{\delta u_x}{\delta x} \right) + \left( \frac{2\mu}{3} \right) \left[ \frac{\delta u_x}{\delta x} + \frac{\delta u_y}{\delta y} + \frac{\delta u_z}{\delta z} \right] \tag{2.48}$$

$$\tau_{yy} = -2\mu \left( \frac{\delta u_y}{\delta y} \right) + \left( \frac{2\mu}{3} \right) \left[ \frac{\delta u_x}{\delta x} + \frac{\delta u_y}{\delta y} + \frac{\delta u_z}{\delta z} \right] \tag{2.49}$$

$$\tau_{zz} = -2\mu \left( \frac{\delta u_z}{\delta z} \right) + \left( \frac{2\mu}{3} \right) \left[ \frac{\delta u_x}{\delta x} + \frac{\delta u_y}{\delta y} + \frac{\delta u_z}{\delta z} \right] \tag{2.50}$$

$$\tau_{xy} = \tau_{yx} = -\mu \left[ \frac{\delta u_x}{\delta y} + \frac{\delta u_y}{\delta x} \right] \tag{2.51}$$

$$\tau_{xz} = \tau_{zx} = -\mu \left[ \frac{\delta u_x}{\delta z} + \frac{\delta u_z}{\delta x} \right] \tag{2.52}$$

$$\tau_{yz} = \tau_{zy} = -\mu \left[ \frac{\delta u_y}{\delta z} + \frac{\delta u_z}{\delta y} \right] \tag{2.53}$$

With the help of the above nine equations [from (2.45) to (2.53)], can eliminate the stress ($\tau$) terms from Eqs. (2.39–2.44) and assuming $\mu$ is constant (i.e. Newtonian fluid). Now by substitute of Eqs. (2.48–2.53) into Eq. (2.43) for the x component of momentum:

$$\rho \left\{ \frac{Du_x}{Dt} \right\} = -\left( \frac{\delta p}{\delta x} \right) - \left[ \left( \frac{\delta \tau_{xx}}{\delta x} \right) + \left( \frac{\delta \tau_{yx}}{\delta y} \right) + \left( \frac{\delta \tau_{zx}}{\delta z} \right) \right] + \rho g_x \tag{2.43}$$

So,

$$\rho \left\{ \frac{Du_x}{Dt} \right\} = -\left( \frac{\delta p}{\delta x} \right) + \frac{\delta}{\delta x} \left[ 2\mu \left( \frac{\delta u_x}{\delta x} \right) - \left( \frac{2\mu}{3} \right) \left\{ \frac{\delta u_x}{\delta x} + \frac{\delta u_y}{\delta y} + \frac{\delta u_z}{\delta z} \right\} \right]$$
$$+ \frac{\delta}{\delta y} \left[ \mu \left\{ \frac{\delta u_x}{\delta y} + \frac{\delta u_y}{\delta x} \right\} \right] + \frac{\delta}{\delta z} \left[ \mu \left\{ \frac{\delta u_x}{\delta z} + \frac{\delta u_z}{\delta x} \right\} \right] + \rho g_x \tag{2.54}$$

Or

$$\rho \left\{ \frac{Du_x}{Dt} \right\} = -\left( \frac{\delta p}{\delta x} \right) + \mu \left[ \frac{\delta^2 u_x}{\delta x^2} + \frac{\delta^2 u_x}{\delta y^2} + \frac{\delta^2 u_x}{\delta z^2} \right]$$
$$+ \left( \frac{\mu}{3} \right) \left( \frac{\delta}{\delta x} \right) \left[ \frac{\delta u_x}{\delta x} + \frac{\delta u_y}{\delta y} + \frac{\delta u_z}{\delta z} \right] + \rho g_x \tag{2.55}$$

Similarly for y and z components of momentum:

$$\rho\left\{\frac{Du_y}{Dt}\right\} = -\left(\frac{\delta p}{\delta y}\right) + \mu\left[\frac{\delta^2 u_y}{\delta x^2} + \frac{\delta^2 u_y}{\delta y^2} + \frac{\delta^2 u_y}{\delta z^2}\right] + \left(\frac{\mu}{3}\right)\left(\frac{\delta}{\delta y}\right)\left[\frac{\delta u_x}{\delta x} + \frac{\delta u_y}{\delta y} + \frac{\delta u_z}{\delta z}\right] + \rho g_y$$

$$\text{(2.56)}$$

and

$$\rho\left\{\frac{Du_z}{Dt}\right\} = -\left(\frac{\delta p}{\delta z}\right) + \mu\left[\frac{\delta^2 u_z}{\delta x^2} + \frac{\delta^2 u_z}{\delta y^2} + \frac{\delta^2 u_z}{\delta z^2}\right] + \left(\frac{\mu}{3}\right)\left(\frac{\delta}{\delta z}\right)\left[\frac{\delta u_x}{\delta x} + \frac{\delta u_y}{\delta y} + \frac{\delta u_z}{\delta z}\right] + \rho g_z$$

$$\text{(2.57)}$$

For constant $\rho$ and $\mu$, Eqs. (2.55)–(2.57) may be simplified in vector-tensor notation form:

$$\rho\left\{\frac{Du}{Dt}\right\} = -\nabla p + \mu\nabla^2 u + \left(\frac{\mu}{3}\right)\left(\frac{\delta}{\delta z}\right)\nabla u + \rho g \qquad \text{(2.58)}$$

For an incompressible fluid (by means of the equation of continuity): u = 0
Then Eq. (2.58) can be rewritten as

$$\rho\left\{\frac{Du}{Dt}\right\} = -\nabla p + \mu\nabla^2 u + \rho g \qquad \text{(2.59)}$$

The above Eq. (2.59) is known as *Navier–Stokes equation*, which first developed by Navier in France on 1922. This Eq. (2.59) represents the equation of motion in the cartesian coordinate system.

Similarly, *Navier–Stokes equations* are also available for cylindrical and spherical coordinates in standard textbooks. *Appendix III* gives the list of equations.

Fluids are classified as (i) ideal and (ii) non-ideal. Ideal fluids are those for which viscosity, $\mu$ is zero. The equation of motion for ideal fluid, Eq. (2.59) reduces to

$$\rho\left\{\frac{Du}{Dt}\right\} = -\nabla p + \rho g \qquad \text{(2.60)}$$

This Eq. (2.60) is the famous *Euler equation*, first derived on 1755. It has been widely used for describing fluid flow systems in which viscous effects are relatively unimportant.

## 2.5  Application of Continuity and Navier–Stokes Equations

### 2.5.1  Flow of Falling Fluid

Considering the flows of a fluid along a flat surface forming an angle $\beta$ with the vertical plane (Fig. 2.3). In the absence of external pressure, gravity is the only force acting on the fluid. Again, considering this is the two-dimensional problem, the direction y is parallel with the plane, i.e. flows of fluid; and the direction x is perpendicular to it towards the gravity.

From equation of continuity (Appendix II, Eq. II. 1):

$$\left(\frac{\delta\rho}{\delta t}\right) = -\left[\left\{\frac{\delta(\rho u_x)}{\delta x}\right\} + \left\{\frac{\delta(\rho u_y)}{\delta y}\right\} + \left\{\frac{\delta(\rho u_z)}{\delta z}\right\}\right] \qquad (2.9)$$

For steady state: $\left(\frac{\delta\rho}{\delta t}\right) = 0$, and for incompressible fluid (i.e. $\rho$ is constant); continuity Eq. (2.9) become

$$\frac{\delta u_x}{\delta x} + \frac{\delta u_y}{\delta y} + \frac{\delta u_z}{\delta z} = 0 \qquad (2.61)$$

Again, for two-dimensional flow of an incompressible fluid (i.e. $u_z = 0$), Eq. (2.61) is further modified to

$$\frac{\delta u_x}{\delta x} + \frac{\delta u_y}{\delta y} = 0 \qquad (2.62)$$

Since there is no velocity at x direction, so $u_x = 0$, but $u_y \neq 0$.
Equation (2.62) becomes

$$\frac{\delta u_y}{\delta y} = 0 \qquad (2.63)$$

Now from Appendix III, Eq. III.11, Navier–Stokes equation, for y component of cartesian coordinates, can be written as :

**Fig. 2.3**  Flow of fluid along an incline flat surface

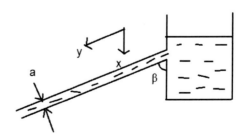

$$\rho\left(\frac{\delta u_y}{\delta t} + u_x\frac{\delta u_y}{\delta x} + u_y\frac{\delta u_y}{\delta y} + u_z\frac{\delta u_y}{\delta z}\right) = -\frac{\delta p}{\delta y} + \mu\left[\frac{\delta^2 u_y}{\delta x^2} + \frac{\delta^2 u_y}{\delta y^2} + \frac{\delta^2 u_y}{\delta z^2}\right] + \rho g_y$$

$$(2.64)$$

Now for the two-dimensional flow: $u_z = 0$, so the term $u_z\frac{\delta u_y}{\delta z} = 0$ and $\frac{\delta^2 u_y}{\delta z^2} = 0$.
Hence, Eq. (2.64) becomes

$$\rho\left(\frac{\delta u_y}{\delta t} + u_x\frac{\delta u_y}{\delta x} + u_y\frac{\delta u_y}{\delta y}\right) = -\frac{\delta p}{\delta y} + \mu\left[\frac{\delta^2 u_y}{\delta x^2} + \frac{\delta^2 u_y}{\delta y^2}\right] + \rho g_y \qquad (2.65)$$

Again $u_x = 0$, and $\frac{\delta u_y}{\delta y} = 0$ (Eq. 2.63), so $\frac{\delta^2 u_y}{\delta y^2} = 0$ and steady-state condition:
$\frac{\delta u_y}{\delta t} = 0$;
   so Eq. (2.65) further becomes

$$\frac{\delta p}{\delta y} = \mu\left[\frac{\delta^2 u_y}{\delta x^2}\right] + \rho g_y = \mu\left[\frac{\delta^2 u_y}{\delta x^2}\right] + \rho g\cos\beta \qquad (2.66)$$

[Since $\sin(90 - \beta) = g/g_y$, therefore, $g_y = g\cos\beta$].
In the absence of external pressure, $\frac{\delta p}{\delta y} = 0$,
So

$$\mu\left[\frac{\delta^2 u_y}{\delta x^2}\right] + \rho g\cos\beta = 0 \text{ or } \frac{d}{dx}\left[\mu\left(\frac{du_y}{dx}\right)\right] = -\rho g\cos\beta \qquad (2.67)$$

Now integrating: $\int d\left[\mu\left(\frac{du_y}{dx}\right)\right] = -\rho g\cos\beta\int dx$.
Or

$$\mu\left(\frac{du_y}{dx}\right) = -\rho g\cos\beta x + I_1 \text{(where } I_1 = \text{Integration constant)} \qquad (2.68)$$

Now putting boundary conditions, $u_y = 0$ at $x = a$, i.e. $\frac{du_y}{dx} = 0$.
So Eq. (2.68) becomes $0 = -\rho g\cos\beta a + I_1$ so $I_1 = \rho g a\cos\beta$.
Again Eq. (2.68) becomes

$$\mu\left(\frac{du_y}{dx}\right) = -\rho g\cos\beta x + \rho g a\cos\beta \qquad (2.69)$$

By integrating Eq. (2.69):

$$\mu.u_y = -\rho g\cos\beta\left(\frac{x^2}{2}\right) + \rho g a\cos\beta.x + I_2 \qquad (2.70)$$

(where $I_2$ = Integration constant).

Now putting boundary conditions, $u_y = 0$ at $x = a$,

$$0 = -\rho g \cos \beta \left( \frac{a^2}{2} \right) + \rho g a \cos \beta \cdot a + I_2$$

Therefore, $I_2 = -\rho g \left( \frac{a^2}{2} \right) \cos \beta$.
Hence, Eq. (2.70) becomes

$$u_y = \frac{1}{\mu} \left[ -\rho g \cos \beta \left( \frac{x^2}{2} \right) + \rho g a x \cos \beta - \rho g \left( \frac{a^2}{2} \right) \cos \beta \right]$$

$$= \frac{1}{2\mu} \left[ -\rho g \cos \beta (a - x)^2 \right] \tag{2.71}$$

### 2.5.2  Axial Flow Through Pipe

Consider the laminar flows of an incompressible fluid of constant viscosity through a straight circular, horizontal cylindrical pipe under steady-state condition (Fig. 2.4).
    Using cylindrical coordinates (r, θ, z), there are three components of velocity vector, e.g. $u_r$, $u_\theta$, and $u_z$.
    From equation of continuity for cylindrical coordinates (Appendix II, Eq. II.2):

$$\left( \frac{\delta \rho}{\delta t} \right) + \frac{1}{r} \frac{\delta (\rho r u_r)}{\delta r} + \frac{1}{r} \frac{\delta (\rho u_\theta)}{\delta \theta} + \frac{\delta (\rho u_z)}{\delta z} = 0 \tag{2.72}$$

Due to cylindrical symmetry, $u_z$ is not a function of θ, i.e. $\frac{du_z}{d\theta} = 0$.
    For steady state: $\frac{\delta \rho}{\delta t} = 0$, from physical reasonings: $u_r = u_\theta = 0$, and incompressible fluid, ρ is constant.
    Now Eq. (2.72) becomes

$$\frac{\delta u_z}{\delta z} = 0, \text{ i.e. } u_z \neq 0 \tag{2.73}$$

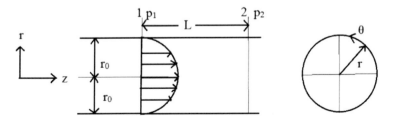

**Fig. 2.4**  Fluid flow through a circular, horizontal cylindrical pipe

Z component of the Navier–Stokes equation for cylindrical coordinate (Appendix III, Eq. III.15):

$$\rho\left(\frac{\delta u_z}{\delta t} + u_r\frac{\delta u_z}{\delta r} + \frac{u_\theta}{r}\frac{\delta u_z}{\delta\theta} + u_z\frac{\delta u_z}{\delta z}\right) = -\frac{\delta p}{\delta z} + \mu\left[\frac{1}{r}\frac{\delta}{\delta r}\left(r\frac{\delta u_z}{\delta r}\right) + \frac{1}{r^2}\frac{\delta^2 u_z}{\delta\theta^2} + \frac{\delta^2 u_z}{\delta z^2}\right] + \rho g_z$$

$$(2.74)$$

For steady state: $\frac{\delta u_z}{\delta t} = 0$, $g_z = g\cos\beta = g\cos 90° = 0$.
Equation (2.74) can be modified as

$$-\frac{\delta p}{\delta z} + \mu\left[\frac{1}{r}\frac{\delta}{\delta r}\left(r\frac{\delta u_z}{\delta r}\right)\right] = 0 \qquad (2.75)$$

Since p is a function of z only and u is a function of r only.
Therefore,

$$\mu\left[\frac{1}{r}\frac{\delta}{\delta r}\left(r\frac{\delta u_z}{\delta r}\right)\right] = \frac{\delta p}{\delta z} = p' = \text{a constant} \qquad (2.76)$$

Integrating Eq. (2.76):

$$\int d\left(r\frac{\delta u_z}{\delta r}\right) = \frac{p'}{\mu}\int r\,dr \quad \text{or} \quad \left(r\frac{\delta u_z}{\delta r}\right) = \frac{p'}{\mu}\frac{r^2}{2} + I_1 \qquad (2.77)$$

where $I_1$ is an integration constant.
Boundary condition: at $r = 0$, $u_z = \infty$, i.e. maximum velocity. So, $I_1 = 0$.
Therefore,

$$\left(r\frac{\delta u_z}{\delta r}\right) = \left(\frac{p'}{\mu}\right)\frac{r^2}{2} \qquad (2.78)$$

Again integrating Eq. (2.78):

$$\int (du_z) = \left(\frac{p'}{2\mu}\right)\int r\,dr \quad \text{or} \quad u_z = \left(\frac{p'}{4\mu}\right)r^2 + I_2 \qquad (2.79)$$

where $I_2$ is another integration constant.
Again, boundary condition: at $r = r_0$, $u_z = 0$: therefore, $I_2 = -\left(\frac{p'r_0^2}{4\mu}\right)$.
Hence, Eq. (2.79) becomes

$$u_z = \left(\frac{p'}{4\mu}\right)r^2 - \left(\frac{p'r_0^2}{4\mu}\right) = \left(\frac{-p'r_0^2}{4\mu}\right)\left\{1 - \left(\frac{r^2}{r_0^2}\right)\right\} \qquad (2.80)$$

This Eq. (2.80) gives rise to a parabolic velocity profile with a maximum velocity $(u_z, max)$ at the axis $(r = 0)$ as shown in Fig. 2.4.

The average velocity,

$$u_{\bar{z}} = \frac{\int_0^{r_0} u_z(2\pi r)dr}{\pi r_0^2} \tag{2.81}$$

where $2\pi r$ is the perimeter of cylinder and $\pi r_0^2$ is the cross-sectional area of cylinder. By putting the value of $u_z$ (from Eq. 2.80) to Eq. (2.81):

$$u_{\bar{z}} = \frac{\int_0^{r_0}\left[\left(\frac{-p'r_0^2}{4\mu}\right)\left\{1-\left(\frac{r^2}{r_0^2}\right)\right\}\right](2\pi r)dr}{\pi r_0^2} = -\left(\frac{p'}{2\mu}\right)\left[\int_0^{r_0} r dr - \int_0^{r_0}\frac{r^3}{r_0^2}dr\right] = -\left(\frac{p'r_0^2}{8\mu}\right) \tag{2.82}$$

Considering the section of pipe from plane 1 to plane 2 having length, L. Therefore, $-p' = \frac{p_1-p_2}{L}$.

Then (From Eq. 2.82):

$$\Delta p = p_1 - p_2 = L(-p') = \frac{L8\mu u_{\bar{z}}}{r_0^2} = \frac{32\mu u_{\bar{z}}L}{d^2} \tag{2.83}$$

where d is diameter of pipe $= 2r_0$.

Total volumetric flow rate,

$$Q = u_{\bar{z}}\pi r_0^2 = \left(\frac{-p'r_0^2}{8\mu}\right)\cdot\pi r_0^2 = \frac{\pi(p_1-p_2)r_0^4}{8\mu L} \tag{2.84}$$

Again, in terms of diameter,

$$Q = \frac{\pi(p_1-p_2)d^4}{128\mu L} \tag{2.85}$$

Equation (2.84) is the well-known *Hagen–Poiseuille equation*. This can be used to determine the viscosity of the fluid:

$$\mu = \frac{\pi(p_1-p_2)r_0^4}{8LQ} \tag{2.86}$$

It may be emphasized that these equations are applicable only under the following conditions:

1. The fluid flow should be continuous and laminar,
2. The viscosity of the fluid does not change with velocity, i.e. flow is Newtonian,

3. The fully developed steady-state flow is maintained in the portion of the pipe. This has been estimated to hold for pipe length above the value of $0.07r_0$ Re, where $r_0$ is the radius of the pipe and Re is Reynolds number (Re $= \frac{\rho u L}{\mu}$, L is characteristic length such as diameter of the pipe).

**Example 2.2** Calculate the viscosity of a fluid flowing with laminar flow at the rate of 400 cm³/min in a capillary of diameter 70 mm given that the pressure drops over a length of 300 m of the capillary is $1.42 \times 10^5$ Pa. Find the velocity of fluid.

**Solution**

Volumetric flow rate, Q = 400 cm³/min = $6.67 \times 10^{-6}$ m³/s.
　　Radius of the capillary, $r_0$ = 70/2 mm = $35 \times 10^{-3}$ m.
　　Length of the capillary, L = 300 m.
　　Pressure drop, $p_1 - p_2 = 1.42 \times 10^5$ Pa.
　　Since Q = $u.\pi r_0^2$.
　　Therefore, u = Q / $(\pi r_0^2)$ = $[(6.67 \times 10^{-6})/\{3.14 \times (35 \times 10^{-3})^2\}]$ = 1.73 $\times$ $10^{-3}$ m/s = **1.73 mm/s**.
　　From Eq. (2.86): $\mu = \frac{\pi(p_1-p_2)r_0^4}{8LQ} = \frac{\pi(1.42\times10^5)(35\times10^{-3})^4}{8\times300\times6.67\times10^{-6}}$ = **41.8 kgm⁻¹s⁻¹**.

## 2.5.3　High Speed Flow of Gases

In a number of metallurgical processes such as the BOF process, flash roasting, BF smelting, there are concerned with the flow of gases or compressible fluids in a reactor [1]. Since these gases enter the reactor at very high speeds. Such a flow is achieved in practice using tapered nozzles. The shapes of these nozzles can be arrived at from a study of the equations of continuity and motion. These high speeds of gases, generally in the supersonic range are usually represented by a dimensionless group known as *Mach number*, which is defined as the ratio of the velocity of gas to that of sound at the same temperature.

The velocity of sound, $u_s$ is related to the properties of the medium in which it travels. Thus, for a gaseous medium, the velocity of sound is expressed by the relation:

$$u_s = \left(\frac{\gamma RT}{M_g}\right)^{1/2} = \left(\frac{\gamma pV}{M_g}\right)^{1/2} = \left(\frac{\gamma p}{\rho_g}\right)^{1/2} \qquad (2.87)$$

where $\gamma$ is ($C_p/C_V$), R is gas constant, T is temperature (in K), $M_g$ is molecular weight of gas, and $\rho_g$ is density of gas (i.e. $M_g/V$).

From the equation of continuity for cartesian coordinates (Appendix II, Eq. II. 1):

$$\left(\frac{\delta\rho}{\delta t}\right) + \left[\left\{\frac{\delta(\rho u_x)}{\delta x}\right\} + \left\{\frac{\delta(\rho u_y)}{\delta y}\right\} + \left\{\frac{\delta(\rho u_z)}{\delta z}\right\}\right] = 0 \qquad (2.9)$$

For steady state $\left(\frac{\delta\rho}{\delta t}\right) = 0$, and again for the two-dimensional flow of a compressible fluid ($u_z = 0$, i.e.; $\frac{\delta(\rho u_z)}{\delta z} = 0$) continuity Eq. (2.9) become

$$\frac{\delta(\rho u_x)}{\delta x} + \frac{\delta(\rho u_y)}{\delta y} = 0 \tag{2.88}$$

Now velocity at x direction, $u_x = 0$, but $u_y \neq 0$;
Therefore,

$$\frac{\delta(\rho \mu_y)}{\delta y} = 0 \tag{2.89}$$

Now from Appendix III, Eq. III.11 (Navier–Stokes equation), for y component of cartesian coordinates, can be written as

$$\rho\left(\frac{\delta u_y}{\delta t} + u_x\frac{\delta u_y}{\delta x} + u_y\frac{\delta u_y}{\delta y} + u_z\frac{\delta u_y}{\delta z}\right) = -\frac{\delta p}{\delta y} + \mu\left[\frac{\delta^2 u_y}{\delta x^2} + \frac{\delta^2 u_y}{\delta y^2} + \frac{\delta^2 u_y}{\delta z^2}\right] + \rho g_y \tag{2.90}$$

Now for the two-dimensional flow: $u_z = 0$, so the term $u_z\frac{\delta u_y}{\delta z} = 0$ and $\frac{\delta^2 u_y}{\delta z^2} = 0$. Hence, Eq. (2.90) becomes

$$\rho\left(\frac{\delta u_y}{\delta t} + u_x\frac{\delta u_y}{\delta x} + u_y\frac{\delta u_y}{\delta y}\right) = -\frac{\delta p}{\delta y} + \mu\left[\frac{\delta^2 u_y}{\delta x^2} + \frac{\delta^2 u_y}{\delta y^2}\right] + \rho g_y \tag{2.91}$$

Again $u_x = 0$, and $\frac{\delta(\rho u_y)}{\delta y} = 0$ (Eq. 2.89), and steady-state condition: $\frac{\delta(\rho u_y)}{\delta t} = 0$; So Eq. (2.91) further becomes

$$\frac{\delta p}{\delta y} = \mu\left[\frac{\delta^2 u_y}{\delta x^2}\right] + \rho g\cos 90^0 = \mu\left[\frac{\delta^2 u_y}{\delta x^2}\right] \quad \left(\text{since } \cos 90^0 = 0\right) \tag{2.92}$$

Therefore,

$$\frac{\delta p}{\delta y} = \frac{d}{dx}\left(\mu\frac{\delta u_y}{\delta x}\right) \tag{2.93}$$

Integrating Eq. (2.93) $\int dp = \int\left\{\frac{d}{dx}\left(\mu\frac{\delta u_y}{\delta x}\right)\right\}dy$
Or,

$$p = \left\{\frac{d}{dx}\left(\mu\frac{\delta u_y}{\delta x}\right)y\right\} + I_1 \tag{2.94}$$

where $I_1$ is an integration constant.

At $y = 0$, $p = p_1$: so $p_1 = I_1$.

Again at

$$y = L, p = p_2; \text{ so } p_2 = \left\{ \frac{d}{dx}\left( \mu \frac{\delta u_y}{\delta x} \right)L \right\} + I_1 = \left\{ \frac{d}{dx}\left( \mu \frac{\delta u_y}{\delta x} \right)L \right\} + p_1 \quad (2.95)$$

Therefore,

$$\frac{p_1 - p_2}{L} = -p' = \frac{d}{dx}\left( \mu \frac{\delta u_y}{\delta x} \right) \quad (2.96)$$

Integrating Eq. (2.96):

$$-p' \int dx = \int d\left( \mu \frac{\delta u_y}{\delta x} \right) \text{ or } -p'x = \mu\left( \frac{\delta u_y}{\delta x} \right) + I_2 \quad (2.97)$$

where $I_2$ is another integration constant.

At $x = 0$, $u_y = u_{y,max}$. So, $I_2 = -\mu\left( \frac{\delta u_{y,max}}{\delta x} \right)$.

Therefore, Eq. (2.97) becomes

$$-p'x = \mu\left( \frac{\delta u_y}{\delta x} \right) - \mu\left( \frac{\delta u_{y,max}}{\delta x} \right) = -\mu\left( \frac{\delta u_y}{\delta x} \right) \quad (2.98)$$

Again integrating Eq. (2.98): $p' \int x dx = \mu \int du_y$  or  $p'\left( \frac{x^2}{2} \right) = \mu u_y$.

Therefore,

$$u_y = \left( \frac{p'x^2}{2\mu} \right) \quad (2.99)$$

The Eq. (2.99) means that gas velocity increased in y direction with increasing the pressure difference per unit length as well as increasing the value of x but decreased with increasing viscosity.

## 2.6  Mechanism of Momentum Transport

Considering the free jet falls of liquid steel from ladle to mold. Two basic equations governing the transport of momentum in fluid, these are Newton's 2nd law of motion and Newton's law of viscosity [3].

(a)  Newton's 2nd law of motion: Fig. 2.5 shows an element of liquid steel of mass m, moving from position 1 to 2 as it accelerates towards the surface of the filling mold. Newton's 2nd law states that the sum of the normal forces acting on this

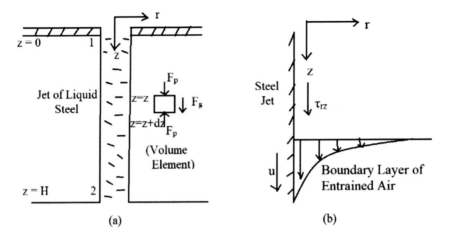

**Fig. 2.5** Liquid steel falls from ladle to mold

body in the vertical z direction is equal to the mass of the body multiplied by its z acceleration, or time rate of z momentum change, i.e.

$$\Sigma F = ma \quad \text{or} \quad \Sigma F = \frac{d(mu)}{dt} \tag{2.100}$$

The atmospheric pressure forces ($F_p$) acting normally to the two horizontal surfaces of the volume element will be equal and opposite over the small height, so that

$$\Sigma F = -F_p + F_g + F_p = F_g \tag{2.101}$$

where $F_g$ is the force due to gravity, mg. (g is acceleration due to gravity = 9.807 m/s$^2$).

Consider the volume element (i.e. control volume) fixed in space, into which liquid steel enters at z and leave at $(z + dz)$. Since the liquid steel flows at z direction. Therefore,

$$a = \frac{du}{dt} = \frac{du}{dz}\frac{dz}{dt} = u\frac{du}{dz} \quad \left(\text{since } \frac{dz}{dt} = u\right) \tag{2.102}$$

From Eq. (2.100):

$$\Sigma F = ma \quad \text{or} \quad F_g = ma = mu\frac{du}{dz} \tag{2.103}$$

[from Eqs. (2.101) and (2.102)].

Integrating Eq. (2.103) from z = 0 to H and u = $u_0$ to $u_f$:

$$F_g \int_0^H dz = m \int_{u_0}^{u_f} u du$$

Or

$$F_g H = \frac{m}{2}\left(u_f^2 - u_0^2\right) \tag{2.104}$$

Since $F_g$ = mg; and m = $\rho$.A.dz (where $\rho$ is density of liquid steel, A is cross-sectional area).

Equation (2.104) becomes $\rho \cdot A \cdot dz \cdot g \cdot H = \frac{1}{2}\rho \cdot A \cdot dz\left(u_f^2 - u_0^2\right)$.

Or

$$\rho.g.H = \frac{1}{2}\rho\left(u_f^2 - u_0^2\right) = \frac{1}{2}\rho u_f^2 - \frac{1}{2}\rho u_0^2 \tag{2.105}$$

The decrease in potential energy of the liquid steel in moving from z = 0 to z = H, is exactly compensated by an increase in its kinetic energy.

Again, net force acting on the control volume, $F_g$ is equal to the rate of momentum change:

From Eq. (2.103):

$$F_g = mu\frac{du}{dz} = \rho \cdot A \cdot dz \cdot u\frac{du}{dz} = \rho \cdot A \cdot u \cdot du = m'du = m'(u_f - u_0) \tag{2.106}$$

where $m'$ is the mass flow rate (kg/s).

(b)  Newton's law of viscosity: Newton's law of viscosity states that a fluid element, in steady parallel flow, will resist the application of a shear stress on one of its surfaces, and this resistance will increase in proportion to the rate of deformation, i.e.

$$F_s = A_s \cdot \mu \cdot \frac{du}{dr} \quad \text{or } \tau_{rz} = -\frac{F_s}{A_s} = -\mu \cdot \frac{du}{dr} \tag{2.107}$$

where $\mu$ is the viscosity of liquid steel, $\tau_{rz}$ is customarily regarded as a shear stress acting in the z direction on a plane perpendicular to the r axis.

The negative sign results from the sign convention normally adopted, which specifies that the direction of increasing z is to be the direction of positive shear stress, $\tau_{rz}$.

Figure 2.5b shows an element of gas (air), adjacent to the liquid steel jet, moving tangentially downwards at a constant velocity, u. The higher velocity of gas, closer to the jet, will transmit a shearing force to this gaseous element, cause a corresponding downward motion. Partly as a result of this shearing action, a boundary layer of air is formed, which can lead to the entrainment of air in the liquid steel.

To interpret Newton's law of viscosity as a momentum interchange phenomenon, one may draw an analogy from kinetic theory of gases. Thus, in the very neighborhoods of the steel surface, the gas molecules will move at the surface velocity of the steel, adhering to surface atoms of iron (zero-slip condition). The gas will be moving much more slowly and with much reduced tangential or z momentum. z momentum will be lost by faster gas molecules when they move away from the surface of the liquid steel jet, and that this flux of z momentum in the r direction will result in the generation of a shearing force $F_s$ transmitted to the steel surface.

## 2.7  Dimensional Analysis

According to Langhaar (1951): Dimensional analysis is a method by which deduce information about a phenomenon from the single premise that the phenomenon can be described by a dimensionally correct equation among certain variables. That means with a little effort, a partial solution is obtained [4].

Every physical quantity has some dimension by virtue of either its definition or its relationship with other quantities [1]. For example, velocity, which is defined as distance traveled per unit time, has the dimensions of length (L) per unit time (t), i.e. $Lt^{-1}$. It has been seen that velocity is expressed in terms of dimensions of length and time.

Thus, physical quantities can be divided into two categories as follows:

1. Those units are expressed in terms of certain others. Such quantities are called *derived quantities.*
2. Those quantities in terms of which all derived quantities can be expressed. Only four quantities namely mass (M), length (L), time (t), and temperature (T); come under this category. These four quantities are called *basic quantities.*

Determination of dimensions of any physical quantity from its relationship with other quantities is based on the principle of dimensional homogeneity. According to the principle of dimensional homogeneity, both right- and left-hand sides of the above equation must have the same dimensions. For example, force (F) is related to mass (m) and acceleration (a), i.e. F = m.a. Since mass and acceleration have dimension, M and $Lt^{-2}$ respectively; hence force (F) will have the dimensions of $MLt^{-2}$.

Considering example: rate of mass of fluid flow (kg/s) through any one of the faces of cube (i.e. volume element) is represented by the product [(density of fluid) x (velocity perpendicular to the face) x (cross-sectional area)], i.e. $(\rho u_x)_x \cdot \delta y \delta z$ (from Eq. 2.5).

Now density of fluid ($\rho$), velocity of fluid ($u_x$), and cross-sectional area ($\delta y\ \delta z$) have the dimensions $ML^{-3}$, $Lt^{-1}$, and $L^2$, respectively; hence rate of mass flow will have the dimension $Mt^{-1}$, i.e. kg/s.

Again, from *Hagen–Poiseuille equation* the viscosity of the fluid can be written as

$$\mu = \frac{\pi(p_1 - p_2)r_0^4}{8LQ} \tag{2.86}$$

where $p_1 - p_2$ is the pressure difference, dimensions $ML^{-1}t^{-2}$; $r_0$ and L are the radius and length having dimensions L; Q is volumetric flow rate, dimension $L^3t^{-1}$.

Now if the dimensions of the different physical quantities are substituted in Eq. (2.86).

$$\text{RHS} = \frac{(ML^{-1}t^{-2})L^4}{L(L^3t^{-1})} = M.L^{-1+4-1-3} \cdot t^{-2+1} = ML^{-1}t^{-1} = \text{this is dimension of}$$

viscosity ($\mu$) = LHS.

It is observed that both the side have the same dimensions, i.e. $ML^{-1}t^{-1}$.

Therefore,

$$\mu = f(\Delta p, r_0, L, Q) \tag{2.108}$$

To determine experimentally the effect of the different variables in Eq. (2.108) within the parentheses on the right-hand side, that need to conduct a large number of experiments on flow of fluids on different radii ($r_0$) and length of pipes, different volumetric flow rate under different pressure differences across the tube to arrive at Eq. (2.99).

Again,

$$m' = Q \cdot \rho = \frac{\pi\rho(p_1 - p_2)r_0^4}{8L\mu} \tag{2.109}$$

where $m'$ is the mass flow rate, dimension $Mt^{-1}$; $(p_1-p_2) = \Delta p$ is the pressure difference across the length, L of the capillary, dimension $ML^{-1}t^{-2}$; $\rho$ is the density of fluid with dimension $ML^{-3}$, $r_0$ and L are the radius and length of the tube, respectively, dimension L; and $\mu$ is viscosity of fluid, dimension $ML^{-1}t^{-1}$.

If the dimensions of the different physical quantities are substituted in Eq. (2.109):
Then

$$\text{RHS} = \frac{(ML^{-3})(ML^{-1}t^{-2})L^4}{L(ML^{-1}t^{-1})} = M^{1+1-1}L^{-3-1+4-1+1}t^{-2+1} = ML^0t^{-1} = Mt^{-1} = \text{LHS}$$

Therefore,

$$m' = f(\Delta p, \rho, r_0, L, \mu) \tag{2.110}$$

To determine experimentally the effect of the different variables in Eq. (2.110) within the parentheses on the right-hand side, that needs to conduct a large number of experiments on flow of fluids on different radii and length of pipes, different viscosity and density under different pressure difference across the tube to arrive at Eq. (2.109).

It is possible to reduce the large number of experiments; it can be done by rearranging Eq. (2.109) in the alternative forms:

$$(m')^2 = \left(\frac{\pi \rho \Delta p r_0^4}{8L\mu}\right)^2 \tag{2.111}$$

Or

$$\left[\frac{m'^2}{\Delta p \rho (\pi r^2)^2}\right] = \left[\left(\frac{1}{64\rho}\right)\left(\frac{\Delta p \rho \pi r^2}{\mu^2}\right)\left(\frac{r_0}{L}\right)^2\right] \tag{2.112}$$

This Eq. (2.112) is written in such a way that each of the terms in different parentheses has zero dimension, i.e. dimensionless numbers. In the form a functional relationship this equation can be rewritten as

$$\pi_1 = f(\pi_2, \pi_3) \tag{2.113}$$

where

$$\pi_1 = \left\{\frac{m'^2}{\Delta p \rho (\pi r^2)^2}\right\} \tag{2.114}$$

$$\pi_2 = \left(\frac{\Delta p \rho \pi r^2}{\mu^2}\right) \tag{2.115}$$

$$\pi_3 = \left(\frac{r_0}{L}\right)^2 \tag{2.116}$$

Equation (2.114): RHS $= \frac{(Mt^{-1})^2}{(ML^{-1}t^{-2})(ML^{-3})L^4} = M^{2-1-1}t^{-2+2}L^{1+3-4} = M^0 t^0 L^0$.

Equation (2.115): $= \frac{(ML^{-1}t^{-2})(ML^{-3})L^2}{(ML^{-1}t^{-1})^2} = M^{1+1-2}L^{-1-3+2+2}t^{-2+2} = M^0 L^0 t^0$.

Equation (2.116): RHS $= \left(\frac{L^2}{L^2}\right) = L^{2-2} = L^0$.

Hence, it is proved that $\pi_1$, $\pi_2$ and $\pi_3$ are the dimensionless numbers.

### 2.7.1   Methods of Dimensional Analysis

The major advantage of dimensional analysis is to reduce the number of independent variables in a problem. With the help of dimensional analysis, variables of the
problem can be combined into dimensionless groups such as Reynolds number (Re),
Prandtl number (Pr) and the experimental data can be very conveniently presented in
terms of these numbers. Dimensional analysis does not give the precise relationship
among these non-dimensional groups; these relations or equations are obtained by
correlating experimental data. Although dimensional analysis is a very powerful tool
yet without experimental data, it is of no value [4].

As mentioned above (Sect. 2.7), there are four primary dimensions namely mass
(M), length (L), time (t), and temperature (T). The dimensions of all other physical
variables of any phenomenon can be obtained in terms of these four basic dimensions.
e.g.

(i)   Reynolds number $(Re) = \left(\dfrac{u\rho d}{\mu}\right) = \left[\dfrac{\left(\frac{L}{t}\right)\left(\frac{M}{L^3}\right)L}{\left(\frac{M}{Lt}\right)}\right] = L^{1-3+1+1}M^{1-1}t^{-1+1}$

$= L^0 M^0 t^0$

where dimensions of u, $\rho$, d, and $\mu$ are L/t, $M/L^3$, L and $ML^{-1}t^{-1}$, respectively.

(ii)   Prandtl number
$(Pr) = \dfrac{\mu C_p}{k} = \dfrac{(ML^{-1}t^{-1})(L^2 t^{-2}T^{-1})}{(MLt^{-3}T^{-1})} = M^{1-1}L^{-1+2-1}t^{-1-2+3}T^{-1+1} = M^0 L^0 t^0 T^0$

where dimensions of $\mu$, $C_p$, and k are $ML^{-1}t^{-1}$, $L^2 t^{-2}T^{-1}$, and $MLt^{-3}T^{-1}$,
respectively.

The different methods used for dimensional analysis can be classified as follows:

(1)   Those based on knowledge of the physical parameters affecting the dependent
variables of the system under study. In this category, a) Rayleigh's method,
and b) Buckingham's pi ($\pi$) theory,
(2)   Those based on knowledge of differential equations governing the system under
study, and
(3)   Those based on similarity criteria among similar systems of different sizes.

#### 2.7.1.1   Rayleigh's Method

In this method, the first step is to write the functional relationship between the
independent physical quantities and the dependent variables in the form of a product.
e.g. the functional relationship of Eq. (2.110) may be rewritten as

$$m' = \alpha(\Delta p)^a(\rho)^b(r_0)^c(L)^d(\mu)^e \tag{2.117}$$

where $\alpha$ is the dimensionless constant.

Now substitute the primary dimensions (i.e. M, L, t) of each of the variables in the functional relationship, Eq. (2.117) becomes

$$\left(Mt^{-1}\right) = \left(ML^{-1}t^{-2}\right)^a \left(ML^{-3}\right)^b (L)^c (L)^d \left(ML^{-1}t^{-1}\right)^e \qquad (2.118)$$

Using the concept of dimensional homogeneity, writing equations for the dimensions of each of the basic quantities, viz., mass (M), length (L), time (t), and temperature (T) on both sides of the Eq. (2.118).

(i)    For mass,

$$M : 1 = a + b + e \qquad (2.119)$$

(ii)    For length,

$$L : 0 = -a - 3b + c + d - e \qquad (2.120)$$

(iii)    For time,

$$t : -1 = -2a - e \qquad (2.121)$$

In Eqs. (2.119)–(2.121), there are five unknown variables namely a, b, c, d, and e; but only three equations interrelating them. To solve the equations, now express any three of them in terms of the rest two with the help of the above three equations. The selection of these two variables are generally arbitrary. In the present case, d and e are these two variables.
From Eq. (2.121):

$$2a = 1 - e, \text{ therefore, } a = \frac{1 - e}{2} \qquad (2.122)$$

From Eq. (2.119):

$$b = 1 - a - e = 1 - e - \left(\frac{1 - e}{2}\right) = \frac{1 - e}{2} \qquad (2.123)$$

From Eq. (2.120):

$$c = a + 3b + e - d = \left(\frac{1 - e}{2}\right) + 3\left(\frac{1 - e}{2}\right) + e - d = 2(1 - e) + e - d = (2 - e - d) \qquad (2.124)$$

Substituting values of a, b, and c in Eq. (2.117):

$$m' = \alpha(\Delta p)^{\left(\frac{1-e}{2}\right)}(\rho)^{\left(\frac{1-e}{2}\right)}(r_0)^{(2-e-d)}(L)^d(\mu)^e \qquad (2.125)$$

$$m'^2 = \alpha^2 (\Delta p)^{(1-e)} (\rho)^{(1-e)} (r_0)^{(4-2e-2d)} (L)^{2d} (\mu)^{2e} \qquad (2.126)$$

Rearrangement of terms, like Eq. (2.112):

$$\left[ \frac{m'^2}{\Delta p \rho r^4} \right] = \left[ \alpha^2 \left( \frac{\Delta p \rho r^2}{\mu^2} \right)^{-e} \left\{ \left( \frac{r_0}{L} \right)^2 \right\}^{-d} \right] \qquad (2.127)$$

Now instead of writing $r_0^4$ or $r_0^2$ to the above equation, the cross-sectional area $(\pi r_0)^2$ will be written.

$$\left[ \frac{m'^2}{\Delta p \rho (\pi r_0^2)^2} \right] = \left( \frac{\alpha^2}{\pi^3} \right) \left( \frac{\Delta p \rho \pi r_0^2}{\mu^2} \right)^{-e} \left\{ \left( \frac{r_0}{L} \right)^2 \right\}^{-d} \qquad (2.128)$$

Now putting the values $\pi_1$, $\pi_2$, and $\pi_3$ from Eqs. [(2.114) to (2.116)] to Eq. (2.128). Therefore,

$$\pi_1 = \left( \frac{\alpha^2}{\pi^3} \right) (\pi_2)^{-e} (\pi_3)^{-d} \qquad (2.129)$$

Or

$$\pi_1 = f(\pi_2, \pi_3) \qquad (2.113)$$

There are three ratios in Eq. (2.128), they are the dimensional groups. In the original empirical correlation [Eq. (2.117)], $m'$ has five independent variables, that has reduced to three variables appearing as dimensional groups.

If the value of the constants $\alpha$, e, d in Eq. (2.128) are found by experiment for one set of these three groups, these values would be valid for only other sets of dimensionless groups within experimental range.

## 2.7.1.2  Buckingham's Pi ($\pi$) Theorem

Buckingham's pi ($\pi$) theorem states that the number of independent dimensionless groups (r) that can be formed from a set of variables (n) having basic dimensions (m); i.e. $r = (n - m)$.

This theorem is used as a rule of thumb for determining the number of independent dimensionless groups that can be obtained from a set of variables.

Buckingham's $\pi$ Theorem can be divided into two parts as follows:

(i)   The solution of any dimensionally physical equation can be expressed in terms of a certain number of dimensionless groups: $\pi_1$, $\pi_2$, $\pi_3$, and so forth, as follows:

$$f(\pi_1, \pi_2, \pi_3 \ldots) = 0 \tag{2.130}$$

(ii)  If an equation containing n, independent variables (i.e. physical quantities) is analyzed dimensionally on the basis of primary dimensions (M, L, t), the number of groups (r) in the resulting dimensionless equation will be equal to

$$r = (n - m) \tag{2.131}$$

These r groups are generally designated by the symbols $\pi_1, \pi_2, \ldots \pi_r$.
Thus, according to Buckingham's $\pi$ Theorem:

$$f(\pi_1, \pi_2, \ldots \pi_r) = 0 \tag{2.132}$$

In Eq. (2.117), there are five unknown variables (i.e. a, b, c, d, and e), so n = 5; there are three primary dimensions (i.e. M, L, t), so m = 3; therefore, r = n − m = 5 − 3 = 2.
Hence

$$\pi_1 = f(\pi_2, \pi_3) \tag{2.113}$$

**Example 2.3**  The mass transfer coefficient ($k_m$) of a component is known to depend upon density ($\rho$), viscosity ($\mu$) and velocity (u) of the fluid, the diffusion coefficient (D) of the component under study in the fluid and a characteristic distance (d). Using Buckingham's $\pi$ Theory, derive the functional relationship among these parameters.

**Solution**

The total numbers of physical quantities or parameters are six (i.e. $k_m$, $\rho$, $\mu$, u, D, and d), so n = 6. Number of primary variables, m = 3 (i.e. M, L, t).
Therefore, r = n − m = 6 − 3 = 3, i.e. there will be three dimensionless groups for the system.
The dimension of $k_m$ is $Lt^{-1}$.

$$\text{Therefore, } k_m = Lt^{-1} = \left(\frac{L^2 t^{-1}}{L}\right) = \left(\frac{D}{d}\right) \tag{a}$$

Or $k_m \left(\frac{d}{D}\right) = Lt^{-1}\left(\frac{L}{L^2 t^{-1}}\right) = L^{1+1-2} t^{-1+1} = L^0 t^0$.
= Dimensionless group known as *Sherwood number* (Sh).
The dimensions of $\rho$, $\mu$, and D are $ML^{-3}$, $ML^{-1}t^{-1}$, and $L^2 t^{-1}$, respectively.
Similarly, a combination of $\rho$, $\mu$, and D with dimensions can be written as

$$\rho = ML^{-3} = \left(\frac{ML^{-1} t^{-1}}{L^2 t^{-1}}\right) = \left(\frac{\mu}{D}\right) \tag{b}$$

Therefore, $\left(\frac{\mu}{D\rho}\right) = \left\{\frac{(ML^{-1}t^{-1})}{(L^2t^{-1})(ML^{-3})}\right\} = L^{-1-2+3}t^{-1+1}M^{1-1} = L^0t^0M^0 = $
Dimensionless group is known as *Schmidt number* (Sc).

Again, the dimensions of u, D, and L are $Lt^{-1}$, $L^2t^{-1}$, and L, respectively. Similarly, a combination of u, D, and L with dimensions can be written as

$$u = Lt^{-1} = \left(\frac{L^2t^{-1}}{L}\right) = \left(\frac{D}{L}\right) \tag{c}$$

Therefore, $\left(\frac{D}{uL}\right) = \left(\frac{L^2t^{-1}}{Lt^{-1}L}\right) = L^{2-1-1}t^{-1+1} = L^0t^0 = $ Dimensionless group.
Hence, functional relationship obtains:

$$\left(\frac{k_m d}{D}\right) = \alpha\left\{\left(\frac{\mu}{D\rho}\right)\left(\frac{D}{uL}\right)\right\} \tag{d}$$

$$Or(Sh) = \alpha\left\{Sc.\left(\frac{D}{uL}\right)\right\} \tag{e}$$

## 2.8 Friction Factor

A fluid exerts force on solid surfaces. The force due to its motion is known as *kinetic force* or *frictional force* ($F_k$). An analysis of the experimental data on fluid flow in various systems has shown that the frictional force ($F_k$) acting at the interface between the fluid and the solid surface is directly proportional to the product of the average kinetic energy (K) per unit volume of fluid and contact area of solid (A).

$$F_k \infty AK \tag{2.133}$$

Or

$$F_k = f_{fr}AK = f_{fr}\left(\frac{\rho u^2}{2}\right)A \tag{2.134}$$

where $f_{fr}$ is the proportionality constant, known as *friction factor*, which is a dimensionless quantity; $\left(\frac{\rho u^2}{2}\right)$ is the kinetic energy and A is the area of contact.

In case the fluid is surrounded by the solid surface (e.g. for flow through a pipe) and as drag coefficient for submerged solids or gas bubbles. The frictional force acting at the interface between the fluid and the solid surfaces was calculated by Eq. (2.134).

Rearranging Eq. (2.134):

$$f_{fr} = \left[ \frac{F_k}{A\left(\frac{\rho u^2}{2}\right)} \right] \tag{2.135}$$

Substituting the dimensions of different terms on the RHS of Eq. (2.135):

$$f_{fr} = \left[ \frac{MLt^{-2}}{L^2(ML^{-3})(L^2t^{-2})} \right] = M^{1-1}L^{1-2+3-2}t^{-2+2} = M^0L^0t^0 \tag{2.136}$$

that means, the friction factor ($f_{fr}$) is a dimensionless parameter.

Based on physical analysis of fluid flow through a pipe, it can be concluded that the frictional force acting between the solid surface and fluid.

Hence,

$$f_{fr} = f(u, \rho, \mu, d, L) \tag{2.137}$$

where $\mu$ is viscosity of fluid, d and L are the diameter and length of the pipe.

By dimension analysis, Rayleigh's method:

$$f_{fr} = \alpha(u)^a(\rho)^b(\mu)^c(d)^d(L)^e \tag{2.138}$$

where $\alpha$ is the dimensionless constant.

Now substituting the dimensions of different terms to Eq. (2.138):

$$M^0L^0t^0 = \left(Lt^{-1}\right)^a\left(ML^{-3}\right)^b\left(ML^{-1}t^{-1}\right)^c(L)^d(L)^e$$

$$\text{Balance for M: } 0 = b + c \text{ or } b = -c \tag{i}$$

$$\text{Balance for L: } 0 = a - 3b - c + d + e \tag{ii}$$

$$\text{Balance for t: } 0 = -a - c \text{ or } a = -c \tag{iii}$$

From Eq. (ii): $0 = a - 3b - c + d + e = -c + 3c - c + d + e$ or $d = -c - e$ (iv)

So Eq. (2.135) becomes

$$f_{fr} = \alpha(u)^{-c}(\rho)^{-c}(\mu)^c(d)^{-c-e}(L)^e \tag{2.139}$$

$$= \alpha\left(\frac{u\rho d}{\mu}\right)^{-c}\left(\frac{L}{d}\right)^{e} = f(\text{Re})^{-c}\left(\frac{L}{d}\right)^{e} \tag{2.140}$$

However, it has been found experimentally that for smooth pipes the frictional factor $(f_{fr})$ is independent of $\left(\frac{L}{d}\right)$ ratio.

Hence,

$$f_{fr} = f(\text{Re}) \tag{2.141}$$

- Laminar flow region, is signified by a value of Re below 2100, the friction factor is related to the Reynolds number by the relation

$$f_{fr} = \left(\frac{16}{Re}\right) \tag{2.142}$$

- Transition flow region, is signified by a value of Re varying between 2100 to 3000, the friction factor in this region is an irregular function of Re.
- Turbulent flow region (Re > 3000): A number of relations have been proposed for this region as follows:

$$\text{Blasius relation:} \quad f_{fr} = 0.079(\text{Re})^{-0.25} \quad \text{for } 3000 < \text{Re} < 10^5 \tag{2.143}$$

$$\text{Colburn's relation :} \quad f_{fr} = 0.023(\text{Re})^{-0.25} \tag{2.144}$$

$$\text{Drew's relation :} \quad f_{fr} = 0.0014 + 0.125(\text{Re})^{-0.32} \tag{2.145}$$

**Example 2.4** Calculate the pressure differences under which 30,000 litees per hour of water flow, under steady-state condition, through a horizontal pipe of diameter 20 cm and length 10 m. Assume viscosity of water to be 1 cp, and its density 1000 kg/m³.

**Solution**

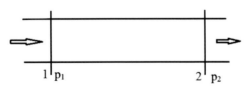

Fig. Ex. 2.4.

Energy balance:

Kinetic energy at point 1 + Work done by the surrounding to the system + Potential energy point at 1 = Kinetic energy at point 2 + Work done by the system to the surrounding + + Potential energy point at 2 + Friction force

$$\frac{m'u_1^2}{2} + \left(\frac{m'}{\rho}\right)p_1 + m'g(z_1 - z_0) = \frac{m'u_2^2}{2} + \left(\frac{m'}{\rho}\right)p_2 + m'g(z_2 - z_0) + F_k$$

Or

$$\frac{m'u_1^2}{2} - \frac{m'u_2^2}{2} + \left(\frac{m'}{\rho}\right)p_1 - \left(\frac{m'}{\rho}\right)p_2 + m'g(z_1 - z_2) - F_k = 0 \qquad (a)$$

The volumetric flow rate, $Q = 30,000$ litees per hour $= 30 \text{ m}^3/\text{hr} = (30/3600) = 8.33 \times 10^{-3} \text{ m}^3/\text{s}$.

Again, $u = \frac{Q}{\frac{\pi d^2}{4}} = \left(\frac{4Q}{\pi d^2}\right) = \frac{4 \times 8.33 \times 10^{-3}}{3.14 \times (0.2)^2} = 0.265 \text{m/s}$

Since pipe is horizontal, so $z_1 = z_2$ and for steady state, $u_1 = u_2 = u$.
Hence, Eq. (2.1) becomes

$$\left(\frac{m'}{\rho}\right)(p_1 - p_2) - F_k = 0 \qquad (b)$$

Again, from Eq. (2.134): $F_k = f_{fr}\left(\frac{\rho u^2}{2}\right)A$.

Or $\left(\frac{m'}{\rho}\right)(p_1 - p_2) = F_k = f_{fr}\left(\frac{\rho u^2}{2}\right)A$.
Therefore,

$$(p_1 - p_2) = f_{fr}\left(\frac{\rho u^2}{2}\right)A\left(\frac{\rho}{m'}\right) = f_{fr}\left(\frac{\rho u^2}{2}\right)A\left(\frac{1}{Q}\right) \quad (\text{Since } m' = Q \cdot \rho) \quad (c)$$

Again, Reynolds number, $R_e = \left(\frac{\rho u d}{\mu}\right) = \frac{10^3 \times 0.265 \times 0.2}{10^{-3}} = 53,000$

$$(\text{Since } \mu = 1cp = 10^{-2}p = 10^{-3}\text{kg/m.s })$$

From Eq. (2.143): Blasius relation: for $3000 < Re < 10^5$, $\quad f_{fr} = 0.079(Re)^{-0.25}$.
Therefore, $f_{ft} = 0.079(53000)^{-0.25} = 5.21 \times 10^{-3}$.
Area, $A = 2\pi r.L = \pi dL = 3.14 \times 0.2 \times 10 = 6.28 \text{ m}^2$.

$$(p_1 - p_2) = f_{fr}\left(\frac{\rho u^2}{2}\right)A\left(\frac{\rho}{m'}\right) = f_{fr}\left(\frac{\rho u^2}{2}\right)A\left(\frac{1}{Q}\right)$$

Now from Eq. (2.3):
$$= \frac{5.21 \times 10^{-3} \times 10^3 \times 0.265^2 \times 6.28}{2 \times 8.33 \times 10^{-3}}$$

$$= 137.92 \text{ Pa}$$

## 2.9 Laminar, Steady Flow Around Submerged Solid Sphere

Figure 2.6 shows the laminar, steady fluid flow passing around a submerged solid sphere. Incompressible fluid is flowing in the vertical upward direction. The fluid exerts the normal due to displacement of fluid by the submerged solid sphere.

Since, normal force, $F_n =$ Buoyant force (which is the mass of displaced fluid times the gravitational acceleration) + Foam drag (which is due to pressure differences in the flow direction resulting in foam formation).

So,

$$F_n = \left[ \left( \frac{4\pi r_0^3 \rho_f g}{3} \right) + (2\pi r_0 \mu u_0) \right] \tag{2.146}$$

Force due to friction,

$$F_k = (4\pi r_0 \mu u_0) \tag{2.147}$$

Therefore, total force,

$$F = F_n + F_k = \left[ \left( \frac{4\pi r_0^3 \rho_f g}{3} \right) + (6\pi r_0 \mu u_0) \right] \tag{2.148}$$

where $u_0$ is the velocity of fluid in the bulk and $r_0$ is the radius of sphere; $\rho_f$ and $\rho_s$ are the density of fluid and solid sphere, respectively.

At steady state, these upward forces are balanced by the force applied by the solid sphere (i.e. weight of solid):

$$\left[ \left( \frac{4\pi r_0^3 \rho_f g}{3} \right) + (6\pi r_0 \mu u_0) \right] = \left( \frac{4\pi r_0^3 \rho_s g}{3} \right) \tag{2.149}$$

Therefore,

$$u_0 = \left[ \frac{4\pi r_0^3 g (\rho_s - \rho_f)}{3(6\pi r_0 \mu)} \right] = \left[ \frac{2 r_0^2 (\rho_s - \rho_f) g}{9\mu} \right] \tag{2.150}$$

**Fig. 2.6** Fluid flows around a submerged solid sphere

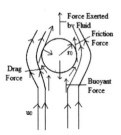

Equation (2.150) is known as *Stokes's law*.

Dimensionless number, the friction factor ($f_{fr}$) is also known as *drag coefficient* [5].
The frictional force ($F_k$) [Eq. (2.134)] can be rewritten as

$$F_k = f_{fr}AK = f_{fr}\left(\frac{\rho_f u_0^2}{2}\right)(\pi r_0^2) \tag{2.151}$$

where A = area and $K = \left(\frac{\rho_f u_0^2}{2}\right)$.

Again, from Eq. (2.148): $F_k = (6\pi r_0 \mu u_0)$].

Putting value of $u_0$, from Eq. (2.150): $F_k = (6\pi r_0 \mu)\left[\frac{2r_0^2(\rho_s-\rho_f)g}{9\mu}\right]$

$$= \left[\frac{4\pi r_0^3(\rho_s - \rho_f)g}{3}\right] \tag{2.152}$$

From Eqs. (2.151) and (2.152): $\left[\left(\frac{\rho_f u_0^2}{2}\right)(\pi r_0^2)f_{fr}\right] = \left[\frac{4\pi r_0^3(\rho_s-\rho_f)g}{3}\right]$

Therefore,

$$f_{fr} = \left[\frac{8r_0(\rho_s - \rho_f)g}{3(\rho_f u_0^2)}\right] = \left(\frac{4dg}{3u_0^2}\right)\left\{\frac{(\rho_s - \rho_f)}{\rho_f}\right\} \tag{2.153}$$

where d is diameter of solid sphere.

Since,

$$Re = \left(\frac{u_0 \rho_f d}{\mu}\right) = \left(\frac{\rho_f d}{\mu}\right)\left[\frac{2r_0^2(\rho_s - \rho_f)g}{9\mu}\right] = \left[\frac{\rho_f d^3(\rho_s - \rho_f)g}{18\mu^2}\right] \tag{2.154}$$

Again, from Eq. (2.153):

$$f_{fr} = \left(\frac{4dg(\rho_s - \rho_f)}{3\rho_f}\right)\left\{\frac{9\mu}{2r_0^2(\rho_s - \rho_f)g}\right\}^2 = \left\{\frac{24 \times 18\mu^2}{\rho_f d^3(\rho_s - \rho_f)g}\right\} = \frac{24}{Re} \tag{2.155}$$

from Eq. [(2.154)].

For laminar flow, Re < 0.1, $f_{fr} = f(Re)$; Fig. 2.7 presents friction factor for solid
particle as a function of Reynolds number.

**Fig. 2.7** Friction factor for
solid particle as a function of
Reynolds number

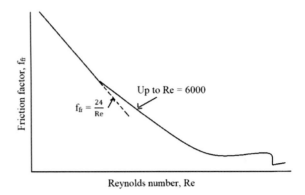

## 2.9.1   Fluid Flow in Circular Pipe

For fluid flowing in circular pipe (Fig. 2.8) whose radius is r and length is L.
    The frictional force ($F_k$) [Eq. (2.134)] can be rewritten as

$$F_k = AKf_{fr} = (2\pi r L)\left(\frac{\rho_f u^2}{2}\right)f_{fr} \qquad (2.156)$$

where contact area, $A = (2\pi r L)$.
    Generally, $F_k$ is not the quantity measured, but rather the pressure drops [i.e. ($p_0 - p_L$)] and the elevation (i.e. height) difference [i.e. ($h_0 - h_L$). A force balance on the fluid between O and L in the direction of flow (Fig. 2.8) gives for fully developed flow:

$$F_k = \left[(p_0 - p_L) + \rho g(h_0 - h_L)\right]\pi r^2 \qquad (2.157)$$

$$= \left[(P_0 - P_L)\pi r^2\right] \qquad (2.158)$$

**Fig. 2.8** Fluid flows in
circular vertical pipe

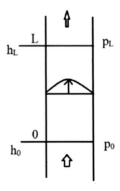

where $P = (p + \rho gh)$, modified pressure $(ML^{-1}t^{-2})$ at constant $\rho$ and g.
  Now eliminating $F_k$ between Eqs. (2.156) and (2.158):

$$(2\pi r L)\left(\frac{\rho_f u^2}{2}\right) f_{fr} = (P_0 - P_L)\pi r^2$$

Or,

$$f_{fr} = \left(\frac{r}{2L}\right)\left\{\frac{(P_0 - P_L)}{\left(\frac{\rho_f u^2}{2}\right)}\right\} = \left(\frac{d}{2L}\right)\left\{\frac{(P_0 - P_L)}{(\rho_f u^2)}\right\} \qquad (2.159)$$

where d is the diameter of pipe.

**Example 2.5** Water is being pumped through a horizontal pipe of 5 cm internal diameter and 60 m length at a flow rate of $2.5 \times 10^{-2}$ m³/s. Friction factor ($f_{fr}$) is 0.0058. Calculate the pressure difference and power required to maintain the water flow.

(Given: $\rho_{H2O} = 10^3$ kg/m³ and $\mu_{H2O} = 0.85 \times 10^{-3}$ kg/m.s.)

**Solution**

Velocity of water, $u = \frac{Q}{\frac{\pi d^2}{4}} = \left(\frac{4Q}{\pi d^2}\right) = \frac{4 \times 2.5 \times 10^{-2}}{3.14 \times (0.05)^2} = 12.74$ m/s.

Reynolds number, $Re = \left(\frac{\rho u d}{\mu}\right) = \frac{10^3 \times 12.74 \times 0.05}{0.85 \times 10^{-3}} = 7.49 \times 10^5$.

Hence flow of water is turbulent.

From Eq. (2.159): $f_{fr} = \left(\frac{d}{2L}\right)\left\{\frac{(P_0 - P_L)}{(\rho_f u^2)}\right\}$.

Or $\Delta p = \frac{2Lf_{fr}}{d}(\rho_f u^2) = \left(\frac{2 \times 60 \times 0.0058}{0.05}\right)\{10^3 \times (12.74)^2] = \mathbf{2.26 \times 10^6\ N/m^2}$.

Again, $-w_s^{\wedge} = \left(\frac{u^2}{2}\right) + \left(\frac{\Delta p}{\rho}\right) = \left(\frac{(12.74)^2}{2}\right) + \left(\frac{2.26 \times 10^6}{10^3}\right) = 2341.15$ J/kg.

Therefore,

$$w_s = m'(-w_s^{\wedge}) = (Q.\rho)(-w_s^{\wedge}) = (2.5 \times 10^{-2} \times 10^3) \times 2341.15 = \mathbf{58.53\ kW}$$

## 2.10 Fluid Flow Through Packed Bed of Solids

A packed bed of solids, through which fluid is flowing, illustrated by blast furnace of ironmaking, sponge iron production, sinter bed, calcination of limestone, etc. The simple representation of a packed bed is that of a vertical tube filled with particles of a uniform size (Fig. 2.9). Gas is passing through difficult path within the packed bed. Bed diameter ($d_T$) is very much larger than diameter of solid particle ($d_p$), i.e. $d_T \gg d_p$.

**Fig. 2.9** Fluid flow through
a packed bed of solids

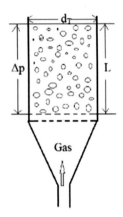

Derivative represents bulk of bed sampling should be from center of bulk than near to the wall of tube. Assuming that fluid flow will be uniform across the bed.

Some terms, which are used frequently, are defined as follows:

(1)   Void fraction $(\varepsilon) = \dfrac{\text{Total volume of bed} - \text{Volume of solid particles}}{\text{Total volume of bed}}$

$$= \dfrac{\text{Volume of empty space (i.e. void) in bed}}{\text{Total volume of bed}} \qquad (2.160)$$

(2)   Superficial velocity (velocity of gas when bed is empty, i.e. no obstacles for gas flow):

$$U_0 = \left\{\dfrac{Q}{\left(\dfrac{\pi d_T^2}{4}\right)}\right\} = \varepsilon U \qquad (2.161)$$

where $Q$ is the volumetric flow rate of gas, $\varepsilon$ is the void fraction, and $U$ is the average velocity of gas.

$$\left(\dfrac{\pi d_T^2}{4}\right) U_0 = Q = \left(\dfrac{\pi d_T^2}{4}\right) \varepsilon U \qquad (2.162)$$

Again assume: (i) packed bed is full of bundle of tubes with different cross sections. (i.e. tube bundle model), (as shown in Fig. 2.10a).
(ii) submerged object like single particle in fluid.

The friction $(f_{fr})$ for the packed bed can be applied [Eq. (2.159)], same analogous as for fluid flowing in circular pipe:

$$f_{fr} = \left(\dfrac{d_e}{2L}\right) \left\{\dfrac{(P_0 - P_L)}{(\rho_f U^2)}\right\} \qquad (2.163)$$

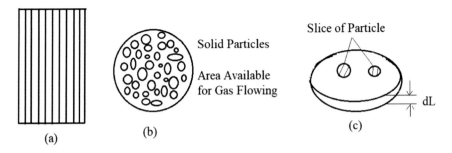

**Fig. 2.10  a** Tube bundle model, **b** cross section of tube, and **c** cross section of a slice

Therefore,

$$\left\{\frac{(P_0 - P_L)}{\rho_f}\right\} = \left(\frac{2L}{d_e}\right) U^2\, f_{fr} = \left(\frac{2L}{d_e}\right) \left(\frac{U_0^2}{\varepsilon^2}\right) f_{fr} \qquad \text{(since } U_0 = \varepsilon U) \qquad (2.164)$$

where $d_e$ is the equivalent diameter of tube.

Since it is not a circular tube, mean hydraulic radius, $r_h$ can be used. The mean hydraulic radius ($r_h$) can be expressed in terms of the void fraction ($\varepsilon$) and the wetted surface (a) per unit volume of bed. $r_h$ is depend on size of particles.

The diameter $d_e$ of the circular pipe can be replaced by $4r_h$. Therefore,

$$d_e = 4r_h \qquad \text{(a)}$$

Therefore, $r_h$

$$= \frac{\text{Cross-section available for flow}}{\text{Wetted perimeter}} = \frac{\text{Cross-section available for flow } x\,dL}{\text{Wetted perimeter } x\,dL}$$

$$= \frac{\text{Volume available for flow/volume of bed}}{\text{Total wetted surface/volume of bed}} = \frac{\text{Volume available for flow in the slice dL}}{\text{surface area of particle in the slice dL}}$$

$$= \frac{\text{Volume available for flow/volume of slice}}{\text{surface area of particle in the slice/volume of slice}} = \frac{\varepsilon}{a} \qquad \text{(b)}$$

Considering cross section of tube like a slice of thickness, dL. The quantity a is related to the specific surface, $a_v$ (i.e. total particle surface per volume of particles).

Therefore, specific surface area, $a_v = \dfrac{\text{Surface area/volume of bed}}{\text{Volume of particle/volume of bed}} = \dfrac{a}{1 - \varepsilon}$ (c)

$$\text{For spherical particle, } a_v = \frac{\frac{4\pi d_p^2}{4}}{\frac{4\pi d_p^3}{3 \times 2^3}} = \frac{6}{d_p} \qquad \text{(d)}$$

$$\text{For lump particle, } a_v = \frac{6}{d_p}\lambda \tag{e}$$

where $\lambda$ is correction factor.

$$\text{From Eq. (c): } a = (1-\varepsilon)a_v = (1-\varepsilon)\left(\frac{6\lambda}{d_p}\right) \tag{f}$$

$$\text{From Eq. (b) } r_h = \frac{\varepsilon}{a} = \frac{\varepsilon d_p}{(1-\varepsilon)6\lambda} \tag{g}$$

Since $d_e = 4r_h$, so Eq. (2.164) becomes

$$\left\{\frac{(P_0-P_L)}{\rho_f}\right\} = \left(\frac{2L}{4r_h}\right)\left(\frac{u_0^2}{\varepsilon^2}\right)f_{fr} = \left(\frac{L(1-\varepsilon)6\lambda}{2\varepsilon d_p}\right)\left(\frac{u_0^2}{\varepsilon^2}\right)f_{fr} = \left\{\frac{L\,u_0^2\,(1-\varepsilon)\lambda}{\varepsilon^3\,d_p}\right\}.3f_{fr} \tag{2.165}$$

If the flow is turbulent, $f_{fr}$ is dependent on relative roughness, not on Reynolds number.

$3f_{fr}$ is constant for all packed bed, since roughness is the same for small as well as large particles (since $d_T \gg d_p$). Hence Reynolds number is used in the Ergun equation ($Re_E$):

$$Re_E = \frac{\rho_f\,u\,d_e}{\mu} \tag{2.166}$$

From Eq. (2.161): $u = \frac{u_0}{\varepsilon}$, from eq. (a): $d_e = 4r_h = \frac{4\varepsilon d_p}{(1-\varepsilon)6\lambda}$ [from Eq. (g)]. Putting the values of $u$ and $d_e$ from the above equations, to Eq. (2.166):

$$Re_E = \left(\frac{\rho_f}{\mu}\right)\left(\frac{u_0}{\varepsilon}\right)\left(\frac{4\,\varepsilon\,d_p}{(1-\varepsilon)6\lambda}\right) = \left(\frac{\rho_f\,u_0 d_p}{(1-\varepsilon)\,\mu}\right)\left(\frac{2}{3\lambda}\right) \tag{2.167}$$

(I) For turbulent flow: $Re_E > 1000$, $3f_{fr} = 1.75$; therefore, Eq. (2.165) becomes

$$\left\{\frac{(P_0-P_L)}{L}\right\} = 1.75\left\{\frac{\rho_f\,u_0^2\,(1-\varepsilon)\lambda}{\varepsilon^3\,d_p}\right\} \tag{2.168}$$

Equation (2.168) is known as *Burke–Plummers equation*.

(II) For laminar flow: $Re_E < 10$ and $f_{fr} = \left(\frac{16}{Re_E}\right) = \left(\frac{16(1-\varepsilon)\mu}{\rho_f\,u_0 d_p}\right)\left(\frac{3\lambda}{2}\right)$ [from Eq. (2.167)]. Therefore, Eq. (2.165) becomes

$$\left\{\frac{(P_0-P_L)}{L}\right\} = \left\{\frac{\rho_f\,u_0^2\,(1-\varepsilon)\lambda}{\varepsilon^3\,d_p}\right\}.\,3\left(\frac{24\,(1-\varepsilon)\,\mu\,\lambda}{\rho_f\,u_0 d_p}\right)$$

$$= 72.\left(\frac{\mu\,u_0}{d_p^2}\right)\left\{\frac{(1-\varepsilon)^2\lambda^2}{\varepsilon^3}\right\} \tag{2.169}$$

This Eq. (2.169) derive by assuming tube is straight, but gas flow in zic-zac ways. In Eq. (2.169), 72 value can be replaced by 150:

$$\{\frac{(P_0-P_L)}{L}\} = 150.(\frac{\mu\,u_0}{d_p^2})\,\{\frac{(1-\varepsilon)^2\lambda^2}{\varepsilon^3}\} \tag{2.170}$$

This Eq. (2.170) is known as *Blake-Kozeny equation*, when $Re_E < 10$ and $\varepsilon < 0.5$.
(III) $10 < Re_E < 1000$, i.e. for transition region: [Combination of Eqs. (2.170) and (2.168)]

$$\{\frac{(P_0-P_L)}{L}\} = 150.(\frac{\mu\,u_0}{d_p^2})\,\{\frac{(1-\varepsilon)^2\lambda^2}{\varepsilon^3}\} + 1.75\{\frac{\rho_f\,u_0^2\,(1-\varepsilon)\lambda}{\varepsilon^3\,d_p}\} \tag{2.171}$$

This Eq. (2.171) is known as *Ergun equation*.

## 2.11  Overall Force Balance for Packed Bed of Solids

Consider packed bed of solids, when fluid goes upward (Fig. 2.11).

Now when there was no flow, weight of bed supported by the force given by the plate:

$$\left[\left(\frac{\pi d_T^2 L}{4}\right)(1-\varepsilon)(\rho_s-\rho_f)g\right] = \left[\Delta p\left(\frac{\pi d_T^2}{4}\right) + F^\rightarrow\right] \tag{2.172}$$

where $\left(\frac{\pi d_T^2 L}{4}\right)$ is the volume of bed, $(1-\varepsilon)$ is the volume occupied by the particles in the bed, $\Delta p$ is the pressure difference in the bed, and $F^\rightarrow$ is force by supporting plate.

By increasing velocity of fluid, value of $\left\{\Delta p\left(\frac{\pi d_T^2}{4}\right)\right\}$ will be increased.

If the rate of flow is increased continuously, a stage will come when the particles remain freely suspended in the fluid stream above the supporting plate. Then the bed is said to be *fluidized* [1].

**Fig. 2.11**  Force balance for packed bed of solids

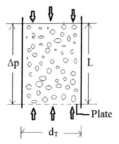

There are some conditions:

(i)   If flow rate of fluid increases, then volume of bed also increases, i.e. fluidized bed is formed.
(ii)  If flow rate of fluid decreases, then volume of bed also decreases.
(iii) Lower limit, i.e. minimum fluidization, when the upward force on the bed due to pressure difference balance the weight of the bed:

$$\left[\left(\frac{\pi d_T^2 L}{4}\right)(1 - \varepsilon_{mf})(\rho_s - \rho_f)g\right] = \Delta p\left(\frac{\pi d_T^2}{4}\right) \tag{2.173}$$

where subscript $mf$ represents by minimum fluidization.

Therefore,

$$\frac{\Delta p}{L} = \left[(1 - \varepsilon_{mf})(\rho_s - \rho_f)g\right] \tag{2.174}$$

$$= 150.\left(\frac{\mu\, u_{0,mf}}{d_p^2}\right)\left\{\frac{(1-\varepsilon_{mf})^2\lambda^2}{\varepsilon_{mf}^3}\right\} + 1.75\left\{\frac{\rho_f\, u_{0,mf}^2\,(1-\varepsilon_{mf})\lambda}{\varepsilon_{mf}^3\, d_p}\right\} \tag{2.175}$$

[from Ergun Eq. (2.171)].

Now multiply by $\left\{\frac{\rho_f d_p^3}{\mu^2(1-\varepsilon_{mf})}\right\}$ to Eq. (2.175):

$$\left\{\frac{(\rho_s - \rho_f)\, g\, \rho_f\, d_p^3}{\mu^2}\right\} = 150.\left(\frac{\rho_f\, d_p\, u_{0,mf}}{\mu}\right)\left\{\frac{(1-\varepsilon_{mf})\lambda^2}{\varepsilon_{mf}^3}\right\} + 1.75\left(\frac{\rho_f\, d_p\, u_{0,mf}}{\mu}\right)^2\left(\frac{\lambda}{\varepsilon_{mf}^3}\right) \tag{2.176}$$

This expression can be rewrite, as alternative form, in terms of the dimensionless *Galileo number* (Ga) and Reynolds number for minimum fluidization ($Re_{mf}$).

$$Ga = 150.\left\{\frac{(1-\varepsilon_{mf})\lambda^2}{\varepsilon_{mf}^3}\right\}Re_{mf} + 1.75 \cdot \left(\frac{\lambda}{\varepsilon_{mf}^3}\right)(Re_{mf})^2 \tag{2.177}$$

where $Ga = \left\{\frac{\rho_f\,(\rho_s - \rho_f)\, g\, d_p^3}{\mu^2}\right\}$ and $Re_{mf} = \left(\frac{\rho_f\, u_{0,mf}\, d_p}{\mu}\right)$

Wen and Yu [6] have suggested, based on their studies, as follows:

- At minimum fluidization: $\left\{\frac{(1-\varepsilon_{mf})\lambda^2}{\varepsilon_{mf}^3}\right\} \approx 11$, and $\left(\frac{\lambda}{\varepsilon_{mf}^3}\right) \approx 14$
- Hence, Eq. (2.177) further modifies as:

$$Ga = (150 \times 11)Re'_{mf} + (1.75 \times 14)(Re'_{mf})^2$$

$$= 1650 Re'_{mf} + 24.5 \left( Re'_{mf} \right)^2 \qquad (2.178)$$

where

$$Re'_{mf} = \left[ \left\{ (33.7)^2 + 0.0408 Ga \right\}^{0.5} - 33.7 \right] \qquad (2.179)$$

In the above discussion, the equation for the minimum fluidization velocity is derived. If the velocity of the fluidizing gas in the bed is increased further, the bed will also expand further. Ultimately when the upward thrust by the fluid and buoyancy effects together become greater than the effect of the terminal velocity of the particles falling under gravity, they are carried away by the stream of the fluid [1].

So

$$U_t = \sqrt{\frac{\frac{4 d_p \left( \rho_s - \rho_f \right) g}{3}}{\rho_f . f_{fr}}} \qquad (2.180)$$

The minimum velocity of the fluid, at which the particles are carried away by the stream of the fluid, is known as *elutriation velocity*, $U_E$ and the equation used for evaluating its value is dependent upon the particle Reynolds number $[Re_p = (\frac{\rho_f . U d_p}{\mu})]$ as given below:

(i) For $Re_p < 0.4$:
$$U_E = \left\{ \frac{(\rho_s - \rho_f) g d_p^2}{18 \mu} \right\} \qquad (2.181)$$

(ii) For $0.4 < Re_p < 500$:
$$U_E = 0.0178 \left\{ \frac{(\rho_s - \rho_f) g d_p}{(\mu \rho_f)^{1/2}} \right\} \qquad (2.182)$$

(iii) For $500 < Re_p < 2 \times 10^5$:
$$U_E = \sqrt{\left\{ \frac{(\rho_s - \rho_f) g d_p}{\rho_f} \right\}} \qquad (2.183)$$

**Example 2.6** Sponge iron powder is produced in fluidized bed reactor by reducing fine particles of iron ore ($Fe_2O_3$) (average particle size is 0.5 mm) at 1000 K by a reducing gas. The gas consist of 50% (by volume) $H_2$ and 50% CO and its viscosity is $2 \times 10^{-5}$ kg/m.s. Calculate (a) critical pressure drop per meter height of the bed at the onset of fluidization, (b) minimum fluidization velocity, and elutriation velocity.

(Given: i) Critical bed void fraction at fluidization point $(\varepsilon_{mf}) = 0.4$ (ii) $\rho_{ore} = 5 \times 10^3$ kg/m³,
(iii) $Re_{mf} = [\{(33.7)^2 + 0.0408 \, Ga\}^{0.5} - 33.7]$.

**Solution**

Since $\rho_{ore} = 5 \times 10^3$ kg/m³, assume that there is no change of volume of the particle upon reduction, then density of resulting iron produced would be $5 \times 10^3 \times (\frac{2 \times 56}{160}) = 3.5 \times 10^3$ kg/m³.

Since 1 kg mol of gas at STP occupies 22.4 m$^3$ volume.

So $[(0.5 \times 2) + (0.5 \times 28)]$ kg mol of gas at STP occupies 22.4 m$^3$ volume.

Therefore, $\rho_{gas} = \frac{[(0.5 \times 2) + (0.5 \times 28)]}{22.4} = 0.6696$ kg/m$^3$ at STP.

Since $\frac{p_1 V_1}{T_1} = \frac{p_2 V_2}{T_2}$. At constant pressure, $p_1 = p_2$.

Therefore $\frac{V_1}{T_1} = \frac{V_2}{T_2}$ or $\frac{V_1}{V_2} = \frac{T_1}{T_2} = \frac{\rho_2}{\rho_1}$, Hence $\rho_2 = \rho_1 \times \frac{T_1}{T_2}$.

[Since $\rho = \frac{M}{V}$ or $V = \frac{M}{\rho}$, therefore $\frac{V_1}{V_2} = \frac{M}{\rho_1} \times \frac{\rho_2}{M} = \frac{\rho_2}{\rho_1}$].

Therefore, $\rho_{gas}$ at 1000 K $= \rho_{gas}$ at STP $\times \frac{273}{1000} = 0.6696 \times \frac{273}{1000} = 0.183$ kg/m$^3$

(a)  From Eq. (2.174): $\frac{\Delta p}{L} = [(1 - \varepsilon_{mf})(\rho_s - \rho_f)g]$

$= [(1 - 0.4)\{(3.5 \times 10^3) - 0.183\} \times 9.807] = \mathbf{20{,}593.62 \ N/m^3}$.

(b)  Since Ga $= \left\{ \frac{\rho_f(\rho_s - \rho_f)g d_p^3}{\mu^2} \right\} = \left\{ \frac{0.183(5000 - 0.183) \times 9.807 \times (5 \times 10^{-4})^3}{(2 \times 10^{-5})^2} \right\} = 2804.09$.

Therefore, Re$_{mf} = [\{(33.7)^2 + 0.0408 \ \text{Ga}\}^{0.5} - 33.7] = 1.657$.

Since Re$_{mf} = \left( \frac{\rho_f u_{mf} d_p}{\mu} \right)$.

Hence u$_{mf} = \frac{Re_{mf} \mu}{\rho_f d_p} = \frac{1.657 \times 2 \times 10^{-5}}{0.183 \times 5 \times 10^{-4}} = 0.362 \text{m/s}$.

(c)  From Eq. (2.182): For $0.4 < Re_{mf} < 500$:

$$U_E = 0.0178 \left\{ \frac{(\rho_s - \rho_f)g \, d_p}{(\mu \rho_f)^{1/2}} \right\} = 0.0178 \left\{ \frac{(5 \times 10^3 - 0.183)9.807 \times (5 \times 10^{-4})}{[(2 \times 10^{-5}) \, 0.183]^{1/2}} \right\} = \mathbf{228.11 \ m/s}.$$

**Example 2.7** Average particle diameter of copper ore is 0.1 mm and is roasted in a fluidized bed roaster at 550 °C. Calculate (i) minimum fluidization velocity, (ii) terminal velocity, and (iii) elutriation velocity. Given: densities of ore and air (at 550 °C) are $2.27 \times 10^3$ kg/m$^3$ and 0.43 kg/m$^3$ respectively; viscosity of air at 550 °C is $3.7 \times 10^{-5}$ kg/ms.

**Solution** Ga $= \left\{ \frac{\rho_f(\rho_s - \rho_f)g d_p^3}{\mu^2} \right\} = \left\{ \frac{0.43 \times (2270 - 0.43) \times 9.81 \times (10^{-4})^3}{(3.7 \times 10^{-5})^2} \right\} = 6.99$.

From Eq. (2.179): $Re'_{mf} = [\{(33.7)^2 + 0.0408 \ \text{Ga}\}^{0.5} - 33.7]$.

$= [\{(33.7)^2 + 0.0408 \times 6.99\}^{0.5} - 33.7] = 0.004$.

(i)  Re$_{mf} = \left( \frac{\rho_f u_{0,mf} d_p}{\mu} \right)$, therefore, u$_{0,mf} = \left( \frac{\mu \, Re_{mf}}{\rho_f \, d_p} \right) = \left( \frac{0.004 \times 3.7 \times 10^{-5}}{0.43 \times 10^{-4}} \right) = 0.0036$ m/s $= \mathbf{3.6}$

So, minimum fluidization velocity $= \mathbf{3.6 \ mm/s}$

(ii)  From Eq. (2.180): $u_t = \sqrt{\frac{4 d_p (\rho_s - \rho_f)g}{3 \rho_f . f_{fr}}} = \sqrt{\frac{4 \times 10^{-4} (2270 - 0.43) \, 9.81}{3 \, 0.43 \times 6000}}$     $[f_{fr} = 24/R_e = 24/0.004 = 6000]$

$= \mathbf{0.034 \ m/s} = \mathbf{34 \ mm/s}$

$= 0.034$ m/s $= 34$ mm/s

(iii)  For Re$_p < 0.4$: $u_E = \left\{ \frac{(\rho_s - \rho_f) g \, d_p^2}{18 \mu} \right\} = \frac{(2270 - 0.43) \, 9.81 \times 10^{-8}}{18 \times 3.7 \times 10^{-5}} = \mathbf{0.334 \ m/s} = \mathbf{334 \ mm/s}$

**Example 2.8** Calculate the velocity of a 1 μm slag particle, which is rising through stagnant liquid steel at 1600 °C. Given: Slag density is 3000 kg/m³ and density of liquid steel is 7000 kg/m³ and viscosity of liquid steel 7cP and 1000 cP = 1 kg/m s.

**Solution** From Stokes's law [Eq. (2.150)]:

$$
u_0 = \left[\frac{2r_0^2(\rho_s - \rho_f)g}{9\mu}\right] = \left[\frac{2 \times 0.5^2 10^{-12}(3000 - 7000)9.81}{9 \times 7 \times 10^{-3}}\right] = -3.1143 \times 10^{-7} \text{m/s}
$$

Minus velocity means rising against gravity.

**Example 2.9** During deoxidation of liquid steel in a furnace, the time taken for deoxidation products of radius $10^{-3}$ cm to rise through 40 cm of the liquid steel bath is 12 min. Calculate the viscosity of liquid steel given that the densities of liquid steel and the deoxidation product was 7000 and 2800 kg/m³, respectively, and g is 9.807 m/s².

**Solution**

Since density of the deoxidation product is less than the liquid steel, so that will rise, i.e. float.

The rate of rise, i.e. velocity, $u_0 = \frac{\text{distance}}{\text{time}} = \frac{0.4}{12 \times 60} = 5.56 \times 10^{-4}$ m/s.
From modified Stokes's law:

$$
\mu = \left[\frac{2r_0^2(\rho_f - \rho_s)g}{9u_0}\right] = \frac{2x10^{-10}x(7000 - 2800) \times 9.807}{9 \times 5.56 \times 10^{-4}} = 1.646 \times 10^{-3} \text{ kg/m.s.}
$$

## 2.12 Exercises

**Problem 2.1** Calculate the viscosity of a fluid flowing with laminar flow at the rate of 500 cm³/min in a capillary of diameter 60 mm given that the pressure drops over a length of 300 m of the capillary is $1.5 \times 10^5$ Pa. Find velocity of fluid. [Ans: 11.26 kg m⁻¹ s⁻¹, 2.95 mm/s].

**Problem 2.2** Water is being pumped through a horizontal pipe of 10 cm internal diameter and 50 m length at a flow rate of 1.0 m³/s. Friction factor ($f_{fr}$) is 0.006. Is the flow laminar or turbulent? Calculate the pressure difference and power required to maintain the water flow. (Given: $\rho_{H2O} = 10^3$ kg/m³ and $\mu_{H2O} = 0.8 \times 10^{-3}$ kg/m.s.) [Ans: Turbulent, $97.37 \times 10^6$ N/m² and 105,484 kw].

**Problem 2.3** Average particle diameter of magnetite ore is 0.1 mm and is roasted in a fluidized bed roaster at 700 °C. Calculate (i) minimum fluidization velocity, (ii) terminal velocity, and (iii) elutriation velocity. Given: densities of ore and air (at 700 °C) are $5.15 \times 10^3$ kg/m$^3$ and 0.363 kg/m$^3$ respectively; viscosity of air at 700 °C is $4.11 \times 10^{-5}$ kg/ms. [Ans: 7.3 mm/s, 68.11 mm/s and 682.8 mm/s].

**Problem 2.4** Calculate the velocity of a 1.5 μm slag particle, which is rising through a stagnant liquid steel at 1600 °C in ladle of 1 m height. Find out the time required for slag particle comes out. Given: Slag density is 3000 kg/m$^3$ and density of liquid steel is 7500 kg/m$^3$ and viscosity of liquid steel 8 cP. [Ans: $689.76 \times 10^{-9}$ m/s and 24.16 min].

## 2.13  Questions

Q1. Explains with example (a) Partial Time Derivative, (b) Total Time Derivative, and (c) Substantial Time Derivative.

Q2. Derive the equation for axial flow of fluid through a circular horizontal pipe by using (i) Equation of Continuity and (ii) Navier–Stokes equation.

Q3. Derive the equation for flow of falling fluid from (i) Equation of Continuity and (ii) Navier–Stokes equation.

Q4. Discuss Buckingham's Pi ($\pi$) Theorem.

Q5. Derive the equation for laminar, steady flow around a submerged solid sphere.

Q6. Derive the equation for fluid flow through a packed bed of solids.

Q7. Derive Ergun equation for fluid flow through a packed bed of solids.

## References

1. Mohanty AK (2012) Rate processes in metallurgy. PHI Learning Pvt Ltd, New Delhi
2. Bird RB, Stewart WE, Lightfoot EN (2006) Transport phenomena, 2nd Ed., Wiley, Singapore
3. Guthrie RIL (1992) Engineering in process metallurgy. Oxford University Press, USA
4. Sachdeva RC (1998) Fundamental of engineering heat and mass transfer. Wiley Eastern Ltd, New Delhi
5. Szekely J, Themalis NJ (1971) Rate phenomena in process metallurgy. Wiley Interscience, New York
6. Wen CY, Yu YH (1966) AIChE J, 12 p 610

# Chapter 3
# Heat Transfer

Basic concept of heat and modes of heat transfer are discussed. Fourier's law, thermal conductivity, concept of driving potential, and combined mechanism of heat transfer are described. General differential equation of heat conduction, equation of one-dimensional heat conduction, equation of systems with variable thermal conductivity, and equation of composite systems are derived. Systems with different heat sources, convective heat transfer, thermal radiation, concept of black body, and radiation from non-black surfaces are also discussed in this chapter.

## 3.1 Basic Concept

First question arises: what is heat?

*Heat* is a thermal energy due to that temperature increases to the materials.

Next question arises: what is heat transfer?

*Heat transfer* is the transfer of energy (i.e. heat) occurring due to temperature difference. *Heat transfer* is done due to a temperature gradient within the materials. Heat transfer must occur, whenever exists a temperature difference in a medium or between media.

There are three *modes of heat transfer*: (i) *conduction*, (ii) *convection*, and (iii) *radiation*.

(i)   Conduction: Conduction of heat is done through a solid or a stationary fluid, i.e. heat transfer is done through solid or liquid body. The term conduction refers to the heat transfer that can occur across the solid or liquid body. Heat can be conducted through solids, liquids, and gases [1]. Heat is conducted by the transfer of energy of motion between adjacent molecules. The kind of motion depends on the molecular state of the system and the range from the vibration of atoms in a crystal lattice to the random motion of molecules of a gas. There are different mechanisms in certain cases, as for example, the transfer of energy by free electrons in metallic solids [2].

(ii)   Convection: Convection of heat is done from a hot surface of a solid or a liquid to a moving fluid, i.e. heat exchange is done from solid or liquid body to moving fluid. The term convection refers to the heat transfer that can be occurred between a hot surface and a moving fluid, when they are at different temperatures [1]. Since motion of fluid is involved, heat transfer by convection is partially governed by the laws of fluid mechanics [2]. Convection of heat is possible only in the presence of a fluid medium [3].

(iii)  Radiation: The third mode of heat transfer is thermal radiation. Radiation of heat is done from a hot surface of a solid or a liquid to surrounding without help of media, i.e. heat exchange is done from solid or liquid body to surrounding without help of any media. All hot surfaces emit energy in the form of electro-magnetic waves [1], and no medium is required for its emission. Although energy can be transferred by radiation through gases, liquids, and solids; these media absorb partially or completely of the energy; so that energy is radiated most efficiently through empty space [2], i.e. under vacuum condition.

## 3.2   Temperature Field

The process of heat transfer can only take place when different points in a body (or a system) have different temperatures. Conduction in solids is generally occurred by variations of temperature in both space and time [3].

In general, temperature field is a function of space and time:

$$T = f(x, \ y, \ z, \ t) \tag{3.1}$$

The temperature fields are either steady or unsteady (i.e. transient) condition. If in a heat flow system, the temperature at each point is constant with time, it is referred as a steady state condition; and then the temperature is only a function of space coordinates.

Therefore,

$$T = f(x, y, z), \text{ and } \frac{\delta T}{\delta t} = 0 \tag{3.2}$$

If the temperature is a function of only two coordinates, and its variation in the third coordinate is negligible; it is called a two-dimensional case. Equation (3.1) becomes:

$$T = f(x, y, t), \text{ and } \frac{\delta T}{\delta z} = 0 \tag{3.3}$$

Similarly, if the temperature is a function of only one coordinate, the system is called one-dimensional:

$$T = f(x, t), \text{ and } \frac{\delta T}{\delta y} = \frac{\delta T}{\delta z} = 0 \tag{3.4}$$

The equation for a one-dimensional, steady state temperature field becomes [3]:

$$T = f(x), \frac{\delta T}{\delta t} = 0 \text{ and } \frac{\delta T}{\delta y} = \frac{\delta T}{\delta z} = 0 \tag{3.5}$$

i.e. $\frac{\delta T}{\delta x} \neq 0$.

## 3.3 Conduction

Conduction may be termed as the transfer of energy from the more energetic particles to the less energetic particles of a substance due to interactions between the particles. Now considering a gas in which there exists a temperature gradient and assuming there is no bulk motion. The gas may occupy the space between two surfaces that are maintained at different temperatures, as shown in Fig. 3.1. The energy of gas molecules is related to the random translational motion, as well as to the internal rotational and vibrational motions.

Higher temperature is associated with higher molecular energies, and when neighboring molecules collide, a transfer of energy from the more energetic molecules to the less energetic molecules must occur. In the presence of a temperature gradient, energy transfer by conduction must occur in the direction of lower temperature. Considering a hypothetical plane at $x_0$ is constantly crossed by molecules from above and below due to their random motion. However, molecules from above are associated with a higher temperature than those from below, in which case there must be a net transfer of energy in the positive x direction. The net transfer of energy by random molecular motion is referred as a *diffusion of energy* [1].

In a solid, conduction may be due to atomic activity in the form of electron movement. The energy transfer to lattice waves is induced by atomic motion. In a non-conductor material, the energy transfer is exclusively via these lattice waves; in

**Fig. 3.1** Heat transfer due to conduction with different temperatures

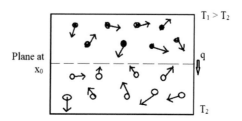

a conductor material, it is also due to the motion of free electrons [1]. As for example, when a metal rod is placed in a hot furnace, metal rod immediately becomes hot due to conduction of heat i.e. due to movement of electrons, through metal rod.

### 3.3.1 Fourier's Law

Law of heat transfer was first proposed by French scientist J. B. J. Fourier (1822). Fourier proposed that the rate at which heat is conducted through a solid is proportional to the cross-sectional area perpendicular to the heat flow, the temperature gradient in the solid at the location and finally, the nature of the material itself (i.e. its thermal conductivity). It is possible to quantify heat transfer processes in term of rate equations. These equations may be used to calculate the amount of energy being transferred per unit time. For heat conduction, the rate equation is known as *Fourier's law*. For one-dimensional plane wall is shown in Fig. 3.2, having a temperature distribution, T(x). The rate of heat conduction (q) is proportional to the area (A) perpendicular to the x direction of heat transfer, and to the temperature gradient $\left(\frac{dT}{dx}\right)$ in x direction, i.e.

$$q_x \infty\, A \left(\frac{dT}{dx}\right) \tag{3.6}$$

The rate equation can be expressed as:

$$q_x = -kA\left(\frac{dT}{dx}\right) \text{ or } q'_x = -k\left(\frac{dT}{dx}\right) \left(\text{since } \frac{q_x}{A} = q'_x\right) \tag{3.7}$$

where $q'_x$ is heat flux (W.m$^{-2}$), and k is the proportionality constant, which is heat transfer property known as the thermal conductivity (W.m$^{-1}$.K$^{-1}$) of the conducting material.

**Fig. 3.2** One-dimensional
heat transfer by conduction

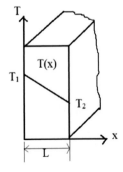

*Heat flux* $q'_x$ is the heat transfer rate in the x direction per unit area perpendicular to the direction of heat transfer. The thermal conductivity is a physical property of the substance and is defined as the ability of a substance to conduct heat [3].

The minus sign [in Eq. (3.7)] means heat is transferred in the direction of lower temperature. Steady-state conduction means that heat transfer rate does not vary with time. Under the steady-state condition shown in Fig. 3.2, where the temperature distribution is linear, the temperature gradient may be expressed as:

$$\frac{dT}{dx} = \frac{(T_2 - T_1)}{L} \qquad (3.8)$$

Now Eq. (3.7) becomes:

$$q'_x = -k\left\{\frac{(T_2 - T_1)}{L}\right\} = k\left\{\frac{(T_1 - T_2)}{L}\right\} = k\left(\frac{\Delta T}{L}\right) \quad \text{(Since } T_1 > T_2) \qquad (3.9)$$

**Example 3.1** A stainless steel plate 3.0 cm thick has a temperature of 550 °C at one face and 50 °C on the other. The thermal conductivity of stainless steel at 300 °C is 19.1 W/m·K. Calculate the heat transfer through the stainless steel per unit area.

**Solution**

According to Eq. (3.9): $q'_x = k\left\{\frac{(T_1 - T_2)}{L}\right\}$.

So the heat transfer$= 19.1 \times \left\{\frac{(823 - 323)}{3.0 \times 10^{-2}}\right\} = \mathbf{320\,kW/m^2}$.

## 3.3.2 Thermal Conductivity

Students must have some knowledge of the conduction mechanisms and the magnitudes of the thermal conductivity in various systems. Hence, thermal conductivity is briefly discussed here. For more information, students can refer textbooks of Modern Physics or Physical Chemistry.

Thermal conductivity is a physical property of a substance, like viscosity of a liquid, it is primarily a function of temperature and pressure. The effect of pressure on the conductivities of solids and liquids has received very little attention from engineers, probably because of their primary concern with applications at one atmospheric pressure [2]. The conductivities of gases have been found to increase with pressure. But the thermal conductivity of an ideal gas is independent of pressure.

The thermal conductivity for most materials can be determined experimentally by measuring the rate of heat flow and temperature gradient of the given substance. Table 3.1 shows the thermal conductivities of the common substances. The metallic solids have higher thermal conductivity values than non-metallic solids. The thermal conductivities of gases have the lowest values.

**Table 3.1** Thermal conductivities of the common substances

| Substance | Thermal conductivity (k), W/mK | |
|---|---|---|
| | at 20 °C | at 300 °C |
| Pure silver | 417.5 | 362.3 |
| Pure copper | 386.0 | 368.7 |
| Pure aluminum | 204.0 | 228.5 |
| Pure zinc | 112.5 | 100.5 |
| Mild steel (0.5% C) | 55.4 | 45.0 |
| Lead | 35.0 | 29.8 |
| Stainless steel | 16.3 | 19.1 |
| Mercury | 8.18 | 8.54 |
| Wood | 0.15 | – |
| Asbestos (fiber) | 0.095 | – |
| Water | 0.659 | 0.613 |
| Air | 0.018 | 0.026 |
| CO | 0.017 | 0.025 |
| $H_2$ | 0.131 | 0.183 |

The thermal conductivity is also a function of temperature, for most pure metals it decreases with increasing temperature, whereas for gases and insulating materials it increases with rise in temperature.

## 3.4  Convection

The convection heat transfer mode is comprised of two motions. In addition to energy transfer, due to random molecular motion (i.e. diffusion), is also transferred by the bulk, i.e. motion of fluid. This fluid motion is associated with large numbers of molecules, which are moving collectively. Such motion, in the presence of temperature gradient, contributes to heat transfer. The total heat transfer is then due to a superposition of energy transport by the random motion of the molecules and by the bulk motion of the fluid [1].

The convection heat transfer occurs between a fluid in motion and a bounding surface when the two are at different temperatures. Figure 3.3 shows fluid flow over the heated surface. A consequence of the fluid–surface interaction is the development of a region in the fluid through which the velocity varies from zero at the surface to a finite value, $u_\infty$ associated with the flow. This region of the fluid is known as velocity boundary layer. If the surface and flow temperatures differ, there will be a region of the fluid through which the temperature varies from $T_s$ at y = 0 (at the surface) to $T_\infty$ at y = y (in the outer flow). This region, called the thermal boundary layer, may be smaller, larger, or the same size as that through which the velocity varies. In any

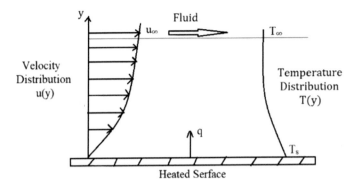

**Fig. 3.3**  Boundary layer development in convection heat transfer

case, if $T_s > T_\infty$, convection heat transfer will occur between the heated surface and the outer flow.

Convection heat transfer is classified according to the nature of the flow as follows:

(i)  *Forced convection*, when the flow is caused by external means, such as by a fan, or a blower etc. As for example, considering the use of a fan provides forced convection air cooling of hot components.

(ii)  *Free or natural convection*, the flow is induced by buoyancy forces, which arise from density differences caused by temperature variations within the fluid. As for example, free or natural convection occurs from hot components in stagnant air.

Convection heat transfer is a mode of energy transfer occurring within a fluid due to the conduction and bulk fluid motion. Typically, the energy, that is being transferred, is the sensible or internal thermal energy of the fluid. However, there are convection processes for which there is *latent heat* exchange. This latent heat exchange is generally associated with a phase change between (a) solid and liquid i.e. melting; or (b) liquid and vapor, i.e. boiling etc. [1].

Convective heat flux is proportional to the difference of temperatures between the surface and bulk fluid:

$$q' \infty (T_s - T_\infty) \tag{3.10}$$

where $q'$ is the convective heat flux $(W.m^{-2})$; $T_s$ and $T_\infty$ are the surface and bulk fluid temperature respectively.

The rate equation for convection heat transfer can be written as:

$$q' = h(T_s - T_\infty) \tag{3.11}$$

where h is the proportionality constant (W.m$^{-2}$.K$^{-1}$), which is termed as *convection heat transfer coefficient*. It depends on conditions in the boundary layer, which are influenced by surface geometry and the nature of the fluid motion etc.

Equation (3.11) is also known as *Newton's law of cooling*. When Eq. (3.11) is used, the convection heat flux is presumed to be positive, if heat is transferred from the surface ($T_s > T_\infty$); and negative if heat is transferred to the surface ($T_s < T_\infty$).

If $T_\infty > T_s$, then the equation of Newton's law of cooling becomes:

$$q' = h\ (T_\infty - T_s) \tag{3.12}$$

**Example 3.2** A hot flat metal plate at 500 °C, length 3 m and width 400 cm, is placed for cooling in blowing air at 25 °C. The convective heat transfer coefficient is 30 W/m$^2$K. Calculate the convective heat transfer of the plate.

**Solution**

The convective heat transfer,

$$
\begin{aligned}
q &= hA(T_s - T_\infty) \\
  &= 30 \times (3 \times 0.4)(773 - 298) \\
  &= \mathbf{17.1\,kW.}
\end{aligned}
$$

## 3.5  Radiation

Thermal radiation is energy emitted by material, which is at a high temperature. Radiation may be occurred from solid surfaces, liquids, and gases. Regardless of the form of material, the emission may be attributed to changes in the electron configurations of the atoms or molecules. The energy of the radiation field is transported by electromagnetic waves (or alternatively photons). While the transfer of energy by conduction or convection requires the presence of a medium, but radiation does not require any medium [1]. In fact, radiation energy transfer occurs most efficiently in a vacuum.

Radiation is emitted by the surface originates from the thermal energy of matter bounded by the surface, and the rate at which energy is released per unit area is termed the surface *emissive power,* E (W.m$^{-2}$). *Stefan-Boltzmann law* helps to find out emissive power of matter:

$$E_b = \sigma\,T_s^4 \tag{3.13}$$

where $T_s$ is the temperature (K) of the surface, and $\sigma$ is the *Stefan-Boltzmann constant* ($\sigma = 5.67 \times 10^{-8}$ W·m$^{-2}$·K$^{-4}$). Such a surface is called an ideal radiator or *blackbody*. So $E_b$ is the emissive power of blackbody.

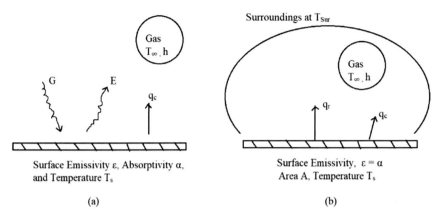

Fig. 3.4 Radiation exchange **a** at a surface and **b** between surface and large surroundings

The heat flux emitted by a real surface is less than that of a blackbody at the same temperature and is given by:

$$E = \varepsilon\, \sigma\, T_s^4 \tag{3.14}$$

where $\varepsilon$ is a radiative property of the surface termed as *emissivity*, which provides a measure of how efficiently a surface emits energy relative to a blackbody; values in the range $0 \le \varepsilon \le 1$.

Radiation may also be incident on a surface from its surroundings. The radiation may originate from a special source, such as sun. Irrespective of the source, the rate at which all such radiation is incident on a unit area of the surface as the *irradiation* G (as shown in Fig. 3.4).

A portion of the irradiation may be absorbed by the surface, thereby increasing the thermal energy of the material. The rate at which radiant energy is absorbed per unit surface area may be calculated from the knowledge of a surface radiative property termed *absorptivity* $\alpha$.

$$G_{abs} = \alpha G \quad (\text{where } 0 \le \alpha \le 1) \tag{3.15}$$

If $\alpha < 1$ and the surface is *opaque*, portions of the irradiation are reflected. If the surface is *semi-transparent*, portion of the irradiation may also be transmitted. However, while absorbed and emitted radiation increase and decrease, respectively, the thermal energy of matter, reflected and transmitted radiation have no effect on this energy. The value of $\alpha$ depends on the nature of the irradiation, as well as on the surface itself.

If radiation exchange between a small surface at $T_s$ and a much larger, isothermal surface completely surrounds the smaller one (Fig. 3.4b). The surroundings could be the walls of a room or a furnace whose temperature $T_{sur}$ differs from that of an enclosed surface ($T_{sur} \neq T_s$). For such a condition, the irradiation may be approximated by emission from a blackbody at $T_{sur}$:

$$G = \sigma\, T_{sur}^4 \tag{3.16}$$

If the surface is assumed to be one for which $\alpha = \varepsilon$ (a gray surface), the net rate of radiation heat transfer from the surface, expressed per unit area of the surface as:

$$q'_{rad} = \frac{q}{A} = \{\varepsilon E_b - \alpha G\} = \varepsilon\, \sigma\, \left(T_s^4 - T_{sur}^4\right) \tag{3.17}$$

This Eq. (3.17) provides the difference between thermal energy that is released due to radiation emission, and which is gained due to radiation absorption.

There are many applications for which it is convenient to express the net radiation heat exchange in the form:

$$q_{rad} = h_r\, A(T_s - T_{sur}) \tag{3.18}$$

where $h_r$ is the *radiation heat transfer coefficient*.

$$h_r = \varepsilon\, \sigma\, (T_s + T_{sur})\left(T_s^2 + T_{sur}^2\right) \tag{3.19}$$

The surface of Fig. 3.4 may also simultaneously transfer heat by convection to an adjoining gas. For the condition Fig. 3.4b, the total rate of heat transfer from the surface is then written as:

$$q = q_{conv} + q_{rad} = hA(T_s - T_\infty) + \varepsilon\, A\sigma\, \left(T_s^4 - T_{sur}^4\right) \tag{3.20}$$

**Example 3.3** A radiator in a heating system operates at a surface temperature of 47 °C. Determine the rate at which it emits radiant heat per unit area if it behaves as a black body.

**Solution**

According to Eq. (3.13):

$$\begin{aligned}
E_b &= \sigma\, T_s^4 \\
&= \left(5.67 \times 10^{-8}\right) \times (273 + 47)^4 = 5.945 \times 10^2\,\text{W/m}^2 \\
&= \mathbf{0.59\,kW/m^2}.
\end{aligned}$$

## 3.6  Concept of Driving Potential

Fourier's law in term of rate of heat conduction [Eq. (3.9)] can be written as:

$$q_x = k\,A\left\{\frac{(T_1 - T_2)}{L}\right\} = kA\left(\frac{\Delta T}{L}\right) = \left(\frac{\Delta T}{L/kA}\right) \tag{3.21}$$

Now considering the heat transfer rate as a flow and (L/kA) as a resistance to this flow, then the temperature becomes the potential, or driving function for heat flow. i.e.

$$\text{Heat flow} = \frac{Thermal\ potential\ difference}{Thermal\ resistance} \quad \text{or} \quad q = \left(\frac{\Delta T}{R_{ther}}\right) \tag{3.22}$$

Equation (3.21) is similar to equation of Ohm's law for electric circuit. The thermal resistance ($R_{ther}$) is (L/kA) with unit of (K/W).

An analogy between the flow of heat through a thermal resistance and the flow of direct current through an electric resistance, then a thermal circuit must be represented by electric circuit and vice versa. The electric analogy may be used to solve more complex problems involving both series and parallel thermal resistance [3].

The thermal resistance for the case of a plane wall (Fig. 3.5a) is represented by electric circuit in Fig. 3.5b. The inverse of thermal resistance is known as *thermal conductance* $(K)$ and is equal to the amount of heat conducted through a solid of area (A) and thickness (L) per temperature difference.

$$K = \frac{kA}{L} \tag{3.23}$$

Thermal conductance and thermal resistance for convection and radiation modes of heat can be written as follows:

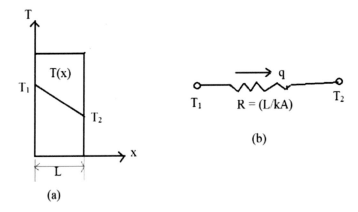

Fig. 3.5  a Plane wall and b its electric circuit

Convection: [From Eq. (3.12)]

$$q = hA \ \Delta T = \left( \frac{\Delta T}{1/hA} \right) \tag{3.24}$$

So thermal resistance for convection,

$$R_{ther,c} = \left( \frac{1}{hA} \right) \tag{3.25}$$

and thermal conductance for convection,

$$K_c = hA \tag{3.26}$$

Radiation: [From Eq. (3.17)]:

$$q = A\varepsilon \ \sigma \ \left( T_s^4 - T_{sur}^4 \right)$$
$$= \frac{\left( T_s^4 - T_{sur}^4 \right)}{(1/A\varepsilon \ \sigma)} \tag{3.27}$$

So thermal resistance for radiation,

$$R_{ther,r} = \left( \frac{1}{A\varepsilon \ \sigma} \right) \tag{3.28}$$

and thermal conductance for radiation,

$$K_r = A\varepsilon \ \sigma \tag{3.29}$$

## 3.7   Combined Mechanism of Heat Transfer

It is not unusual that the heat transfer is taking place due to two, or all three modes. The most frequently encountered instance is one in which a solid wall (usually plane) separates two fluids, e.g. the tubes of a heat exchanger [3].

The overall heat transfer by combined modes is usually expressed in terms of an *overall conductance* or *overall heat transfer coefficient* (U), defined by the relation:

$$q = UA \ (\Delta T) \tag{3.30}$$

The overall heat transfer coefficient is a quantity such that the rate of heat flow through a configuration is given by taking a product of U, the surface area and overall temperature difference.

### 3.7.1 Heat Transfer by Combined Convection and Conduction Modes

Figure 3.6 shows the overall heat transfer through a plane wall, which is heated one side by a hot fluid $F_1$ and cooled on the other side by a cold fluid $F_2$. The heat transfer by hot fluid $F_1$ to plane wall is by convection mode, the heat transfer within the wall is by conduction mode, and finally the heat transfer by cold fluid $F_2$ is again by convection mode.

So, the heat transfer rate is given by:

$$q = h_1 A(T_1 - T_2) = \frac{kA}{L}(T_2 - T_3) = h_2 A(T_3 - T_4) \qquad (3.31)$$

Therefore, for convection due to hot fluid:
$(T_1 - T_2) = \frac{q}{h_1 A}$ (a)
For conduction within plane wall:
$(T_2 - T_3) = \frac{q}{kA/L}$ (b)
Again, for convection due to cold fluid:
$(T_3 - T_4) = \frac{q}{h_2 A}$ (c)
By adding Eqs. (a) to (c) to eliminate the unknown temperatures $T_2$ and $T_3$:
$(T_1 - T_4) = q\left\{ \frac{1}{h_1 A} + \frac{1}{kA/L} + \frac{1}{h_2 A} \right\}$ (d)
Therefore,

$$q = \left\{ \frac{(T_1 - T_4)}{\left( \frac{1}{h_1 A} + \frac{L}{kA} + \frac{1}{h_2 A} \right)} \right\} \qquad (3.32)$$

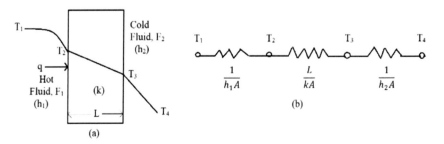

**Fig. 3.6 a** Overall heat transfer through plane wall and **b** its electric circuit

where $\left(\frac{1}{hA}\right)$ and $\left(\frac{L}{kA}\right)$ are the convection resistance and conduction resistance, respectively.

Equation (3.32) is compared with Eq. (3.30), the overall heat transfer coefficient can be written as:

$$U = \left\{ \frac{1}{\left(\frac{1}{h_1} + \frac{L}{k} + \frac{1}{h_2}\right)} \right\} = \frac{1}{\Sigma R_{ther}} \qquad (3.33)$$

The overall heat transfer coefficient depends upon the geometry of the separating wall, its thermal properties, and the convective coefficients at the two surfaces. The overall heat transfer coefficient is particularly useful in the case of composite walls, such as in the design of walls for boiler, refrigerators, air-conditioned buildings, etc.

### 3.7.2  Heat Transfer by Combination of Three Modes

Suppose the inner surface of a combustion chamber wall is received heat from the hot products of combustion. The other outer side of wall is being cooled by a cold fluid (Fig. 3.7). The heat transfer is taking place in three ways, viz. (i) heat is transferred from the hot gases of combustion to the inner surface of wall by both convection and radiation simultaneously, i.e. the convection and radiation resistances are parallel in electric circuit; (ii) heat is conducted through the wall by conduction, and (iii) heat flows by convection from the outer surface of the wall to the cold fluid, assuming there is negligible radiation.

Step 1:

$$q = q_c + q_r = h_c A \, (T_1 - T_2) + h_r A \, (T_1 - T_2)$$

$$= \left\{ \frac{(T_1 - T_2)}{\left(\frac{1}{h_c A} + \frac{1}{h_r A}\right)} \right\} = \left\{ \frac{(T_1 - T_2)}{R_1} \right\} \qquad (3.34)$$

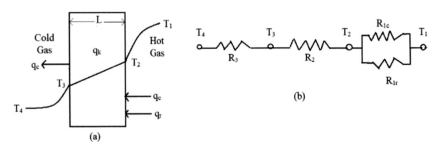

Fig. 3.7  a Overall heat transfer through plane wall and b it's electric circuit

Step 2:

$$q = q_k = \frac{kA}{L}(T_2 - T_3) = \left\{ \frac{(T_2 - T_3)}{(L/kA)} \right\} = \left\{ \frac{(T_2 - T_3)}{R_2} \right\} \quad (3.35)$$

Sep 3:

$$q = q_c = h_C A\,(T_3 - T_4) = \left\{ \frac{(T_3 - T_4)}{\left(\frac{1}{h_C A}\right)} \right\} = \left\{ \frac{(T_3 - T_4)}{R_3} \right\} \quad (3.36)$$

where $q_c$, $q_r$, and $q_k$ are the heat transfer for convection, radiation, and conduction, respectively.

$h_c$ and $h_r$ are the heat transfer coefficient for convection and radiation, respectively.

A is the area of plane and L is the thickness of wall.

k is the thermal conductivity.

$T_1$, $T_2$, $T_3$, and $T_4$ are the temperature of hot gases, temperature of inner surface of wall, temperature of outer surface of wall on the cooling side, and temperature of cold fluid, respectively.

R is the thermal resistance for electric circuit.

$R_1$ is combined thermal resistance due to convection ($R_{1c}$) and radiation ($R_{1r}$) by the hot gases $= \left( \frac{1}{h_c A} + \frac{1}{h_r A} \right)$.

$R_2$ is the thermal resistance of wall due to conduction $= \frac{L}{k\,A}$

$R_3$ is the thermal resistance due to convection between wall and cold fluid $= \frac{1}{h_c\,A}$.

By adding Eqs. (3.34) to (3.36):

$$q = \left\{ \frac{(T_1 - T_4)}{R_1 + R_2 + R_3} \right\} \quad (3.37)$$

Equation (3.37) is for heat flow, which is an analogous to Ohm's law of direct current flow:

$$i = \frac{\Delta E}{R} \quad (3.38)$$

where i is the current, $\Delta E$ is the electrical potential, and R is the resistance.

The overall heat transfer coefficient,

$$U = \left\{ \frac{1}{R_1 + R_2 + R_3} \right\} \quad (3.39)$$

## 3.8   General Differential Equation of Heat Conduction

### 3.8.1   Cartesian Coordinates System

Considering a differential element, i.e. small control volume in Cartesian coordinates have sides δx, δy, and δz as shown in Fig. 3.8. The energy balance, for this control volume, can be obtained from the first law of thermodynamics as follows:

[{I. Net heat conducted into control volume per unit time} + {II. Internal heat generated per unit time}]

$$= [\{III.\ Increase\ in\ internal\ energy\ per\ unit\ time\} + \{IV.\ Work\ done\ by\ control\ volume\ per\ unit\ time\}]$$

$$(3.40)$$

If $q_x$ is the heat transfer in x direction at x, face ABCD and $q_{x+\delta x}$ is the heat transfer in x direction at x + δx, face EFGH. The rate of heat flow into the control volume in the x direction through face ABCD:

$$q_x = -k \left( \frac{\delta T}{\delta x} \right) \delta y\ \delta z \qquad\qquad (3.41)$$

where k is the thermal conductivity of material, which is independent of position and temperature.

$\left( \frac{\delta T}{\delta x} \right)$ is the temperature gradient in x direction.

The rate of heat flow out from the control volume in the x direction through face EFGH at x + δx:

$$q_{x+\delta x} = q_x + \left( \frac{\delta q_x}{\delta x} \right) \delta x = -k \left( \frac{\delta T}{\delta x} \right) \delta y\ \delta z - \frac{\delta}{\delta x} \left\{ k \left( \frac{\delta T}{\delta x} \right) \right\} \delta x \delta y \delta z \qquad (3.42)$$

Thus, net rate of heating entering the element in x direction is the difference between the entering and leaving heat flow rates:

**Fig. 3.8** Differential element in Cartesian coordinates system

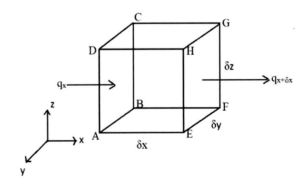

$$q_x - q_{x+\delta x} = \frac{\delta}{\delta x}\left\{k\left(\frac{\delta T}{\delta x}\right)\right\}\delta x\ \delta y\ \delta z \tag{3.43}$$

Similarly, net rate of heating entering the element in y direction and z direction are the difference between the entering and leaving heat flow rates:

$$q_y - q_{y+\delta y} = \frac{\delta}{\delta y}\left\{k\left(\frac{\delta T}{\delta y}\right)\right\}\delta x\ \delta y\ \delta z \tag{3.44}$$

$$q_z - q_{z+\delta z} = \frac{\delta}{\delta z}\left\{k\left(\frac{\delta T}{\delta z}\right)\right\}\delta x\ \delta y\ \delta z \tag{3.45}$$

Hence, the net heat conducted into control volume per unit time, term I in Eq. (3.40):

$$I = \left[\frac{\delta}{\delta x}\left\{k\left(\frac{\delta T}{\delta x}\right)\right\} + \frac{\delta}{\delta y}\left\{k\left(\frac{\delta T}{\delta y}\right)\right\} + \frac{\delta}{\delta z}\left\{k\left(\frac{\delta T}{\delta z}\right)\right\}\right]\delta x\ \delta y\ \delta z \tag{3.46}$$

Suppose $q\cdot$ is the internal heat generation per unit volume and per unit time (W/m$^3$.s), so internal heat generated per unit time, term II in Eq. (3.40):

$$II = q\cdot\delta x\ \delta y\ \delta z \tag{3.47}$$

The change in internal energy for the control volume over a period of time, dt: (mass of control volume) × (specific heat) × (change in temperature of control volume in time dt)

$$= (\rho\ \delta x\ \delta y\ \delta z)\left(C_p\right)dT = \left(\rho\ C_p\ dT\right)\delta x\ \delta y\ \delta z \tag{3.48}$$

where $\rho$ and $C_p$ are the density and specific heat of the material of the control volume. Then the change in internal energy for the control volume per unit time, term III in Eq. (3.40):

$$III = \rho\ C_p\left(\frac{\delta T}{\delta t}\right)\delta x\ \delta y\ \delta z \tag{3.49}$$

The last term (IV) of Eq. (3.40) is very small because the work done by solids due to temperature changes is negligible.

Substituting of Eqs. (3.46), (3.47) and (3.49) into Eq. (3.40) leads to the general three-dimensional equation for heat conduction:

$$\left[\frac{\delta}{\delta x}\left\{k\left(\frac{\delta T}{\delta x}\right)\right\} + \frac{\delta}{\delta y}\left\{k\left(\frac{\delta T}{\delta y}\right)\right\} + \frac{\delta}{\delta z}\left\{k\left(\frac{\delta T}{\delta z}\right)\right\}\right] + q\cdot = \rho\ C_p\left(\frac{\delta T}{\delta t}\right) \tag{3.50}$$

Equation (3.50) is the general form, in Cartesian coordinates, of the *heat diffusion equation*; or simply the *heat conduction equation*.

Special forms of the heat conduction Eq. (3.50) as follows:

(i)   Uniform thermal conductivity:

If the thermal conductivity (k) is a constant, then the heat conduction Eq. (3.50) becomes:

$$\left(\frac{\delta^2 T}{\delta x^2} + \frac{\delta^2 T}{\delta y^2} + \frac{\delta^2 T}{\delta z^2}\right) + \frac{q^{\cdot}}{k} = \frac{\rho C_p}{k}\left(\frac{\delta T}{\delta t}\right) = \frac{1}{\alpha}\left(\frac{\delta T}{\delta t}\right) \tag{3.51}$$

where $\alpha$ is the thermal diffusivity of the material (m$^2$/s) is equal to $\frac{k}{\rho C_p}$.

The physical significance of thermal diffusivity is that how fast heat is propagated or diffusion through a material during changes of temperature with time. The larger the thermal diffusivity, the shorter is the time required for the applied heat to penetrate deeper into the solid material.

Vectoral form:

$$\nabla^2 T + \frac{q^{\cdot}}{k} = \frac{1}{\alpha}\left(\frac{\delta T}{\delta t}\right) \tag{3.52}$$

where $\nabla^2 T = \frac{\delta^2 T}{\delta x^2} + \frac{\delta^2 T}{\delta y^2} + \frac{\delta^2 T}{\delta z^2}$, is called the *Laplacian* (operator) of T.

(ii)   Steady-state conditions:

The temperature at any point in the control volume does not change with time, i.e. $\frac{\delta T}{\delta t} = 0$.

The heat conduction Eq. (3.51) becomes:

$$\left(\frac{\delta^2 T}{\delta x^2} + \frac{\delta^2 T}{\delta y^2} + \frac{\delta^2 T}{\delta z^2}\right) + \frac{q^{\cdot}}{k} = 0 \tag{3.53}$$

or

$$\nabla^2 T + \frac{q^{\cdot}}{k} = 0 \tag{3.54}$$

Equation (3.54) is called *Poisson equation*.

(iii)   No heat generation:

In the absence of any heat generation or release of energy (i.e. $q^{\cdot} = 0$) within the control volume, Eq. (3.51) becomes:

$$\frac{\delta^2 T}{\delta x^2} + \frac{\delta^2 T}{\delta y^2} + \frac{\delta^2 T}{\delta z^2} = \frac{1}{\alpha}\left(\frac{\delta T}{\delta t}\right) \tag{3.55}$$

In addition, if the process is in a steady state, then Eq. (3.55) further modified to:

$$\frac{\delta^2 T}{\delta x^2} + \frac{\delta^2 T}{\delta y^2} + \frac{\delta^2 T}{\delta z^2} = 0 \qquad\qquad (3.56)$$

or

$$\nabla^2 T = 0 \qquad\qquad (3.57)$$

Equation (3.57) is called Laplace *equation*.

(iv)   One-dimensional heat conduction equation:

If the temperature varies only in the x direction, with steady state (i.e. $\frac{\delta T}{\delta t} = 0$) and no heat generation (i.e. $q^{\cdot} = 0$), one-dimensional heat conduction Eq. (3.51) becomes:

$$\frac{\delta^2 T}{\delta x^2} = 0 \qquad (\text{since } \frac{\delta^2 T}{\delta y^2} = \frac{\delta^2 T}{\delta z^2} = 0 \text{ for one-dimensional heat conduction})$$

$$(3.58)$$

## 3.8.2   *Cylindrical Coordinates System*

The heat conduction equation derived in the previous section can be used for solids with rectangular shape like slabs, cubes etc. But there are other shapes of solids like cylinders, tubes, cones, spheres for which the Cartesian coordinate system is not applicable. For cylindrical solids, a cylindrical coordinate (r, θ, z) should be used. The heat conduction equation in cylindrical coordinate can be obtained by doing an energy balance over a differential element. As described previously similar procedure is followed. The following general form of the heat conduction equation is obtained:

$$\frac{1}{r}\frac{\delta}{\delta r}\left(kr\frac{\delta T}{\delta r}\right) + \frac{1}{r^2}\frac{\delta}{\delta\theta}\left(k\frac{\delta T}{\delta\theta}\right) + \frac{\delta}{\delta z}\left(k\frac{\delta T}{\delta z}\right) + q^{\cdot} = \rho\, C_p\left(\frac{\delta T}{\delta t}\right) \qquad (3.59)$$

## 3.8.3   *Spherical Coordinates System*

The heat conduction equation in spherical coordinate (r, θ, φ) can be obtained by doing an energy balance over a differential element. As described previously, similar procedure is followed. The following general form of the heat conduction equation is obtained:

$$\frac{1}{r^2}\frac{\delta}{\delta r}\left(kr^2\frac{\delta T}{\delta r}\right) + \frac{1}{r^2\sin^2\theta}\frac{\delta}{\delta\varphi}\left(k\frac{\delta T}{\delta\varphi}\right) + \frac{1}{r^2\sin\theta}\frac{\delta}{\delta\theta}\left(k\sin\theta\frac{\delta T}{\delta\theta}\right) + q^{\cdot} = \rho\,C_p\left(\frac{\delta T}{\delta t}\right)$$

$$(3.60)$$

Equations for diffusion of heat, i.e. heat conduction equations are shown in *Appendix IV.*

## 3.9  One-Dimensional Heat Conduction

If the temperature field in a system can be described in terms of only one space coordinate, the system is called one-dimensional. In addition to the plane walls, many different shapes can be described by one-dimensional systems, viz., the cylindrical and spherical, if the temperature is a function of radial distance and is independent of axial distance or azimuth angle. A two-dimensional system can also be a one-dimensional system of the effect of the second space coordinate is negligible. Similarly, multi-dimensional problems can also be initially tackled by a one-dimensional analysis because, in these cases, the differential equations are mostly simplified and yield easier solutions [3].

### 3.9.1  Plane Wall

Considering a plane wall, of a material of uniform thermal conductivity (k), are extended to infinity in y and z directions (as shown in Fig. 3.9). The walls of a furnace may be considered as a plane, if the heat lost through the edges is negligible. The heat transfer occurs by conduction through the wall and there is no heat transfer by convection inside and outside of the furnace. The temperature is only a function of x.

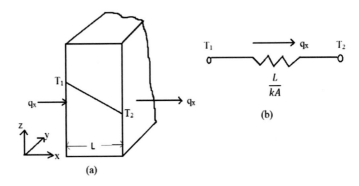

**Fig. 3.9  a** Heat conduction through plane wall and **b** its electric circuit

(i)   Starting with the general form of the heat conduction equation in Cartesian coordinates [from Appendix IV Eq. (IV.1)]:

$$\left[\frac{\delta}{\delta x}\left\{k\left(\frac{\delta T}{\delta x}\right)\right\} + \frac{\delta}{\delta y}\left\{k\left(\frac{\delta T}{\delta y}\right)\right\} + \frac{\delta}{\delta z}\left\{k\left(\frac{\delta T}{\delta z}\right)\right\}\right] + q^{\cdot} = \rho\, C_p\left(\frac{\delta T}{\delta t}\right)$$

or

$$\left(\frac{\delta^2 T}{\delta x^2} + \frac{\delta^2 T}{\delta y^2} + \frac{\delta^2 T}{\delta z^2}\right) + \frac{q^{\cdot}}{k} = \frac{1}{\alpha}\left(\frac{\delta T}{\delta t}\right) \tag{3.51}$$

where $\alpha$ is the thermal diffusivity of the material (m$^2$/s) is equal to $\frac{k}{\rho C_p}$.

Under steady-state condition: $\frac{\delta T}{\delta t} = 0$,

Without heat generation: $\frac{q^{\cdot}}{k} = 0$,

Since the heat conduction is through x direction only: $\frac{\delta^2 T}{\delta y^2} = \frac{\delta^2 T}{\delta z^2} = 0$.

The conduction equation is simplified to:

$$\frac{\delta^2 T}{\delta x^2} = 0 \text{ or } \frac{d^2 T}{dx^2} = 0 \tag{3.61}$$

Equation (3.61) is a second-order differential equation that required two boundary conditions:

(a)   at x = 0, T = $T_1$ , and
(b)   at x = L, T = $T_2$ .

By integration Eq. (3.61): $\frac{dT}{dx} = C_1$

and further integration:

$$T = C_1\, x + C_2 \tag{3.62}$$

where C is integration constant.

Now putting the boundary condition to Eq. (3.62):

(i).   at x = 0, T = $T_1$: $T_1 = C_2$; and
(ii).  at x = L, T = $T_2$: $T_2 = C_1 L + C_2 = C_1 L + T_1$

or

$$C_1 = \left(\frac{T_2 - T_1}{L}\right)$$

Putting the values of $C_1$ and $C_2$ in Eq. (3.62), the equation for temperature distribution becomes:

$$T = \left(\frac{T_2 - T_1}{L}\right) x + T_1 \tag{3.63}$$

or

$$\frac{dT}{dx} = \left(\frac{T_2 - T_1}{L}\right)$$

Therefore, the rate of heat transfer in the wall:

$$q_x = -k A \frac{dT}{dx} = \left\{\frac{kA(T_1 - T_2)}{L}\right\} \tag{3.64}$$

This quantity of heat ($q_x$) must be supplied to the left side of the wall (i.e. inside of the furnace) to maintain a temperature difference ($T_1 - T_2$) across the wall.

The conductive thermal resistance:

$$R_{con} = \frac{L}{kA} \tag{3.65}$$

Equation (3.64) can also be derived alternatively starting with Fourier's equation (3.7):

$$q_x = -k A \left(\frac{dT}{dx}\right)$$

Integrating this equation between the boundaries of the plane wall:

$$\int_0^L q_x \, dx = -kA \int_{T_1}^{T_2} dT \quad \text{or} \quad q_x L = -kA (T_2 - T_1)$$

or

$$q_x = \left\{\frac{kA(T_1 - T_2)}{L}\right\} \tag{3.64}$$

The temperature at any point x along the wall can be obtained by integrating equation (3.7) between 0 and x:

$$q_x \cdot x = -kA (T - T_1)$$

and comparing with Eq. (3.64): $\left\{\frac{kA(T_1 - T)}{x}\right\} = \left\{\frac{kA(T_1 - T_2)}{L}\right\}$

or

$$\left\{\frac{(T_1 - T)}{x}\right\} = \left\{\frac{(T_1 - T_2)}{L}\right\}$$

or

$$T = \left(\frac{T_2 - T_1}{L}\right)x + T_1 \tag{3.63}$$

**Example 3.4** Calculate the rate of heat loss for a brick wall of pre-heating furnace; wall has 5 m length, 4 m height, and 0.5 m thickness. The temperature of the inner wall is 500 °C and that of the outer wall is 30 °C. The thermal conductivity of brick is 0.7 W/mK. Also calculate the temperature at an interior point of the wall, 25 cm from the inner wall.

**Solution**

Consider Fig. 3.9 is the brick wall. Since the thickness is much less than width and height, so one-dimensional equation, Eq. (3.63) can be applied here.

$$q_x = \left\{\frac{kA(T_1 - T_2)}{L}\right\} = \left\{\frac{0.7(5x4)(773 - 303)}{0.5}\right\} = 13160 \, W = \mathbf{13.16 \, kW}$$

At $x = 0.25$, $T = ?$ From Eq. (3.63):

$$T = \left(\frac{T_2 - T_1}{L}\right)x + T_1 = \left(\frac{(303 - 773)}{0.5}\right).(0.25) + 773 = -235 + 773 = 538 \, K = \mathbf{273 \, °C}$$

## 3.9.2 Heat Conduction Through Cylindrical System

Consider a long hollow cylinder has inside radius, $r_i$; outer radius, $r_o$ and length, L (Fig. 3.10). Hot fluid at constant temperature ($T_i$) is passed through the hollow

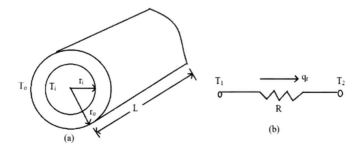

**Fig. 3.10 a** Heat conduction through hollow cylinder and **b** its electric circuit

cylinder. Heat conduction is through radial direction only. Outside surface of cylinder is at constant temperature ($T_o$). Considering the cylinder is very long, so that the heat losses at the end are negligible.

The general form of the heat conduction equation in cylindrical coordinates [from Appendix IV Eq. (IV.2)]:

$$\frac{1}{r}\frac{\delta}{\delta r}\left(kr\frac{\delta T}{\delta r}\right) + \frac{1}{r^2}\frac{\delta}{\delta\theta}\left(k\frac{\delta T}{\delta\theta}\right) + \frac{\delta}{\delta z}\left(k\frac{\delta T}{\delta z}\right) + q^{\cdot} = \rho C_p\left(\frac{\delta T}{\delta t}\right)$$

or

$$\frac{\delta^2 T}{\delta r^2} + \frac{1}{r}\left(\frac{\delta T}{\delta r}\right) + \frac{1}{r^2}\left(\frac{\delta^2 T}{\delta\theta^2}\right) + \frac{\delta^2 T}{\delta z^2} + \frac{q^{\cdot}}{k} = \left(\frac{\rho C_p}{k}\right)\left(\frac{\delta T}{\delta t}\right) = \frac{1}{\alpha}\left(\frac{\delta T}{\delta t}\right)$$

$$(3.66)$$

where $\alpha$ is the thermal of the material (m²/s) is equal to

$$\frac{k}{\rho C_p}.$$

Under steady-state condition:

$$\frac{\delta T}{\delta t} = 0,$$

Without heat generation:

$$\frac{q^{\cdot}}{k} = 0.$$

Since the heat conduction is through radial direction only:

$$\frac{\delta^2 T}{\delta\theta^2} = \frac{\delta^2 T}{\delta z^2} = 0.$$

Hence Eq. (3.66) becomes:

$$\frac{\delta^2 T}{\delta r^2} + \frac{1}{r}\left(\frac{\delta T}{\delta r}\right) = 0$$

or

$$\frac{1}{r}\left[\frac{d}{dr}\left(r\frac{dT}{dr}\right)\right] = 0 \quad \text{or} \quad \frac{d}{dr}\left(r\frac{dT}{dr}\right) = 0 \quad (\text{since } r \neq 0) \qquad (3.67)$$

Boundary condition:

$$\text{at } r = r_i, \ T = T_i; \text{ and at } r = r_o, T = T_o.$$

Integrating Eq. (3.67):
$$\left(r\frac{dT}{dr}\right) = C_1 \text{ or } dT = C_1\frac{dr}{r} \quad \text{(a)}$$
Further integration of above equation:

$$T = C_1 \ln r + C_2 \quad (3.68)$$

where C is integration constant.

Using the boundary conditions to Eq. (3.68):

At $r = r_i$, $T = T_i$; $\quad T_i = C_1 \ln r_i + C_2$ (b)

and

at $r = r_o$, $T = T_o$; $\quad T_o = C_1 \ln r_o + C_2$ (c)

Therefore, from (b) and (c):
$$C_1 = \frac{T_i - T_o}{\ln\frac{r_i}{r_o}} = \frac{T_o - T_i}{\ln\frac{r_o}{r_i}} \quad \text{(d)}$$

and from (b):
$$C_2 = T_i - \frac{T_o - T_i}{\ln\frac{r_o}{r_i}}\ln r_i = \frac{T_i \ln r_o - T_o \ln r_i}{\ln\frac{r_o}{r_i}} \quad \text{(e)}$$

Substituting the values of $C_1$ and $C_2$ in Eq. (3.68):

$$T = \frac{(T_o - T_i)}{\ln\frac{r_o}{r_i}}\ln r + \frac{(T_i \ln r_o - T_o \ln r_i)}{\ln\frac{r_o}{r_i}} \quad (3.69)$$

The rate at which heat is conducted across any cylindrical surface in the solid may be expressed as:
$$q_r = -kA_r\frac{dT}{dr} = -k(2\pi rL)\frac{dT}{dr} \quad \text{(f)}$$
From (a):

$$\frac{dT}{dr} = \frac{C_1}{r} = \frac{1}{r}\cdot\frac{T_o - T_i}{\ln\frac{r_o}{r_i}} \quad \text{[from (d)]}$$

Now from (f):

$$q_r = \frac{2\pi Lk(T_i - T_o)}{\ln\frac{r_o}{r_i}} \quad (3.70)$$

The thermal resistance for the hollow cylinder is:

$$R_{con} = \frac{\ln\frac{r_o}{r_i}}{2\pi Lk} \quad (3.71)$$

Equation (3.70) can also be derived alternatively starting with Fourier's Eq. (3.7):

$$q_r = -k\,A\left(\frac{dT}{dr}\right) = -k(2\pi rL)\left(\frac{dT}{dr}\right) \qquad (\text{since } A = 2\pi rL)$$

or

$$q_r\frac{dr}{r} = -2\pi Lk\,dT \tag{3.72}$$

Integration of Eq. (3.72) with boundary:

$$q_r\int_{r_i}^{r_0}\frac{dr}{r} = -2\pi Lk\int_{T_i}^{T_0}dT$$

or

$$q_r\ln\left(\frac{r_0}{r_i}\right) = -2\pi Lk(T_o - T_i) = 2\pi Lk(T_i - T_o)$$

Hence,

$$q_r = \frac{2\pi Lk(T_i - T_o)}{\ln\frac{r_o}{r_i}} \tag{3.70}$$

**Example 3.5** A hollow cylinder has 4 cm inner diameter and 8 cm outer diameter, and it has an inner surface temperature of 250 °C and outer surface temperature 50 °C. Determine the temperature of the point halfway between the inner and the outer surfaces. Also, determine the heat flow through the cylinder per linear meter. Given: the thermal conductivity of cylinder material is 70 W/mK.

**Solution**

From Eq. (3.70):

$$q_r = \frac{2\pi Lk(T_i - T_o)}{\ln\frac{r_o}{r_i}} = \left\{\frac{2\times 3.14\times 1\times 70(523 - 323)}{\ln\left(\frac{4}{2}\right)}\right\} = 87920/0.693 = \mathbf{126.87\,kW/m}$$

At halfway between the inner and the outer surfaces,

$$r_h = (4 + 2)/2 = 3\,cm$$

Since the heat flow remains the same:

$$q_r = \frac{2\pi Lk(T_i - T_o)}{\ln\frac{r_o}{r_i}} = \frac{2\pi Lk(T_i - T_h)}{\ln\frac{r_h}{r_i}}$$

or

$$\frac{(T_i - T_o)}{\ln \frac{r_o}{r_i}} = \frac{(T_i - T_h)}{\ln \frac{r_h}{r_i}}$$

or

$$(T_i - T_h) = \frac{(T_i - T_o)}{\ln \frac{r_o}{r_i}} \cdot \ln \frac{r_h}{r_i} = \frac{(200)}{\ln \frac{4}{2}} \cdot \ln \frac{3}{2} = \frac{(200)}{0.693} \times 0.405 = 116.9$$

Therefore, $T_h = T_i - 116.9 = 523 - 116.9 = 406.1K = \mathbf{133.1\,°C}$

### 3.9.3 Heat Conduction Through Spherical System

Consider a hollow sphere (Fig. 3.11) whose inside and outside surfaces are at constant temperatures $T_i$ and $T_o$, respectively. Heat conduction is through radial direction only.

The general form of the heat conduction equation in spherical coordinate [from Appendix IV Eq. (IV.3)]]:

$$\frac{1}{r^2}\frac{\delta}{\delta r}\left(kr^2\frac{\delta T}{\delta r}\right) + \frac{1}{r^2\sin^2\theta}\frac{\delta}{\delta\varphi}\left(k\frac{\delta T}{\delta\varphi}\right) + \frac{1}{r^2\sin\theta}\frac{\delta}{\delta\theta}\left(k\sin\theta\frac{\delta T}{\delta\theta}\right) + q^{\cdot} = \rho C_p\left(\frac{\delta T}{\delta t}\right)$$

or

$$\frac{1}{r^2}\frac{\delta}{\delta r}\left(r^2\frac{\delta T}{\delta r}\right) + \frac{1}{r^2\sin^2\theta}\frac{\delta}{\delta\varphi}\left(\frac{\delta T}{\delta\varphi}\right) + \frac{1}{r^2\sin\theta}\frac{\delta}{\delta\theta}\left(\sin\theta\frac{\delta T}{\delta\theta}\right) + \frac{q^{\cdot}}{k} = \left(\frac{\rho C_p}{k}\right)\left(\frac{\delta T}{\delta t}\right)$$
$$= \frac{1}{\alpha}\left(\frac{\delta T}{\delta t}\right) \tag{3.73}$$

Under steady-state condition:

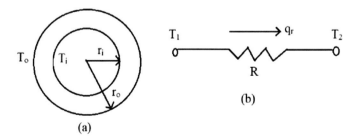

Fig. 3.11  a Heat conduction through hollow sphere and b its electric circuit

$$\frac{\delta T}{\delta t} = 0,$$

Without heat generation:

$$q' = 0,$$

Since the heat conduction is through radial direction only:

$$\frac{\delta^2 T}{\delta \theta^2} = \frac{\delta^2 T}{\delta \varphi^2} = 0.$$

Hence Eq. (3.73) becomes:

$$\frac{1}{r^2}\frac{\delta}{\delta r}\left(r^2 \frac{\delta T}{\delta r}\right) = 0 \quad \text{or} \quad \frac{d}{dr}\left(r^2 \frac{dT}{dr}\right) = 0 \tag{3.74}$$

Integration Eq. (3.74):

$$\left(r^2 \frac{dT}{dr}\right) = C_1 \quad \text{or} \quad dT = C_1\left(\frac{dr}{r^2}\right)$$

Again integrating the above equation:

$$T = \left(\frac{-C_1}{r}\right) + C_2 \tag{3.75}$$

Applying the boundary conditions:

At $r = r_i$, $T = T_i$;   $T_i = \left(\frac{-C_1}{r_i}\right) + C_2$ (a)

at $r = r_o$, $T = T_o$;   $T_o = \left(\frac{-C_1}{r_o}\right) + C_2$ (b)

So, from (a) and (b):

$T_i - T_o = \left(\frac{-C_1}{r_i}\right) - \left(\frac{-C_1}{r_o}\right) = C_1\left(\frac{1}{r_o} - \frac{1}{r_i}\right)$ (c)

Therefore,

$C_1 = \left\{\frac{(T_i - T_o)}{\left(\frac{1}{r_o} - \frac{1}{r_i}\right)}\right\}$ (d)

and from (a):

$C_2 = T_i + \left(\frac{C_1}{r_i}\right) = T_i + \left(\frac{1}{r_i}\right)\left\{\frac{(T_i - T_o)}{\left(\frac{1}{r_o} - \frac{1}{r_i}\right)}\right\}$ (e)

Again from Eq. (3.75):

$$T = \left(\frac{-C_1}{r}\right) + C_2 = \left(\frac{-1}{r}\right)\left\{\frac{(T_i - T_o)}{\left(\frac{1}{r_o} - \frac{1}{r_i}\right)}\right\} + \left[T_i + \left(\frac{1}{r_i}\right)\left\{\frac{(T_i - T_o)}{\left(\frac{1}{r_o} - \frac{1}{r_i}\right)}\right\}\right]$$

$$= T_i + \left\{ \frac{(T_i - T_o)}{\left(\frac{1}{r_o} - \frac{1}{r_i}\right)} \right\} \left(\frac{1}{r_i} - \frac{1}{r}\right) = T_i + \left[ \frac{r_o}{r} \left\{ \frac{(r - r_i)}{(r_o - r_i)} \right\} (T_o - T_i) \right] \quad (3.76)$$

The rate of heat is conducted across any spherical surface in the solid may be expressed as:

$$q_r = -kA_r \frac{dT}{dr} = -k(4\pi r^2) \left(\frac{C_1}{r^2}\right) = \frac{4\pi \, r_o \, r_i \, k \, (T_i - T_o)}{(r_o - r_i)} \quad (3.77)$$

where $A_r = 4\pi r^2$, and $\frac{dT}{dr} = \left(\frac{C_1}{r^2}\right) = \left(\frac{1}{r^2}\right) \left\{ \frac{(T_i - T_o)}{\left(\frac{1}{r_o} - \frac{1}{r_i}\right)} \right\} = \left(\frac{r_o r_i}{r^2}\right) \left\{ \frac{(T_i - T_o)}{(r_i - r_o)} \right\}$

**Example 3.6** A hollow sphere used as a container of liquid chemical, has 20 cm inner diameter and 40 cm outer diameter, the thermal conductivity of sphere material is 50 W/mK. Its inner and outer surface temperatures are 250 °C and 50 °C, respectively. Calculate the heat flow rate through the sphere. Also, estimate the temperature at a point a quarter of the way between inner and outer surfaces.

**Solution**

From Eq. (3.77):

$$q_r = \frac{4\pi \, r_o \, r_i \, k(T_i - T_o)}{(r_o - r_i)} = \frac{4 \times 3.14 \times 0.2 \times 0.1 \times 50(523 - 323)}{(0.2 - 0.1)} = \mathbf{25.12 \, kW}$$

The value of r at one-fourth way of the way between inner and outer surfaces is

$$\{10 + (20 - 10) / 4$$
$$= 10 + 2.5 = 12.5 \, cm$$

According to Eq. (3.76), temperature at r = 12.5 cm:

$$T = T_i + \left[ \frac{r_o}{r} \left\{ \frac{(r - r_i)}{(r_o - r_i)} \right\} (T_o - T_i) \right]$$
$$= 523 + \left[ \left(\frac{0.2}{0.125}\right) \left\{ \frac{(0.125 - 0.1)}{(0.2 - 0.1)} \right\} (323 - 523) \right]$$
$$= 523 - 80 = 443K = \mathbf{170\,°C}$$

## 3.10   Systems with Variable Thermal Conductivity

In earlier Sect. 3.9, the thermal conductivity (k) has been assumed as constant. This assumption is probably satisfactory for materials involving small temperature differences across them. In practice, the thermal conductivity of most material's temperature dependent, and it would be necessary to include in the analysis the variation of thermal conductivity with temperature. Under normal circumstances, it is sufficiently accurate [3] to use the linear expression for k, i.e.

$$k = k_0(1 + \beta T) \tag{3.78}$$

where $k_0$ is the thermal conductivity at $T = 0$, and $\beta$ is the temperature coefficient of thermal conductivity.

### 3.10.1   Plane Wall (Slab)

Consider a plane wall (Fig. 3.9) in the region $0 \le x \le L$ having boundary surface at $x = 0$, the surface temperature $T_1$ and $x = L$, the surface temperature $T_2$. The problem can be formulated as:

$$\frac{d}{dx}\left[\left\{k(T)\frac{dT}{dx}\right\}\right] = 0 \quad \text{in } 0 \le x \le L \tag{3.79}$$

where
  $k(T) = k_0(1 + \beta T)$ [from Eq. (3.78)]
  Now Eq. (3.79) becomes:

$$\frac{d}{dx}\left[\left\{k_0(1 + \beta T)\frac{dT}{dx}\right\}\right] = 0 \tag{3.80}$$

Integration of equation (3.80):

$$k_0(1 + \beta T)\frac{dT}{dx} = C_1 \tag{3.81}$$

Again integration equation (3.81):

$$k_0\left(T + \frac{\beta}{2}T^2\right) = C_1 x + C_2 \tag{3.82}$$

Boundary conditions :
(i)

$$\text{at } x = 0, T = T_1,$$

$$C_2 = k_0\left(T_1 + \frac{\beta}{2}T_1^2\right)$$

and
(ii)

$$\text{at } x = L, T = T_2$$

$$C_1 L + C_2 = k_0\left(T_2 + \frac{\beta}{2}T_2^2\right)$$

Therefore,

$$
\begin{aligned}
C_1 &= \frac{k_0}{L}\left[\left(T_2 + \frac{\beta}{2}T_2^2\right) - \left(T_1 + \frac{\beta}{2}T_1^2\right)\right] = \frac{k_0}{L}\left[(T_2 - T_1) + \frac{\beta}{2}\left(T_2^2 - T_1^2\right)\right] \\
&= \frac{k_0}{L}\left[(T_2 - T_1) + \frac{\beta}{2}\{(T_2 - T_1)(T_2 + T_1)\}\right] = \frac{k_0}{L}\left[(T_2 - T_1)\left\{1 + \frac{\beta}{2}(T_2 + T_1)\right\}\right] \\
&= \frac{k_0}{L}\left[(T_2 - T_1)(1 + \beta T_m)\right] \qquad\qquad\qquad\qquad (3.83)
\end{aligned}
$$

where $T_m$ is the arithmetic mean of the boundary surface temperatures $= \frac{T_1 + T_2}{2}$.
Putting the values $C_1$ and $C_2$ in Eq. (3.82):

$$k_0\left(T + \frac{\beta}{2}T^2\right) = \frac{k_0\, x}{L}\left[(T_2 - T_1)(1 + \beta T_m)\right] + k_0\left(T_1 + \frac{\beta}{2}T_1^2\right)$$

or

$$\left(T + \frac{\beta}{2}T^2\right) - \frac{x}{L}\left[(T_2 - T_1)(1 + \beta T_m)\right] - \left(T_1 + \frac{\beta}{2}T_1^2\right) = 0$$

or

$$\left(T + \frac{\beta}{2}T^2\right) + \frac{x}{L}\left[(T_1 - T_2)(1 + \beta T_m)\right] - T_1\left(1 + \frac{\beta}{2}T_1\right) = 0 \qquad (3.84)$$

Equation (3.84) is a quadratic equation of T, solution is given by:

$$T = \frac{-1 \pm \left[1 - 4\frac{\beta}{2}\left\{\frac{x}{L}(T_1 - T_2)(1 + \beta T_m) - T_1\left(1 + \frac{\beta}{2}T_1\right)\right\}\right]^{1/2}}{2\frac{\beta}{2}}$$

$$= \frac{-1}{\beta} \pm \left[\frac{1}{\beta^2} + \frac{2}{\beta}\left\{T_1\left(1 + \frac{\beta}{2}T_1\right) - \frac{x}{L}(T_1 - T_2)(1 + \beta T_m)\right\}\right]^{1/2}$$

$$\tag{3.85}$$

The heat flow rate,

$$q_x = -kA\frac{dT}{dx} = -k_0(1+\beta T) A \frac{dT}{dx} \tag{3.86}$$

From the Eq. (3.81):

$$\frac{dT}{dx} = \frac{C_1}{k_0(1 + \beta T)} = \frac{k_0[(T_2 - T_1)(1 + \beta T_m)]}{Lk_0(1 + \beta T)} = \frac{[(T_2 - T_1)(1 + \beta T_m)]}{L(1 + \beta T)}$$

Now putting the value of $\frac{dT}{dx}$ in Eq. (3.86):

$$q_x = -k_0(1+\beta T) A. \frac{[(T_2 - T_1)(1 + \beta T_m)]}{L(1 + \beta T)} = \frac{Ak_0[(T_1 - T_2)(1 + \beta T_m)]}{L}$$

$$= \frac{Ak_m(T_1 - T_2)}{L} \tag{3.87}$$

where $k_m$ is the mean thermal conductivity

$$= k_0(1 + \beta T_m) = k_0\{1 + \beta\left(\frac{T_1 + T_2}{2}\right)\} \tag{3.88}$$

The linear variation of k with temperature, the heat flow rate through the slab can be calculated by Eq. (3.87); if the thermal conductivity $k_m$ is evaluated at the mean surface temperature.

**Example 3.7** Calculate the heat loss per square meter surface area of a 50 cm thick furnace wall having surface temperature of 500 °C and 50 °C, if the thermal conductivity k of the wall material is given by: $k = 0.005 \, T–5 \times 10^{-6} \, T^2$; where T is the temperature in °C.

**Solution**

Heat flow,

$$q = -kA\frac{dT}{dx}$$

Therefore, heat flux i.e. heat loss $= \frac{q}{A}$ $= -k\frac{dT}{dx}$ $=$
$-\left(0.005\,T - 5\times 10^{-6}\,T^2\right)\frac{dT}{dx}$

Integrating:

$$\int_0^L \frac{q}{A}\,dx = -\int_{T_1}^{T_2}\left(0.005\,T - 5\times 10^{-6}T^2\right)dT$$

or

$$\frac{q}{A} = -\frac{1}{L}\left[0.005\,\frac{T^2}{2} - 5\times 10^{-6}\,\frac{T^3}{3}\right]_{50}^{500}$$

$$= +\frac{1}{0.5}\,[\frac{0.005}{2}(500^2 - 50^2) - \frac{5\times 10^{-6}}{3}(500^3 - 50^3)]$$

$$= 2\,[\,0.0025\,(250,000-2500)-2.5\times 10^{-6}(\,125,000,000-125,000)$$

$$= 2\,(618.75 - 312.19) = \mathbf{613.12\,W/m^2}$$

**Example 3.8** A furnace wall 30 cm thick has its two-surface maintained at 1200 °C and 200 °C. Calculate the rate of heat flow. The thermal conductivity of the furnace material varies with temperature as: $k = 0.813 + 5.82 \times 10^{-4}\,T$ (where T in °C).

**Solution**

According to Eq. (3.87):

$$q = \frac{Ak_m\,(T_1 - T_2)}{L}$$

So

$$\frac{q}{A} = \frac{k_m\,(T_1 - T_2)}{L}$$

From Eq. (3.88):

$$k_m = k_0\left\{1+\beta\left(\frac{T_1 + T_2}{2}\right)\right\}$$

Since

$$k = 0.813 + 5.82 \times 10^{-4}\,T;\ \text{so } k_0 = 0.813 \text{ and } \beta = 5.82 \times 10^{-4}$$

Therefore,

$$k_m = 0.813\left\{1 + 5.82 \times 10^{-4}(700)\right\}\,K = 1.144\,W/m$$

Hence,

$$\frac{q}{A} = \frac{1.144}{0.3}(1200 - 200) = \textbf{3813.33 W/m}^2$$

### 3.10.2  Hollow Cylinder

Consider a hollow cylinder (Fig. 3.10) with boundary surfaces at $r_i$ and $r_o$ kept at uniform temperatures $T_i$ and $T_o$ respectively. The thermal conductivity of the material of cylinder is temperature dependent and is given by Eq. 3.78 [$k = k_0 (1 + \beta T)$]. The mathematical formulation of the problem is as follows:

$$\frac{d}{dr}\left[\left\{r\,k(T)\frac{dT}{dr}\right\}\right] = 0$$

or

$$\frac{d}{dr}\left[\left\{r\,k_0(1 + \beta T)\frac{dT}{dr}\right\}\right] = 0 \qquad (3.89)$$

where $k = k_0(1 + \beta T)$ [According to Eq. (3.78)]
  Integration of Eq. (3.89):

$$k_0 (1 + \beta T)\frac{dT}{dr} = \frac{C_1}{r} \qquad (3.90)$$

Further integrating the above equation:

$$k_0\left(T + \frac{\beta}{2}T^2\right) = C_1 \ln r + C_2 \qquad (3.91)$$

Now using the boundary conditions
(i) at $r = r_i$ , $T = T_i$ ; and (ii) at $r = r_o, T = T_o$ .
Equation (3.91) becomes:

$$k_0\left(T_i + \frac{\beta}{2}T_i^2\right) = C_1 \ln r_i + C_2 \text{ and } k_0\left(T_o + \frac{\beta}{2}T_o^2\right) = C_1 \ln r_o + C_2$$

Solving these two equations for $C_1$ and $C_2$:

$$C_1 \ln\frac{r_1}{r_o} = k_0\left\{(T_i - T_o) + \frac{\beta}{2}(T_i^2 - T_o^2)\right\}$$

$$= k_o \left\{ (T_i - T_o) \left\{ 1 + \frac{\beta}{2}(T_i + T_o) \right\} \right.$$

$$= k_o (T_i - T_o)(1 + \beta T_m) \quad [\text{ where } T_m = (T_i + T_o)/2]$$

Therefore,

$$C_1 = \left[ k_o \left\{ (T_i - T_o)(1 + \beta T_m) \right\} \right] / \ln \frac{r_1}{r_o}$$

and

$$C_2 = \left\{ k_o \left( T_i + \frac{\beta}{2} T_i^2 \right) \right\} - \left[ \left\{ k_o (T_i - T_o)(1 + \beta T_m) \ln r_i \right\} / \left( \ln \frac{r_1}{r_o} \right) \right]$$

Putting values of $C_1$ and $C_2$ in Eq. (3.91):

$$k_o \left( T + \frac{\beta}{2} \right) T^2 = C_1 \ln r + C_2$$

$$= \left[ k_o \left\{ (T_i - T_0)(1 + \beta T_m) \ln r \right\} / \ln \frac{r_1}{r_0} + \left\{ k_o \left( T_i + \frac{\beta}{2} T_i^2 \right) \right\} \right]$$

$$- \left[ \left\{ k_o T_i - T_0)(1 + \beta T_m) \ln r_i \right\} / \left( \ln \frac{r_1}{r_0} \right) \right]$$

or

$$\frac{\beta}{2} T^2 + T - \left[ \left\{ (T_i - T_o)(1 + \beta T_m) \right\} / \left( \ln \frac{r_1}{r_o} \right) \right] \cdot \ln \frac{r}{r_i} - T_i \left( 1 + \frac{\beta}{2} T_i \right) = 0$$

Therefore,

$$T(r) = \left[ -1 \pm \left\{ 1 + \frac{4\beta}{2}(T_i - T_o)(1 + \beta T_m) \ln \frac{r}{r_i} / \left( \ln \frac{r_1}{r_o} \right) + T_i \left( 1 + \frac{\beta}{2} T_i \right) \right\}^{1/2} \right] / \beta$$

$$(3.92)$$

The heat flow,

$$q = - \left[ kA \frac{dT}{dr} \right]_{r=r_i} = -k_o \left[ (1 + \beta T) \cdot 2\pi r L \cdot \frac{dT}{dr} \right]_{r=r_i} \quad (3.93)$$

Now from Eq. (3.90):

$$\frac{dT}{dr} = \frac{C_1}{r} \frac{1}{k_o(1 + \beta T)} = \frac{k_o\{(T_i - T_o)(1 + \beta T_m)}{r \ln \frac{r_1}{r_o}} \frac{1}{k_o(1 + \beta T)}$$

Hence,

$$q = -k_o (1 + \beta T) \cdot 2\pi rL \cdot \frac{k_o\{(T_i - T_o)(1 + \beta T_m)}{r \ln \frac{r_1}{r_o}} \frac{1}{k_o(1 + \beta T)}$$

$$= \frac{k_o(1 + \beta T_m) 2\pi L(T_i - T_o)}{\ln \frac{r_o}{r_i}} = \frac{k_m 2\pi L(T_i - T_o)}{\ln \frac{r_o}{r_i}} \tag{3.94}$$

Since

$$k_m = k_0 (1 + \beta T_m)$$

**Example 3.9** A thick-walled copper cylinder has an inside radius of 0.5 cm and an outside radius of 2 cm. The inner and outer surfaces are held at 310 °C and 290 °C, respectively. If the thermal conductivity varies with temperature as: k (W/mK) = 371.9 [1–9.25 × 10$^{-5}$ (T–150)].

Calculate the heat loss per unit length.

**Solution**

According to Eq. (3.94):

$$\frac{q}{L} = \frac{k_m 2\pi(T_i - T_o)}{\ln \frac{r_o}{r_i}}$$

Since

$$k_m = k_0(1 + \beta T_m)$$

So

$$T_m = [(310 - 150) + (290 - 150)]/2 = 150\,°C$$

Therefore,

$$k_m = 371.9\left[1 - 9.25 \times 10^{-5} (150)\right] = 371.9\,[1 - 0.0139] = 366.73\,W/mK.$$

Hence,

$$\frac{q}{L} = \frac{k_m 2\pi(T_i - T_o)}{\ln \frac{r_o}{r_i}} = \frac{366.73 \times 2 \times 3.14 \times (310 - 290)}{\ln\left(\frac{2}{0.5}\right)} = \mathbf{33.14\,W/m}$$

### 3.10.3  Hollow Sphere

Consider a hollow sphere (Fig. 3.11) with boundary surfaces at $r_i$ and $r_o$ kept at uniform temperatures $T_i$ and $T_o$, respectively. The thermal conductivity of the material of sphere is temperature dependent and is given by Eq. 3.78 [$k = k_0 (1 + \beta T)$]. The mathematical formulation of the problem can be solved as follows:

For a one-dimensional heat flow in the radial direction:

$$q_r = -k(T)A_r\frac{dT}{dr} = -k_0(1 + \beta T).4\pi r^2\frac{dT}{dr} \tag{3.95}$$

Integrating Eq. (3.95):

$$\int \frac{dr}{r^2} = \frac{-4\pi k_o}{q_r}\int (1 + \beta T)\,dT$$

or

$$\frac{-1}{r} = \frac{-4\pi k_o}{q_r}\left[T + \frac{\beta T^2}{2}\right] + C_1 \tag{3.96}$$

The value of constant $C_1$ is evaluated with boundary condition: at $r = r_i$, $T = T_i$. Now

$$C_1 = \frac{-1}{r_i} + \frac{4\pi k_o}{q_r}\left[T_i + \frac{\beta T_i^2}{2}\right]$$

Substitute the value of $C_1$ into Eq. (3.96):

$$\frac{4\pi k_o}{q_r}\left[T_i + \frac{\beta T_i^2}{2}\right] - \frac{1}{r_i} - \frac{4\pi k_O}{q_r}\left[T + \frac{\beta T^2}{2}\right] + \frac{1}{r} = 0$$

or

$$\left(\frac{\beta T^2}{2} + T\right) - \left[T_i + \frac{\beta T_i^2}{2}\right] + \frac{q_r}{4\pi k_o}\left(\frac{1}{r_i} - \frac{1}{r}\right) = 0$$

Therefore,

$$T = -\left[-1 \pm \left\{1 + \frac{4\beta}{2}\left\{\left(T_i + \frac{\beta T_i^2}{2}\right) - \frac{q_r}{4\pi k_o}\left(\frac{1}{r_i} - \frac{1}{r}\right)\right\}\right]^{1/2}\right]/\beta$$

$$= -\frac{1}{\beta} \pm \frac{1}{\beta}\left[(1 + \beta T_i)^2 - \frac{q_r}{4\pi k_o}\left(\frac{1}{r_i} - \frac{1}{r}\right)\right]^{1/2} \tag{3.97}$$

Equation (3.97) is an expression for temperature distribution when the heat flow through the sphere and the inner surface temperature are known.

**Example 3.10** The inside and outside surfaces of a hollow sphere a $\leq r \leq b$ at r = a and r = b are maintained at temperatures $T_1$ and $T_2$, respectively. The thermal conductivity varies with temperature as: $k(T) = k_o (1 + \alpha T + \beta T^2)$. Derive an expression for total heat flow rate through the sphere.

**Solution**

$q_r = -k(T) A_r \frac{dT}{dr} = -k_0 (1 + \alpha T + \beta T^2) 4\pi r^2 \frac{dT}{dr}$ (i)

Integrating Eq. (i) with boundary:

$$\int \frac{dr}{r^2} = \frac{-4\pi k_o}{q_r} \int (1 + \beta T)\, dT$$

$$q_r \int_a^b \frac{dr}{r^2} = -4\pi k_o \int_{T_1}^{T_2} (1 + \alpha T + \beta T2)\, dT$$

or

$$-q_r \left[\frac{1}{r}\right]_a^b = -4\pi k_0 \left[T + \alpha \frac{T^2}{2} + \beta \frac{T^3}{3}\right]_{T_1}^{T_2}$$

or

$$q_r \left[\frac{(b-a)}{ab}\right] = 4\pi k_o \left[(T_1 - T_2) + \frac{\alpha}{2}(T_1^2 - T_2^2) + \frac{\beta}{3}(T_1^3 - T_2^3)\right]$$

Therefore,

$q_r = \frac{4\pi k_o a b}{(b-a)}(T_1 - T_2)\left[1 + \frac{\alpha}{2}(T_1 + T_2) + \frac{\beta}{3}(T_1^2 + T_1 T_2 + T_2^2)\right]$ (ii)

## 3.11  Composite Systems

Many engineering applications, where involve heat transfer through a medium composed of two or more materials of different thermal conductivities arranged in series or parallel. Composite systems involve a number of series and parallel thermal resistances due to layers of different materials. For examples, furnace wall, tanks of hot gases, etc. which always have insulating materials between the inner and outer walls. A hot fluid flowing inside a tube covered with a layer of thermal insulation is another example of a composite system because, in this case, the thermal conductivities of tube metal and insulation are different [3]. The problem of heat transfer through the composite systems can be solved by the application of thermal resistance concept (as described in Sect. 3.6). To solve the problems, assuming one-dimensional, steady-state heat conduction for composite systems comprising parallel plates, coaxial cylinders, or concentric spheres are considered. It will be further assumed that the parallel layers in the composite system are in perfect thermal contact, or the resistance due to interface loose contact is negligible. That is the heat transfer is continuous at the interface of the two layers in contact.

### 3.11.1 Series Composite Plane Walls

Consider a multilayered wall as shown in Fig. 3.12. Multilayered wall consists of different materials with different thermal conductivities. They make series thermal resistances (Fig. 3.12b). The Fourier equation may be applied directly to get the rate of heat transfer. The heat transfer is taking place in five steps, viz. (i) heat is transferred from the hot fluid to the inner surface of wall A by convection mode, (ii) heat is conducted through the wall A by conduction mode, (iii) heat is conducted through the wall B by conduction, (iv) heat is conducted through the wall C by conduction, and (v) heat flows by convection from the outer surface of the wall C to the cold fluid, assuming that there is negligible radiation.

Step 1:

$$q = q_c = h_1 A (T_1 - T_2) = \left\{ \frac{(T_1 - T_2)}{\left( \frac{1}{h_1 A} \right)} \right\} = \left\{ \frac{(T_1 - T_2)}{R_1} \right\} \qquad (3.98)$$

Step 2:

$$q = q_k = \frac{k_A A}{L_A} (T_2 - T_3) = \left\{ \frac{(T_2 - T_3)}{\left( \frac{L_A}{k_A A} \right)} \right\} = \left\{ \frac{(T_2 - T_3)}{R_2} \right\} \qquad (3.99)$$

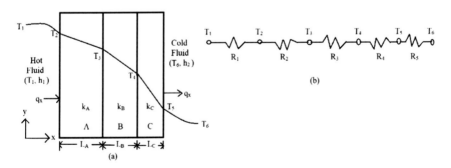

Fig. 3.12 a Heat transfer in composite walls and b its electric circuit

Step 3:

$$q = q_k = \frac{k_B A}{L_B}(T_3 - T_4) = \left\{ \frac{(T_3 - T_4)}{\left(\frac{L_B}{k_B A}\right)} \right\} = \left\{ \frac{(T_3 - T_4)}{R_3} \right\} \tag{3.100}$$

Step 4:

$$q = q_k = \frac{k_C A}{L_C}(T_4 - T_5) = \left\{ \frac{(T_4 - T_5)}{\left(\frac{L_C}{k_C A}\right)} \right\} = \left\{ \frac{(T_4 - T_5)}{R_4} \right\} \tag{3.101}$$

Sep 5:

$$q = q_c = h_2 A (T_5 - T_6) = \left\{ \frac{(T_5 - T_6)}{\left(\frac{1}{h_2 A}\right)} \right\} = \left\{ \frac{(T_5 - T_6)}{R_5} \right\} \tag{3.102}$$

where $q_c$ and $q_k$ are the heat transfer for convection and conduction, respectively.

$h_1$, $h_2$ are the heat transfer coefficient for convection for hot fluid and cold fluid, respectively.

A is the area of plane perpendicular to the heat flow.

$L_A$, $L_B$ and $L_C$ are the thickness of walls A, B, C, respectively.

$k_A$, $k_B$, $k_C$ are the thermal conductivities of walls A, B, C, respectively.

$T_1$ is the temperature of hot fluid inside of the furnace,

$T_2$ is the temperature of inner surface of wall A,

$T_3$ is the temperature of interface of walls A and B,

$T_4$ is the temperature of interface of walls B and C,

$T_5$ is the temperature of outer surface of wall C on the cooling side, and

$T_6$ is the temperature of cold fluid outside of the furnace.

$R_1$ is thermal resistance due to convection ($R_{1c}$) by the hot fluid $= \frac{1}{h_1 A}$,

$R_2$ is the thermal resistance of wall A due to conduction $= \frac{L_A}{k_A A}$,

$R_3$ is the thermal resistance of wall B due to conduction $= \frac{L_B}{k_B A}$,

$R_4$ is the thermal resistance of wall C due to conduction $= \frac{L_C}{k_C A}$, and

$R_5$ is the thermal resistance due to convection between wall C and cold fluid $= \frac{1}{h_2 A}$.

By adding Eqs. (3.98) to (3.102):

$$q = \left\{ \frac{(T_1 - T_6)}{R_1 + R_2 + R_3 + R_4 + R_5} \right\} = \left\{ \frac{(T_1 - T_6)}{\Sigma R_{ther}} \right\} \tag{3.103}$$

where

$$\Sigma R_{ther} = R_1 + R_2 + R_3 + R_4 + R_5 = \frac{1}{A}\left\{\frac{1}{h_1} + \frac{L_A}{k_A} + \frac{L_B}{k_B} + \frac{L_C}{k_C} + \frac{1}{h_2}\right\}$$

Hence,

$$q = \left[\frac{(T_1 - T_6)}{\frac{1}{A}\left\{\frac{1}{h_1} + \frac{L_A}{k_A} + \frac{L_B}{k_B} + \frac{L_C}{k_C} + \frac{1}{h_2}\right\}}\right] \tag{3.104}$$

The temperature at any intermediate point of this composite system can be obtained from Eqs. (3.98) to (3.103). For example, the interface temperature of inner surface of wall A, $T_2$ can be obtained from Eqs. (3.98) and (3.103):

$$\left\{\frac{(T_1 - T_2)}{R_1}\right\} = \left\{\frac{(T_1 - T_6)}{\Sigma R_{ther}}\right\}$$

or

$$(T_1 - T_2) = (T_1 - T_6)\left(\frac{R_1}{\Sigma R_{ther}}\right)$$

or

$$T_2 = T_1 - \left[(T_1 - T_6)\left(\frac{R_1}{\Sigma R_{ther}}\right)\right] \tag{3.105}$$

Again, temperature of outer surface of wall C on the cooling side, $T_5$ can be obtained from Eqs. (3.98) to (3.102) and (3.103):

$$\left\{\frac{(T_1 - T_5)}{R_1 + R_2 + R_3 + R_4}\right\} = \left\{\frac{(T_1 - T_6)}{\Sigma R_{ther}}\right\}$$

or

$$(T_1 - T_5) = (T_1 - T_6)\left(\frac{R_1 + R_2 + R_3 + R_4}{\Sigma R_{ther}}\right)$$

or

$$T_5 = T_1 - \left[(T_1 - T_6)\left(\frac{R_1 + R_2 + R_3 + R_4}{\Sigma R_{ther}}\right)\right] \tag{3.106}$$

Equation (3.104) can be generalized for n-layers wall as:

$$q = \left[ \frac{(T_i - T_o)}{\frac{1}{A}\left\{\left(\frac{1}{h_1} + \frac{1}{h_2}\right) + \Sigma\left(\frac{L_n}{k_n}\right)\right\}} \right] \qquad (3.107)$$

where $T_i$ and $T_o$ are the temperatures of the inside and outside of the furnace.

When the heat transfers by convection, on the two sides of the composite walls, are absent, then Eq. (3.107) becomes:

$$q = \left[ \frac{(T_i - T_o)}{\frac{1}{A}\left\{\Sigma\left(\frac{L_n}{k_n}\right)\right\}} \right] \qquad (3.108)$$

**Example 3.11** A furnace is heated by hot gas (at 2000 °C). Heat flow by radiation from hot gas to inside surface of the wall is 23.26 kW/m² and convective heat transfer coefficient at the interior surface is 11.63 W/m²K. Thermal conductance of the wall material is 58 W/m²K. Interior wall surface temperature is 1000 °C, and temperature of outside of the furnace is 45 °C. Heat flow by radiation from external surface to surroundings is 9.3 kW/m². Find out the surface temperature of external wall and conductive conductance at the outside of the furnace.

**Solution**

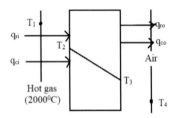

Fig. Ex.3.11($T_1 = 2000^0C$, $T_2 = 1000^0C$, $T_3 = ?$ and $T_4 = 45^0C$)

Figure Ex.3.11 shows the configuration of furnace wall.

Heat flow by radiation from hot gas to inside surface of the wall = $q_{ri}$ = 23.26 kW/m²

Heat flow by convection from hot gas to inside surface of the wall = $q_{ci}$

$$q_{ci} = h_i(T_1 - T_2) = 11.63\ (2000 - 1000) = 11.63\ \text{kW/m}^2.$$

Therefore, total heat entering into the furnace wall = $q_{ri} + q_{ci}$ = 23.26 + 11.63 = 34.89 kW/m².

Thermal resistance of the wall, $R = (1/\text{conductance}) = (1/58) = 0.0172 \ \text{m}^2\text{K/W}$.

Now heat transfer through the wall, $q = \left[\frac{(T_2 - T_3)}{R}\right]$

Therefore, temperature of external wall,

$$T_3 = T_2 - qR$$
$$= 1000 - \{(34.89 \times 10^3)\,0.0172\} = 1000 - 602 = \textbf{398 °C}$$

Now heat loss due to radiation at the external wall, $q_{ro} = 9.3 \ \text{kW/m}^2$.

Therefore, heat loss due to convection, $q_{co} = 34.89 - 9.3 = 25.59 \ \text{kW/m}^2 = h_o$ $(T_3 - T_4)$.

Hence, the conductive conductance at the outside of the furnace

$$h_o = \left[\frac{q_{co}}{(T_3 - T_4)}\right] = \frac{25.59 \times 10^3}{(398 - 45)} = \textbf{72.5 W/m}^2\textbf{K}$$

**Example 3.12** The door of an air condition room is made from two 6 mm thick glass sheets separated by a uniform air gap of 3 mm. The temperature of the inside room is –20 °C, and the outside air temperature is 30 °C. Assuming the heat transfer coefficient between glass and air is 23.2 W/m²K, calculate the rate of heat loss from the room per unit area of the door.

Given: $k_{glass} = 0.75 \ \text{W/mK}$ and $k_{air} = 0.02 \ \text{W/mK}$.

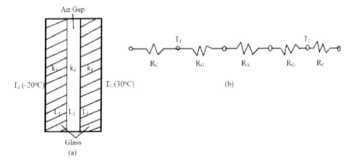

Fig. Ex. 3.12

**Solution**

Figure Ex.3.12 shows the configuration of composite door.

According to Eq. (3.104):

$$q = \left[\frac{(T_1 - T_6)}{\frac{1}{A}\left\{\frac{1}{h_1} + \frac{L_A}{k_A} + \frac{L_B}{k_B} + \frac{L_C}{k_C} + \frac{1}{h_2}\right\}}\right]$$

So

$$\text{Heat loss} = \frac{q}{A} = \left[\frac{(T_1-T_2)}{\left\{\frac{1}{h_1}+\frac{L_1}{k_1}+\frac{L_2}{k_2}+\frac{L_1}{k_1}+\frac{1}{h_2}\right\}}\right] = \left[\frac{\{30-(-20)\}}{\frac{1}{23.2}+\frac{0.006}{0.75}+\frac{0.003}{0.02}+\frac{0.006}{0.75}+\frac{1}{23.2}}\right].$$

$$= \frac{50}{0.252} = 198.4\,\text{W/m}^2$$

**Example 3.13** Calculate the heat transfer through the composite wall shown in Fig. Ex. 3.13. The thermal conductivities of A, B, C, D, and E are 50, 10, 6.7, 20, and 30 W/mK, respectively. Assume one-dimensional heat transfer and unit area.

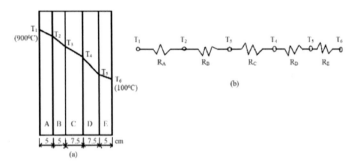

Fig. Ex. 3.13

**Solution**

$$q_x = \frac{k_A A}{L_A}(T_1 - T_2) = \left\{\frac{(T_1-T_2)}{\left(\frac{L_A}{k_A A}\right)}\right\} = \left\{\frac{(T_1-T_2)}{R_A}\right\},\quad R_A = \frac{L_A}{k_A A} = \frac{0.05}{50\times1} = 1\times10^{-3}\ \ (\text{W/K})^{-1}$$

$$q_x = \frac{k_B A}{L_B}(T_2 - T_3) = \left\{\frac{(T_2-T_3)}{\left(\frac{L_B}{k_B A}\right)}\right\} = \left\{\frac{(T_2-T_3)}{R_B}\right\}\quad R_B = \frac{L_B}{k_B A} = \frac{0.05}{10\times1} = 5\times10^{-3}\ \ (\text{W/K})^{-1}$$

$$q_x = \frac{k_C A}{L_C}(T_3 - T_4) = \left\{\frac{(T_3-T_4)}{\left(\frac{L_C}{k_C A}\right)}\right\} = \left\{\frac{(T_3-T_4)}{R_C}\right\}\quad R_C = \frac{L_C}{k_C A} = \frac{0.075}{6.7\times1} = 11.2\times10^{-3}\ \ (\text{W/K})^{-1}$$

$$q_x = \frac{k_D A}{L_D}(T_4 - T_5) = \left\{\frac{(T_4-T_5)}{\left(\frac{L_D}{k_D A}\right)}\right\} = \left\{\frac{(T_4-T_5)}{R_D}\right\}\quad R_D = \frac{L_D}{k_D A} = \frac{0.075}{20\times1} = 3.75\times10^{-3}\ \ (\text{W/K})^{-1}$$

$$q_x = \frac{k_E A}{L_E}(T_5 - T_6) = \left\{\frac{(T_5 - T_6)}{\left(\frac{L_E}{k_E A}\right)}\right\} = \left\{\frac{(T_5 - T_6)}{R_E}\right\}\quad R_E = \frac{L_E}{k_E A} = \frac{0.05}{30\times1} = 1.67\times10^{-3}\ \ (\text{W/K})^{-1}$$

$$\sum R_i = R_A + R_B + R_C + R_D + R_E = (1+5+11.2+3.75+1.67)\times10^{-3} = 22.62\times10^{-3}$$

Heat transfer,

$$q_x = \left\{ \frac{(T_1 - T_6)}{\Sigma R_i} \right\} = \left\{ \frac{(T_1 - T_6)}{R_A + R_B + R_C + R_D + R_E} \right\} = \frac{(900 - 100)}{22.62 \times 10^{-3}} = \textbf{35.37 kW}$$

**Example 3.14**

(i)   A gas fired furnace is made of a layer (12.5 cm) of fireclay brick and a 50 cm layer of red brick. If the inside wall temperature of the furnace is 1100 °C, and the outside wall temperature is 50 °C. Determine the amount of heat loss per square meter of the wall and temperature at the interface of fireclay-red brick.

(ii)  To reduce the thickness of the red brick layer of the furnace to half by filling the layers by dolomite brick. Calculate the thickness of the dolomite brick to ensure an identical loss of heat for the same inside and outside temperatures.

Given: k for fireclay = 0.533 W/mK, k for red brick = 0.7 W/mK and k for dolomite = $0.113 + 2.3 \times 10^{-4}$ T W/mK (Fig. 3.16).

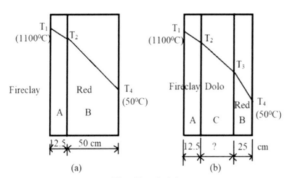

Fig. Ex. 3.14

**Solution**

(i)   Resistance of fireclay,

$$R_A = \frac{L_A}{k_A A} = \frac{0.125}{0.533 \times 1} = 0.235 (W/K)^{-1}$$

Resistance of red brick,

$$R_B = \frac{L_B}{k_B A} = \frac{0.5}{0.7 \times 1} = 0.714 \ (W/K)^{-1}$$

Heat loss,

$$q_x = \left\{ \frac{(T_1 - T_3)}{R_A + R_B} \right\} = \frac{(1100 - 50)}{(0.235 + 0.714)} = \textbf{1106.43W/m}^2$$

The temperature at the interface between fireclay and red brick is $T_2$;

$$q_x = \left\{ \frac{(T_1 - T_2)}{R_A} \right\}$$

or

$$T_2 = T_1 - q_x \cdot R_A = 1100 - 1106.43 \times 0.235 = 1100 - 260 = \textbf{840}\,^\circ\textbf{C}$$

(ii)   Since the heat loss remains same, the temperature at the interface between dolomite and red brick is $T_3$:

$$q_x = \left\{ \frac{(T_3 - T_4)}{R_{B2}} \right\}$$

or

$$T_3 = T_4 + q_x\, R_{B2} = 50 + 1106.43 \left( \frac{0.25}{0.7 \times 1} \right) = 445.2\,^\circ\text{C}$$

Mean thermal conductivity of dolomite layer,

$$k_m = 0.113 + 2.3 \times 10^{-4} \left( \frac{840 + 445.2}{2} \right) = 0.261 \text{ W/mK}$$

$$q_x = \frac{k_m A}{L_C}(T_2 - T_3)$$

or

$$L_C = \frac{k_m A}{q_x}(T_2 - T_3) = \left( \frac{0.261 \times 1}{1106.43} \right)(840 - 445.2) = 0.093 \text{ m} = \textbf{9.3\,cm}$$

## 3.11.2 Series–Parallel Composite Walls

Composite walls may also be characterized by series–parallel configurations, as shown in Fig. 3.13a. It is presumed that surfaces parallel to the x direction are isothermal. The equivalent thermal circuit for a series-parallel composite wall is shown in Fig. 3.13b. The heat transfer is taking place in three steps by conduction, viz. (i) heat is transferred from the hot outer to the inner surface of wall A, (ii) heat is conducted through the wall A to the parallel walls B and C, and (iii) heat is conducted through the wall B and C to wall D.

Step1:

$$q = q_k = \frac{k_A A}{L_A}(T_i - T_1) = \left\{ \frac{(T_i - T_1)}{\left(\frac{L_A}{k_A A}\right)} \right\} = \left\{ \frac{(T_i - T_1)}{R_1} \right\} \tag{3.109}$$

Sep 2:

$$q = q_k = \left[ \frac{(T_1 - T_2)}{\left\{ \frac{1}{\frac{1}{R_{2a}} + \frac{1}{R_{2b}}} \right\}} \right] = \left\{ \frac{(T_1 - T_2)}{R_2} \right\} \tag{3.110}$$

Step3:

$$q = q_k = \frac{k_D A}{L_D}(T_2 - T_o) = \left\{ \frac{(T_2 - T_0)}{\left(\frac{L_D}{k_D A}\right)} \right\} = \left\{ \frac{(T_2 - T_o)}{R_3} \right\} \tag{3.111}$$

Thermal resistance:

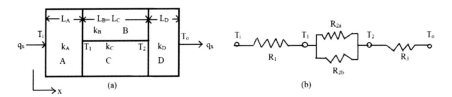

(a)                                              (b)

**Fig. 3.13**  **a** Series–parallel composite wall and **b** its electric circuit

$$R_1 = \frac{L_A}{k_A A},$$

$$R_2 = \left\{ \frac{1}{\frac{1}{R_{2a}} + \frac{1}{R_{2b}}} \right\} = \left\{ \frac{1}{\left( \frac{L_B}{k_B \left( \frac{A}{2} \right)} \right)} + \frac{1}{\left( \frac{L_C}{k_C \left( \frac{A}{2} \right)} \right)} \right\} = \left\{ \frac{1}{\frac{k_B (A/2)}{L_B} + \frac{k_C (A/2)}{L_C}} \right\},$$

$$R_3 = \frac{L_D}{k_D A}.$$

Therefore,

$$q = \left\{ \frac{(T_i - T_o)}{R_1 + R_2 + R_3} \right\} = \left\{ \frac{(T_i - T_o)}{\Sigma R} \right\} \tag{3.112}$$

### 3.11.3  Coaxial Cylinder

A composite cylinder consisting of two coaxial layers of different materials of different thermal conductivities like $k_1$ and $k_2$ has perfect thermal contact is shown in Fig. 3.14. A hot fluid at temperature $T_i$ is flowing inside the cylinder. The heat transfer steps are: (i) heat is transferred by convection with heat transfer coefficient $h_i$ to the inner wall of layer 1. (ii) and (iii) Then heat is transferred through layers 1 and 2 by conduction. (iv) On the outer surface, heat is transferred by convection to a cold fluid at a temperature $T_o$ with a heat transfer coefficient $h_o$. The heat transfer is only radial direction. If L is the length of the cylinder, the total heat transfer rate $q_r$ from the hot fluid to cold fluid over length L is the same through each layer.

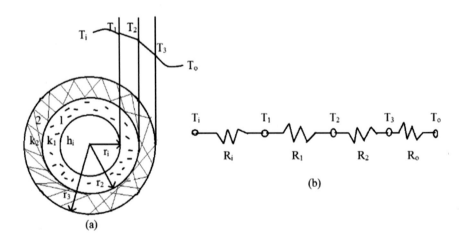

(a)

(b)

**Fig. 3.14  a** Composite cylinder and **b** its electric circuit

Step 1:

$$q_r = h_i\, A(T_i - T_1) = h_i 2\pi r_1 L(T_i - T_1) \quad \text{(since } A = 2\pi r_1 L)$$

$$= \left\{ \frac{(T_i - T_1)}{\left(\frac{1}{h_i(2\pi r_1 L)}\right)} \right\} = \left\{ \frac{(T_i - T_1)}{R_i} \right\} \tag{3.113}$$

Step 2:

$$q_r = \frac{2\pi\, L k_1 (T_1 - T_2)}{\ln \frac{r_2}{r_1}} \quad \{\text{from Eq. (3.70)}\}$$

$$= \left\{ \frac{(T_1 - T_2)}{\left(\frac{1}{k_1\{2\pi L\}}\right)\ln \frac{r_2}{r_1}} \right\} = \left\{ \frac{(T_1 - T_2)}{R_1} \right\} \tag{3.114}$$

Step 3:

$$q_r = \frac{2\pi\, L k_2 (T_2 - T_3)}{\ln \frac{r_3}{r_2}} = \left\{ \frac{(T_2 - T_3)}{\left(\frac{1}{k_2\{2\pi L\}}\right)\ln \frac{r_3}{r_2}} \right\} = \left\{ \frac{(T_2 - T_3)}{R_2} \right\} \tag{3.115}$$

Sep 4:

$$q_r = h_0 A(T_3 - T_0) = h_0 2\pi r_3 L(T_3 - T_0)$$

$$= \left\{ \frac{(T_3 - T_0)}{\frac{1}{h_o(2\pi r_3 L)}} \right\} = \left\{ \frac{T_3 - T_0}{R_0} \right\} \tag{3.116}$$

By adding Eqs. (3.113) to (3.116):

$$q_r = \left\{ \frac{(T_i - T_o)}{R_i + R_1 + R_2 + R_o} \right\} = \left\{ \frac{(T_i - T_o)}{\Sigma R} \right\} \tag{3.117}$$

where

$$R_i = \left( \frac{1}{h_i(2\pi r_1 L)} \right), R_1 = \left( \frac{\ln \frac{r_2}{r_1}}{k_1\{2\pi L\}} \right), R_2 = \left( \frac{\ln \frac{r_3}{r_2}}{k_2\{2\pi L\}} \right) \text{ and } R_o = \left( \frac{1}{h_o(2\pi r_3 L)} \right)$$

so

$$q_r = \left\{ \frac{2\pi L(T_i - T_o)}{\left[ \frac{1}{h_i r_1} + \frac{1}{h_o r_3} + \frac{\ln \frac{r_2}{r_1}}{k_1} + \frac{\ln \frac{r_3}{r_2}}{k_2} \right]} \right\} \tag{3.118}$$

If there are n concentric cylinders, then Eq. (3.118) becomes:

$$q_r = \left\{ \frac{2\pi L(T_i - T_o)}{\left[ \frac{1}{h_i r_1} + \frac{1}{h_o r_{n+1}} + \Sigma \left\{ \frac{1}{k_n} \ln \left( \frac{r_{n+1}}{r_n} \right) \right\} \right]} \right\} \tag{3.119}$$

The overall coefficient of heat transfer, U, can be calculated for this system. The area on which it is based must be specified because the area of a cylinder varies in the radial direction. $U_i$ based on the interior surface area $A_i$ of the cylinder:

$$U_i A_i = \frac{1}{\Sigma R} = \frac{1}{R_i + R_1 + R_2 + R_O} \tag{3.120}$$

and $U_o$ based on the exterior surface area $A_o$ of the cylinder:

$$U_o A_o = \frac{1}{\Sigma R} \tag{3.121}$$

where
$A_i = 2\pi r_1 L$ and $A_o = 2\pi r_3 L$ for the two-layer composite cylinder as shown in Fig. 3.14.

**Example 3.15** Calculate the heat loss from an insulated steel pipe, carrying a hot liquid, to the surroundings per meter length of the pipe. The data are given the following:

Internal diameter of the pipe = 10 cm, wall thickness = 1 cm, thickness of insulation = 3 cm, temperatures of hot liquid and surroundings are 120 °C and 25 °C. k for steel = 58 W/mK, k for insulating material = 0.2 W/mK, Heat transfer coefficient for inside and outside are 720 W/m²K and 9 W/m²K, respectively.

**Solution**

According to Eq. (3.118):
    The heat loss per unit length:

$$(q_r/L) = \left\{ \frac{2\pi(T_i - T_o)}{\left[ \frac{1}{h_i r_1} + \frac{1}{h_o r_3} + \frac{\ln \frac{r_2}{r_1}}{k_1} + \frac{\ln \frac{r_3}{r_2}}{k_2} \right]} \right\}$$

$$= \left\{ \frac{2 \times 3.14\,(120 - 25)}{\left[ \frac{1}{720 \times 0.05} + \frac{1}{9 \times 0.09} + \frac{\ln \frac{6}{5}}{58} + \frac{\ln \frac{9}{6}}{0.2} \right]} \right\} = \mathbf{181.34\ W/m}$$

**Example 3.16**  A steel pipe (k = 50 W/mK) is covered with two layers of insulation each having a thickness of 50 mm. The pipe has internal diameter 100 mm and external diameter 110 mm. k for first insulation material is 0.06 W/mK and for second is 0.12 W/mK. Calculate the loss of heat per meter length of pipe and the interface temperatures between the two layers of insulation when the temperature of the inside pipe surface is 250 °C and that of the outside surface of the insulation is 50 °C.

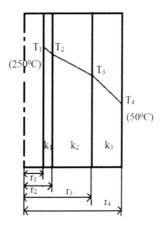

Fig. Ex. 3.16

**Solution**

The insulation pipe is shown in Fig. Ex. 3.16.
    Here $r_1 = 50$ mm $= 0.05$ m, $r_2 = 55$ mm $= 0.055$ m, $r_3 = 105$ mm $= 0.105$ m and

$r_4 = 155$ mm $= 0.155$ m
$k_1 = 50$, $k_2 = 0.06$ and $k_3 = 0.12$ W/mK; $T_1 = 250$ °C, $T_4 = 50$ °C, $T_3 = ?$
According to Eq. (3.119) with n = 3:
The heat loss per unit length:

$$\left( q_r / L \right) = \left\{ \frac{2\pi\,(T_1 - T_4)}{\left[ \frac{\ln \frac{r_2}{r_1}}{k_1} + \frac{\ln \frac{r_3}{r_2}}{k_2} + \frac{\ln \frac{r_4}{r_3}}{k_3} \right]} \right\}$$

$$= \left\{ \frac{2 \times 3.14\,(250 - 50)}{\left[ \frac{\ln \frac{55}{50}}{50} + \frac{\ln \frac{105}{55}}{0.06} + \frac{\ln \frac{155}{105}}{0.12} \right]} \right\} = \mathbf{89.52\,W/m}$$

The interface temperature, $T_3$ can be obtained from the equation:

$$(q_r/L) = \left\{ \frac{2\pi(T_3 - T_4)}{\frac{\ln \frac{r_4}{r_3}}{k_3}} \right\} = \left\{ \frac{2 \times 3.14\,(T_3 - 50)}{\frac{\ln \frac{155}{05}}{0.12}} \right\} = 1.93\,(T_3 - 50) = 89.52$$

Therefore,

$$T_3 = (89.52/1.93) + 50 = \mathbf{96.4\,°C}$$

**Example 3.17** If the order of insulation materials for steel pipe of Ex. 3.16 was reversed, that is, the insulation with a higher value of thermal conductivity was put first, now calculate the change of heat loss with all other conditions remaining unchanged. Comment also on the result.

**Solution**

Now new values of $k_2$ and $k_3$ are 0.12 and 0.06 W/mK, respectively.
    Therefore,

$$(q_r/L) = \left\{ \frac{2\pi\,(T_1 - T_4)}{\left[ \frac{\ln \frac{r_2}{r_1}}{k_1} + \frac{\ln \frac{r_3}{r_2}}{k_2} + \frac{\ln \frac{r_4}{r_3}}{k_3} \right]} \right\} = \left\{ \frac{2 \times 3.14\,(250 - 50)}{\left[ \frac{\ln \frac{55}{50}}{50} + \frac{\ln \frac{105}{55}}{0.12} + \frac{\ln \frac{155}{105}}{0.06} \right]} \right\} = \mathbf{105.72\,W/m}$$

It is seen that the loss of heat is increased by about 18.1% compared with Ex. 3.16.
    Since the purpose of insulation is to reduce the loss of heat, it is always better to place the insulating material with low thermal conductivity on the surface of the tube first, then insulating material with higher thermal conductivity.

### 3.11.4  Coaxial Sphere

A composite hollow sphere consisting of two coaxial layers of different materials of different thermal conductivities like $k_1$ and $k_2$ has perfect thermal contact,, which is shown in Fig. 3.14. A hot fluid at temperature $T_i$ is flowing inside the sphere. The heat transfer steps are: (i) heat is transferred by convection with heat transfer coefficient $h_i$ to the inner wall of layer 1. (ii) and (iii) Then heat is transferred through layers 1 and 2 by conduction. (iv) On the outer surface heat is transferred by convection to a cold fluid at a temperature $T_o$ with a heat transfer coefficient $h_o$. The heat transfer is only radial direction.

Step 1:

$$q_r = h_i A(T_i - T_1) = h_i 4\pi r_1^2 (T_i - T_1) \quad \left( \text{since } A = 4\pi r_1^2 \right)$$

$$= \left\{ \frac{(T_i - T_1)}{\left( \frac{1}{h_i 4\pi r_1^2} \right)} \right\} = \left\{ \frac{(T_i - T_1)}{R_i} \right\} \tag{3.122}$$

Step 2:

$$q_r = \frac{4\pi r_1 r_2 k_1 (T_1 - T_2)}{(r_2 - r_1)} \quad \text{[from Eq. (3.77)]}$$

$$= \left\{ \frac{(T_1 - T_2)}{\left( \frac{(r_2 - r_1)}{k_1 \{4\pi r_1 r_2\}} \right)} \right\} = \left\{ \frac{(T_1 - T_2)}{R_1} \right\} \tag{3.123}$$

Step 3:

$$q_r = \frac{4\pi r_2 r_3 k_2 (T_2 - T_3)}{(r_3 - r_2)} = \left\{ \frac{(T_2 - T_3)}{\frac{(r_3 - r_2)}{k_2 \{4\pi r_2 r_3\}}} \right\} = \frac{(T_2 - T_3)}{R_2} \tag{3.124}$$

Step 4:

$$q_r = h_o A(T_3 - T_o) = h_o 4\pi r_3^2 (T_3 - T_o)$$

$$= \left\{ \frac{(T_3 - T_o)}{\left( \frac{1}{h_o 4\pi r_3^2} \right)} \right\} = \left\{ \frac{(T_3 - T_o)}{R_o} \right\} \tag{3.125}$$

By adding Eqs. (3.122) to (3.125):

$$q_r = \left\{ \frac{(T_i - T_o)}{R_i + R_1 + R_2 + R_o} \right\} = \left\{ \frac{(T_i - T_o)}{\Sigma R} \right\} \tag{3.126}$$

where

$$R_i = \frac{1}{h_i 4\pi r_1^2}, R_1 = \left( \frac{(r_2 - r_1)}{k_1 \{4\pi r_1 r_2\}} \right), R_2 = \left( \frac{(r_3 - r_2)}{k_2 \{4\pi r_2 r_3\}} \right) \quad \text{and } R_o = \left( \frac{1}{h_o 4\pi r_3^2} \right)$$

so

$$q_r = \left\{ \frac{4\pi (T_i - T_o)}{\left[ \frac{1}{h_i r_i^2} + \frac{1}{h_o r_3^2} + \frac{(r_2 - r_1)}{k_1 r_1 r_2} + \frac{(r_3 - r_2)}{k_2 r_2 r_3} \right]} \right\} \tag{3.127}$$

**Example 3.18** A hollow sphere is made of two materials, first with k = 70 W/mK is having an internal diameter of 10 cm and outer diameter of 30 cm and the second with k = 15 W/mK forms the outer layer with outer diameter of 40 cm. The inside and outside temperatures are 300 °C and 30 °C, respectively. Estimate the rate of heat flow through this sphere assuming perfect contact between two materials.

**Solution**

According to Fig. 3.14 and Eq. (3.127):

$$q_r = \left\{ \frac{4\pi(T_i - T_o)}{\left[ \frac{(r_2 - r_1)}{k_1 r_1 r_2} + \frac{(r_3 - r_2)}{k_2 r_2 r_3} \right]} \right\} = \frac{4 \times 3.14(300 - 30)}{\left[ \frac{(0.15 - 0.05)}{70 \times 0.05 \times 0.15} + \frac{(0.2 - 0.15)}{15 \times 0.15 \times 0.2} \right]} = \mathbf{11.3\,kW}$$

### 3.11.5  Critical Radius of Insulation

The addition of insulation to the outside surface of cylindrical or spherical walls does not reduce heat loss. In fact, under certain circumstances, it increases the heat loss up to a certain thickness of insulation. Consider an insulating layer in the form of a hollow cylinder of length L. $r_i$ and $r_o$ are the inner and outer radii of insulation. The temperatures of inside and outside surfaces of insulation are $T_i$ and $T_o$, respectively; outside surface temperature is dissipating heat by convection to the surroundings at temperature $T_\infty$ with heat transfer coefficient, h. If radiation is also taking place, then h can be taken as the combined convection and radiation coefficient.

The rate of heat transfer, q, through this insulation layer can be written from Eq. (3.117):

$$q_r = \left\{ \frac{(T_i - T_\infty)}{R_{ins} + R_o} \right\} \tag{3.128}$$

where

$$R_0 = \left( \frac{1}{h(2\pi r_o L)} \right), \quad R_{ins} = \left( \frac{\ln \frac{r_o}{r_i}}{k\{2\pi L\}} \right)$$

With increasing $r_o$, $R_{ins}$ will be increased, but $R_o$ will be decreased. It is possible that $R_o$ may decrease faster than the increase of $R_{ins}$, causing an increase in q as given by Eq. (3.126). It is also a well-known fact that q would approach zero if an infinite amount of insulation was added. This means that there must be a value of $r_o$ for which q is maximum. This value of $r_o$ is known as the critical radius of insulation, $r_c$. Figure 3.15 shows the effect of increasing the radius $r_o$ on q. It is seen that if $r_o$ is less than $r_c$ and insulation is added to the cylinder heat losses, which will be increased and go through a maximum at $r_c$ and then decrease. If $r_o$ is greater than $r_c$, and insulation is added, the losses will continually decrease.

The concept of critical radius of insulation can be applied to advantage in different situations. For example, the pipelines carrying hot gas must be lagged with insulation so that $r_o$ is always greater than $r_c$. But in the case of electrical wires, $r_o$ should be less than or equal to $r_c$. The proper thickness of insulation will not only provide protection from hazards but will also increase the dissipation of heat (generated within the conducting wire) from that of the bare wire.

Putting the values of $R_o$ and $R_{ins}$ to the Eq. (3.128):

$$q_r = \left\{ \frac{(T_i - T_\infty)}{R_{ins} + R_o} \right\} = q_r = \left\{ \frac{(T_i - T_\infty)}{\left( \frac{\ln \frac{r_o}{r_i}}{k\{2\pi L\}} \right) + \left( \frac{1}{h(2\pi r_o L)} \right)} \right\} \tag{3.129}$$

Therefore,

$$\frac{dq}{dr_o} = \left\{ \frac{-(T_i - T_\infty)\left[ \frac{1}{k(2\pi L)r_o} - \frac{1}{h(2\pi L)r_o^2} \right]}{\left( \frac{\ln \frac{r_o}{r_i}}{k\{2\pi L\}} \right) + \left( \frac{1}{h(2\pi L)r_o^2} \right)} \right\} \tag{3.130}$$

The value of $r_c$, that is, $r_o$ for which q is a maximum may be obtained by equating $(\frac{dq}{dr_o})$ to zero.

Since $(T_i - T_\infty) \neq 0$,

Therefore,

**Fig. 3.15** Critical radius of insulation for cylinder

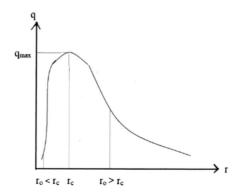

$$\left[\frac{1}{k(2\pi L)r_o} - \frac{1}{h(2\pi L)r_o^2}\right] = 0$$

or

$$r_o = \frac{k}{h} = r_c \qquad\qquad (3.131)$$

To prevention of heat flow by insulation in practice always necessitates a compromise between the effectiveness of insulation and the minimization of cost. By employing a multi-layer aluminum foil technique in vacuum, it is possible to achieve nearly perfect insulation; but the cost involved is prohibitive for general engineering purposes.

Gases with their low conductivity values offer the greatest resistance to heat flow by conduction but allow appreciable heat transfer by convection and radiation. However, if the gas, usually air, is trapped in a porous or fibrous material, its capacity to transfer heat by convection and radiation is tremendously reduced.

**Example 3.19** An electrical conducting copper wire with a diameter of 1 mm is covered with a plastic insulation of thickness 1 mm. The temperature of its surroundings is 25 °C. Find the maximum current carried by the copper wire so that no part of the plastic is above 80 °C.

Given: k for copper = 400 W/mK, k for plastic = 0.5 W/mK and h = 8 W/m²K. Specific electric resistance of copper, $\rho = 3 \times 10^{-8}$ $\Omega$ m.

**Solution**

The electric resistance/meter length

$$= \frac{\text{specific resistance}}{\text{cross sectional area}} = \frac{3 \times 10^{-8}}{\pi\left(0.05 \times 10^{-3}\right)^2} = 3.82\,\Omega\,\text{m}$$

Since

$$q = I^2 R = 3.82\,I^2\,\text{W}$$

Thermal resistance of the convective film insulation per meter length:

$$R_{\text{ther}} = \left(\frac{\ln\frac{r_o}{r_i}}{2\pi k}\right) + \left(\frac{1}{2\pi r_o h}\right) = \left(\frac{\ln\frac{0.015}{0.005}}{2 \times 3.14 \times 0.5}\right) + \left(\frac{1}{2 \times 3.14 \times 0.015 \times 8}\right) = 1.68$$

$$q = \left\{\frac{(T_i - T_\infty)}{R_{\text{ther}}}\right\} = \left\{\frac{(80 - 25)}{1.68}\right\} = 32.74\,\text{W} = 3.82 I^2\,\text{W}$$

Therefore, $I = \left\{\frac{32.74}{3.82}\right\}^{1/2} = 2.93$ A

The maximum safe current limit is 2.93 A for the plastic temperature not to exceed 80 °C.

The critical radius,

$$r_c = \frac{k}{h} = \frac{0.5}{8} = 0.0625 \, m = 6.25 \, cm$$

As the critical value of insulation is much greater than that provided in this problem, the current capacity of the conductor can be raised considerably by increasing the radius of plastic covering upto 6.25 cm.

## 3.12 Systems with Different Heat Sources

There are many instances of generation or absorption of heat taking place within the body itself, e.g. chemical and combustion processes, in electrical conductors. In all these cases, the temperature distribution is of more interest. Some simple systems with uniform heat sources like plane walls, solid and hollow cylinders, spheres etc. will be considered in this section. One-dimensional steady-state conditions and constant conductivity of materials are considered.

### 3.12.1 Plane Wall with Internal Heat Generation

Considering a slab (plane wall) has thickness L, in the region $0 \leq x \leq L$, and uniform thermal conductivity (k), to be extended to infinity in y and z directions (as shown in Fig. 3.16). The electrical current is passed through the slab causing a uniform heat generation of $q$ (W/m³). The temperature on the two faces of the slab, $T_w$, will be the same because it loses the same amount of heat by convection on the two sides. Consider cross-sectional area of slab is A.

**Fig. 3.16** Slab with uniform heat generation

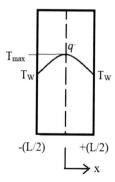

The walls of a furnace may be considered as a plane, if the heat lost through the edges is negligible.

Starting with the general form of the heat conduction equation in Cartesian coordinates [from Eq. (3.51)]:

$$\left( \frac{\delta^2 T}{\delta x^2} + \frac{\delta^2 T}{\delta y^2} + \frac{\delta^2 T}{\delta z^2} \right) + \frac{q^\cdot}{k} = \frac{1}{\alpha} \left( \frac{\delta T}{\delta t} \right) \tag{3.51}$$

where $\alpha$ is the thermal of the material (m$^2$/s) is equal to

$$\frac{k}{\rho C_p}.$$

Under steady-state condition:

$$\frac{\delta T}{\delta t} = 0,$$

Since the heat conduction is through x direction only:

$$\frac{\delta^2 T}{\delta y^2} = \frac{\delta^2 T}{\delta z^2} = 0.$$

The conduction equation [Eq. (3.51)] is simplified to:

$$\frac{\delta^2 T}{\delta x^2} + \frac{q^\cdot}{k} = 0$$

or

$$\frac{d^2 T}{dx^2} = -\frac{q^\cdot}{k} \tag{3.132}$$

Equation (3.132) is a second-order differential equation required two boundary conditions:

(i) at $x = -\frac{L}{2}, T = T_w$;

and (ii) at $x = \frac{L}{2}, T = T_w$

By integration Eq. (3.132):

$$\frac{dT}{dx} = -\frac{q^\cdot}{k} x + C_1 \tag{3.133}$$

Again further integration:

$$T = -\frac{q\cdot}{2k}\,x^2 + C_1 x + C_2 \tag{3.134}$$

The boundary conditions of the problems can be derived as follows:

Since $T = T_w$ at both $x = -\frac{L}{2}$, and $x = \frac{L}{2}$, the temperature distribution about the center plane of the wall would be symmetrical. Under steady-state conditions, the heat generated within must be converted away from the surfaces of the slab. In other words, the maximum temperature must occur at the center line of the slab with one half of the total heat generated going to each face.

That is:

$x = 0, \frac{dT}{dx} = 0$ (a)

Equation (3.133) becomes:

$C_1 = 0$ (b)

And Eq. (3.134) becomes:

$$T = -\frac{q\cdot}{2k}x^2 + C_2 \tag{3.135}$$

Now

at $x = \frac{L}{2}$, $T = T_w$ (c)

Equation (3.135) becomes:

$C_2 = T_w + \frac{q\cdot}{8k}L^2$

Therefore, the temperature distribution:

$$T = -\frac{q\cdot}{2k}x^2 + T_w + \frac{q\cdot}{8k}L^2 = T_w + \frac{q\cdot}{8k}\left(L^2 - 4x^2\right) \tag{3.136}$$

The maximum temperature, $T_{max}$, at the center line is obtained by putting $x = 0$ in Eq. (3.136):

$$T_{max} = T_w + \frac{q\cdot}{8k}L^2 \tag{3.137}$$

The heat flow rate:

$q = -kA\frac{dT}{dx}\Big|_{x=L/2} = -kA\left(-\frac{q\cdot}{k}x + C_1\right)$ [from Eq. (3.133)]

Again

$$C_1 = 0, \text{ and } x = \frac{L}{2}; \text{ so } q = \frac{1}{2}q\cdot AL \tag{3.138}$$

If the ambient temperature is $T_\infty$ and the heat transfer coefficient is h, then at face $x = \frac{L}{2}$:

$$q = \frac{1}{2}q\cdot AL = hA(T_w - T_\infty)$$

or

$$T_w = T_\infty + \frac{1}{2h} q \cdot L \qquad (3.139)$$

Substituting Eq. (3.139) in Eq. (3.136) to get temperature distribution in the slab in terms of the ambient temperature and the heat transfer coefficient:

$$T = T_\infty + \frac{1}{2h} q \cdot L + \frac{q}{8k}(L^2 - 4x^2) = T_\infty + \frac{q}{2}\left\{ \frac{L}{h} + \frac{1}{4k}(L^2 - 4x^2) \right\} \qquad (3.140)$$

**Example 3.20** A plane wall (10 cm thick) generates heat at the rate of $4 \times 10^4$ W/m$^3$, when electric current is passed through it. The convective heat transfer coefficient between each face of the wall and the ambient air is 50 W/m$^2$K. Calculate: (i) the surface temperature, (ii) the maximum temperature in the wall. Given: thermal conductivity of the wall material is 15 W/mK.

**Solution**

Assume the ambient air temperature is 20 °C.
 Here L = 10 cm = 0.1 m, $q = 4 \times 10^4$ W/m$^3$, h = 50 W/m$^2$K, k = 15 W/mK, T = 20 °C = 293 K
 From Eq. (3.139): (i) $T_w = T_\infty + \frac{1}{2h} q \cdot L = 293 + \left(\frac{1}{2 \times 50}\right)\left(4 \times 10^4\right) \times 0.1 =$ 333 K = **60 °C**
 From Eq. (3.137): (ii) $T_{max} = T_w + \frac{q}{8k}L^2 = 333 + \left(\frac{1}{8 \times 15}\right)(4 \times 10^4) \times (0.1)^2 =$ 336.33 K = **63.33 °C**.

### 3.12.2  Hollow Cylinder with Internal Heat Generation

Consider a long hollow cylinder of length, L and inside and outside radius are $r_i$ and $r_o$, respectively. A constant rate of heat ($q$, W/m$^3$) is generated within the cylinder while the boundary surfaces at $r = r_i$ and $r = r_o$ are kept at uniform temperatures $T_i$ and $T_o$, respectively.
 The general form of the heat conduction equation in cylindrical coordinates [from Eq. (3.59)]:

$$\frac{1}{r}\frac{\delta}{\delta r}\left(kr\frac{\delta T}{\delta r}\right) + \frac{1}{r^2}\frac{\delta}{\delta\theta}\left(k\frac{\delta T}{\delta\theta}\right) + \frac{\delta}{\delta z}\left(k\frac{\delta T}{\delta z}\right) + q = \rho C_p\left(\frac{\delta T}{\delta t}\right)$$

Under steady-state condition:

$$\frac{\delta T}{\delta t} = 0.$$

Since the heat conduction is through x direction only:

$$\frac{\delta^2 T}{\delta y^2} = \frac{\delta^2 T}{\delta z^2} = 0.$$

The conduction equation is simplified to:

$$\frac{d}{dr}\left(r\frac{dT}{dr}\right) + \frac{\dot{q}\,r}{k} = 0 \quad (\text{in } r_i \leq r \leq r_o) \tag{3.141}$$

Integrating Eq. (3.141):

$$\left(r\frac{dT}{dr}\right) + \frac{\dot{q}\,r^2}{2k} = C_1 \quad or \quad \frac{dT}{dr} + \frac{\dot{q}\,r}{2k} = \frac{C_1}{r} \tag{3.142}$$

Again integrating (Eq. 3.142):

$$T = -\frac{\dot{q}\,r^2}{4k} + C_1 \ln r + C_2 \tag{3.143}$$

Boundary conditions:
at $r = r_i$, $T = T_i$;  and at $r = r_o$, $T = T_o$ .
Applying the boundary conditions to Eq. (3.143):
$T_i = -\frac{\dot{q}\,r_i^2}{4k} + C_1 \ln r_i + C_2$ (a)
$T_o = -\frac{\dot{q}\,r_o^2}{4k} + C_1 \ln r_o + C_2$ (b)
Substituting (a) from (b):

$$C_1 \ln \frac{r_o}{r_i} = (T_o - T_i) + \frac{\dot{q}\left(r_o^2 - r_i^2\right)}{4\,k}$$

Therefore,
$$C_1 = \frac{(T_o - T_i) + \frac{\dot{q}\left(r_o^2 - r_i^2\right)}{4k}}{\ln \frac{r_o}{r_i}} \quad (c)$$

and

$$C_2 = T_i + \frac{\dot{q}\,r_i^2}{4k} - \left[\frac{(T_o - T_i) + \frac{\dot{q}\left(r_o^2 - r_i^2\right)}{4k}}{\ln \frac{r_o}{r_i}}\right] \ln r_i \quad (d)$$

The temperature distribution in the cylinder, Eq. (3.143) becomes:
$T = -\frac{\dot{q}\,r^2}{4k} + C_1 \ln r + C_2$

$$T = -\frac{\dot{q}\,r^2}{4k} + \left[\frac{(T_o - T_i) + \frac{\dot{q}(r_o^2 - r_i^2)}{4k}}{\ln \frac{r_o}{r_i}}\right] \ln r + T_i + \frac{\dot{q}\,r_i^2}{4k} - \left[\frac{(T_o - T_i) + \frac{\dot{q}(r_o^2 - r_i^2)}{4k}}{\ln \frac{r_o}{r_i}}\right] \ln r_i$$

or

$$T\text{-}T_i = \frac{q^{\cdot}\left(r_i^2 - r^2\right)}{4k} + \left\{(T_o - T_i) + \frac{q^{\cdot}(r_o^2 - r_i^2)}{4k}\right\}\left[\frac{\ln\left(\frac{r}{r_i}\right)}{\ln\left(\frac{r_o}{r_i}\right)}\right]$$

or

$$\frac{T - T_i}{T_o - T_i} = \frac{q^{\cdot}}{4k}\left[\frac{\left(r_i^2 - r^2\right)}{(T_o - T_i)} + \frac{(r_o^2 - r_i^2)}{(T_o - T_i)}\left\{\frac{\ln\left(\frac{r}{r_i}\right)}{\ln\left(\frac{r_o}{r_i}\right)}\right\}\right] + \left[\frac{\ln\left(\frac{r}{r_i}\right)}{\ln\left(\frac{r_o}{r_i}\right)}\right]$$

$$= \left[\frac{\ln\left(\frac{r}{r_i}\right)}{\ln\left(\frac{r_o}{r_i}\right)}\right] + \frac{q^{\cdot}}{4k}\frac{\left(r_i^2 - r^2\right)}{(T_o - T_i)}\left[1 + \left\{\frac{\ln\left(\frac{r}{r_i}\right)}{\ln\left(\frac{r_o}{r_i}\right)}\right\}\right] \qquad (3.144)$$

The radial heat flow,

$$q_r = -kA\frac{dT}{dr}$$

$$= -k\,(2\pi r\,L)\frac{dT}{dr}$$

$$= -k\,(2\pi r\,L)\left\{\frac{C_1}{r} - \frac{q^{\cdot} r}{2k}\right\}$$

[from Eq. (3.142)]

$$= 2\pi L\left\{\frac{q^{\cdot} r^2}{2} - C_1 k\right\} = 2\pi L\left\{\frac{q^{\cdot} r^2}{2} - k\left[\frac{(T_o - T_i) + \frac{q^{\cdot}(r_o^2 - r_i^2)}{4k}}{\ln\left(\frac{r_o}{r_i}\right)}\right]\right\} \qquad (3.145)$$

In case no heat generation, $q^{\cdot} = 0$, Eq. (3.145) becomes:

$$q_r = \frac{2\pi k L(T_o - T_i)}{\ln\left(\frac{r_o}{r_i}\right)} \qquad (3.146)$$

Equation (3.146) is same as Eq. (3.70) obtained previously for a hollow cylinder without heat generation.

The temperature distribution in the cylinder, Eq. (3.144) becomes:

$$\frac{T - T_i}{T_o - T_i} = \left[\frac{\ln\left(\frac{r}{r_i}\right)}{\ln\left(\frac{r_o}{r_i}\right)}\right] \qquad (3.147)$$

### 3.12.3 Solid Cylinder with Internal Heat Generation

For solid cylinder with internal heat generation or current carrying electrical conductor, if the heat generating within a conductor is due to the passage of electric current, $q_g$ can be expressed as:

$$q_g = I^2 R, \quad \text{again} \quad R = \frac{\rho L}{A} \quad \text{and} \quad q^{\cdot} = \frac{q_g}{AL}. \tag{3.148}$$

where i is the current, R is the electrical resistance, $\rho$ is the resistivity, L is the length, and A is the cross-sectional area of the conductor.

Combining these relationships:

$$q^{\cdot} = \frac{q_g}{AL} = \frac{I^2 R}{AL} = \frac{I^2}{AL}\frac{\rho L}{A} = \left(\frac{I}{A}\right)^2 \rho = i^2 \rho \tag{3.149}$$

where i is the current density, A/m$^2$.

Now consider a circular solid rod of length, L and radius, $r_o$ conducting electric current. The surface of the wire is maintained at a temperature $T_0$. This is the case of a solid cylinder with uniformly distributed heat generation.

For steady-state radial condition, the differential equation will be the same as Eq. (3.141):

$$\frac{d}{dr}\left(r\frac{dT}{dr}\right) + \frac{q^{\cdot}r}{k} = 0 \quad (in\ 0 \leq r \leq r_o) \tag{3.150}$$

Boundary conditions: at $r = 0$, $\frac{dT}{dr} = 0$ (for symmetry) (a)
at $r = r_o$, $T = T_o$ (b)

The boundary condition (b) is obtained from the fact that heat generation ($q^{\cdot}$) is uniform throughout the cylinder, and $T_o$ is constant over the entire surface of the cylinder. It is expected that the temperature distribution to be symmetrical around the center of the cylinder.

Integrating Eq. (3.150):

$$\left(r\frac{dT}{dr}\right) + \frac{q^{\cdot}r^2}{2k} = C_1 \quad or \quad \frac{dT}{dr} + \frac{q^{\cdot}r}{2k} = \frac{C_1}{r} \tag{3.151}$$

Again integrating Eq. (3.151):

$$T = -\frac{q^{\cdot}r^2}{4k} + C_1 \ln r + C_2 \tag{3.152}$$

Boundary conditions:
at $r = 0$, $dT/dr = 0$; and at $r = r_o$, $T = T_o$ .

Applying the boundary conditions to Eq. (3.152):

$C_1 = 0$ and $C_2 = T_0 + \frac{q \cdot r_o^2}{4k}$

Therefore,

$$T = -\frac{q \cdot r^2}{4k} + T_0 + \frac{q \cdot r_o^2}{4k} = T_o + \frac{q}{4k}(r_o^2 - r^2) \qquad (3.153)$$

The temperature at the center of the cylinder (i.e. $r = 0$):

$$T_c = T_0 + \frac{q \cdot r_o^2}{4k} \qquad (3.154)$$

Substituting value of $q \cdot$ From Eq. (3.149):

So

$$T_c = T_0 + \frac{i^2 \rho \, r_o^2}{4k} \qquad (3.155)$$

Combining Eqs. (3.153) and (3.154):

$$\frac{(T - T_o)}{(T_c - T_o)} = \frac{\frac{q}{4k}(r_o^2 - r^2)}{\frac{q \cdot r_o^2}{4k}} = \frac{(r_o^2 - r^2)}{r_o^2} = 1 - \left(\frac{r}{r_o}\right)^2 \qquad (3.156)$$

which shows that the temperature distribution in the solid cylinder is parabolic.

Heat flow through the cylinder:

$$
\begin{aligned}
q_r &= -k A \frac{dT}{dr}\bigg|_{(r=ro)} = -k(2\pi r_o L) \frac{d}{dr}[T_o + \frac{q}{4k}(r_o^2 - r^2)]\big|_{r=ro} \\
&= -k(2\pi r_o L)\left\{\frac{q}{4k}(-2r_o)\right\} \\
&= \pi r_o^2 L q \cdot
\end{aligned}
\qquad (3.157)
$$

Now heat Conducted = heat generation = heat converted

$$\pi r_o^2 L q \cdot = 2\pi r_o L \, h(T_o - T_\infty)$$

or

$$T_o = T_\infty + \frac{q \cdot r_o}{2h} \qquad (3.158)$$

where $T_\infty$ is the ambient temperature.

Substitute this value in Eq. (3.153):

$$T = T_o + \frac{q}{4k}(r_o^2 - r^2) = T_\infty + \frac{q \cdot r_o}{2h} + \frac{q}{4k}(r_o^2 - r^2) \qquad (3.159)$$

**Example 3.21** An electrical transmission line made of 25 mm copper wire carries 200 A and has a resistance of $0.4 \times 10^{-4}$ $\Omega$/cm length. If the surface temperature is 200 °C and the ambient air temperature is 10 °C, determine the heat transfer coefficient between the wire surface and the ambient air, and the maximum temperature in the wire. Given: k = 15 W/mK.

**Solution**

The heat loss in the conductor, $q = I^2R = (200)^2 \times (0.4 \times 10^{-4}) = 1.6$ W/cm $= 160$ W/m

$$q' = q/AL = 160/(\pi r_o^2 L) = \frac{160}{3.14 \times (12.5 \times 10^{-3})^2 \times 1} = 326.1 \, \text{kW/m}^3$$

The maximum temperature in the wire occurs at the center:

$$T_c = T_0 + \frac{q' \cdot r_o^2}{4k} = 200 + \frac{(3.261 \times 10^5) \times (12.5 \times 10^{-3})^2}{4 \times 15} = 0.849 = \mathbf{200.85\,°C}$$

Since

$$T_o = T_\infty + \frac{q' \, r_o}{2h}$$

or

$$h = \frac{q' \, r_o}{2(T_o - T_\infty)} = \frac{(326.1 \times 10^3) \times (12.5 \times 10^{-3})}{2 \times (200 - 10)} = \mathbf{10.73 \, W/m^2 K}$$

**Example 3.22** Nichrome wire is used as a heating element in a 10 kW heater. The nichrome surface temperature should not exceed 1220 °C. Other data given: Resistivity of nichrome is 100 $\mu\Omega$.cm, surrounding air temperature is 20 °C, outside surface heat transfer coefficient is 1.15 W/m$^2$K, k for nichrome is 17 W/mK.

Find out the diameter of nichrome wire for making 1m long heater, and also find out the rate of current flow.

**Solution**

Let $d_o$ is the diameter of the nichrome wire.
   Here q = 10 kW, L = 1m

$$q' = q/AL = \frac{10 \times 1000}{\pi d_o^2 \times (1/4)} = \frac{4 \times 10^4}{\pi d_o^2} W/m^3$$

The surface temperature:

$$T_0 = T_\infty + \frac{q\,r_o}{2h} = T_\infty + \frac{q\,d_o}{4h} = 20 + \frac{4 \times 10^4}{\pi d_o^2}\left[\frac{d_o}{4 \times (1.15 \times 10^3)}\right]$$

Therefore,

$$d_o = \frac{2.77}{1200} = 2.31 \times 10^{-3}\text{m} = \mathbf{2.31mm}$$

Since

$$R = \frac{\rho L}{A} = \frac{100 \times 10^{-6} \times 10^{-2} \times 1}{\pi(d_o^2/4)} = \frac{4 \times 10^{-6}}{3.14 \times (2.31 \times 10^{-3})^2} = 0.24\Omega$$

Again

$$q = I^2 R \text{ or } I = \sqrt{\frac{q}{R}} = \sqrt{\frac{10^4}{0.24}} = \mathbf{204A}$$

### 3.12.4  Solid Sphere with Internal Heat Generation

Consider a solid sphere with a uniform heat source ($q^{\cdot}$, W/m³). The outside surface (at $r = r_o$) is maintained at a constant temperature $T_0$. The differential equation describing the temperature distribution can be determined by making an energy balance on a spherical shell of thickness, dr, and cross-sectional area, $4\pi r^2$ as shown in Fig. 3.17.

The general form of the heat conduction equation in spherical coordinate [from Eq. (3.60)]:

$$\frac{1}{r^2}\frac{\delta}{\delta r}\left(kr^2\frac{\delta T}{\delta r}\right) + \frac{1}{r^2\sin^2\theta}\frac{\delta}{\delta\varphi}\left(k\frac{\delta T}{\delta\varphi}\right) + \frac{1}{r^2\sin\theta}\frac{\delta}{\delta\theta}\left(k\sin\theta\frac{\delta T}{\delta\theta}\right) + q^{\cdot} = \rho\,C_p\left(\frac{\delta T}{\delta t}\right)$$

**Fig. 3.17** Solid sphere with uniform heat generation

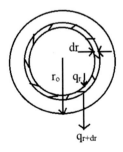

Under steady-state condition: $\frac{\delta T}{\delta t} = 0$,

Since the heat conduction is through r direction only:

$$\frac{\delta^2 T}{\delta \varphi^2} = \frac{\delta^2 T}{\delta \theta^2} = 0.$$

The conduction equation is simplified to:

$$\frac{1}{r^2}\frac{d}{dr}\left(kr^2\frac{dT}{dr}\right) + q\dot{} = 0 \quad (\text{in } 0 \leq r \leq r_0)$$

or

$$\frac{d}{dr}\left(r^2\frac{dT}{dr}\right) + \frac{q\dot{}r^2}{k} = 0 \tag{3.160}$$

$$At\ r = 0,\ \frac{dT}{dr} = 0 \quad (\text{for symmetry}) \quad \text{(a)}$$
$$At\ r = r_0,\ T = T_0 \qquad\qquad\qquad \text{(b)}$$

Integration Eq. (3.160):

$$r^2\frac{dT}{dr} = -\frac{q\dot{}r^3}{3k} + C_1 \ \text{or}\ \frac{dT}{dr} = -\frac{q\dot{}r}{3k} + \frac{C_1}{r^2} \tag{3.161}$$

Again integration Eq. (3.161):

$$T = -\frac{q\dot{}r^2}{6k} - \frac{C_1}{r} + C_2 \tag{3.162}$$

Putting boundary conditions (a and b) in Eqs. (3.161) and (3.162):

$$C_1 = 0$$

and

$$T_0 = -\frac{q\dot{}r_o^2}{6k} + C_2 \ \text{or}\ C_2 = T_0 + \frac{q\dot{}r_o^2}{6k}$$

Now Eq. (3.162) becomes:

$$T = -\frac{q\dot{}r^2}{6k} + T_0 + \frac{q\dot{}r_0^2}{6k} = T_0 + \frac{q\dot{}}{6k}\left(r_0^2 - r^2\right). \tag{3.163}$$

The temperature at the center of the sphere (i.e. r = 0):

$$T_c = T_0 + \frac{q^{\cdot} r_o^2}{6k} \tag{3.164}$$

which is also the maximum temperature in the sphere.

The heat flow rate can be written as:

$$q_r = -k \, 4\pi r_o^2 \frac{dT}{dr}\Big|_{r=ro} = -k \, 4\pi r_o^2 \left(-\frac{q^{\cdot} r_o}{3k}\right) = \frac{4}{3}\pi r_o^3 q^{\cdot}$$

$$= Volume \, of \, sphere \times q^{\cdot} \tag{3.165}$$

Thus, heat generated is equal to the conducted to the surface, which must be equal to the heat convection from the surface under steady-state condition.

Therefore,

$$\frac{4}{3}\pi r_o^3 q^{\cdot} = 4\pi r_o^2 h(T_0 - T_\infty) \text{ or } T_0 = T_\infty + \frac{q^{\cdot} r_o}{3h} \tag{3.166}$$

where $T_\infty$ is the ambient temperature.

The temperature distribution, Eq. (3.166), in term of the heat transfer coefficient, h, and the ambient temperature, $T_\infty$, can be written as:

$$T = T_o + \frac{q^{\cdot}}{6k}\left(r_o^2 - r^2\right) = T_\infty + \frac{q^{\cdot} r_o}{3h} + \frac{q^{\cdot}}{6k}\left(r_o^2 - r^2\right) \tag{3.167}$$

**Example 3.23** The average heat produced by oranges ripening is estimated to be 300 W/m².

Assume the oranges are spherical shape and average size of an orange is 8 cm. Calculate the temperature at the center of the orange. Given: k for orange is 0.15 W/mK and outer temperature is 20 °C.

**Solution**

Since

$$q/A = 300 = (Vol \times q^{\cdot})/A = \left(\frac{4}{3}\pi r_o^3 q^{\cdot}\right)/4\pi r_o^2 = \frac{q^{\cdot}}{3} r_o$$

Therefore,

$$q^{\cdot} = (300 \times 3)/r_0 = 900/(4 \times 10^{-2}) = 225 \times 10^2 \, W/m^2$$

The temperature at the center of the sphere (i.e. r = 0)

$$T_c = T_0 + \frac{q^{\cdot} r_o^2}{6k} = 293 + \frac{(225 \times 10^2)(4 \times 10^{-2})^2}{6 \times 0.15} = 293 + 40 = 333K = \mathbf{60\,°C}$$

## 3.13  Convective Heat Transfer

So far discussed of heat transfer based on conduction and have referred to convection only it provides one of the boundary conditions for conduction problems. Convection is the mode of heat transfer between a surface and a fluid flowing over it. The energy transfer in convection is predominantly due to the bulk motion of the fluid particles, through the molecular conduction within the fluid itself also contributes to some extent. If this motion is mainly due to the density variations associated with temperature gradients within the fluid, the mode of heat transfer is said to be due to *free or natural convection*, when the motion of fluid arises only from the density differences associated with the temperature field. On the other hand, if this fluid motion is produced by some forced velocity field (like a fan, a blower or a pump), the energy transport is said to be due to *forced convection*, i.e. when the motion of fluid arises principally from a pressure gradient caused by a pump or blower.

The heat flux, $q'$, for an arbitrarily shaped surface area A at temperature, $T_s$ (Fig. 3.18a), over which a fluid of velocity, u and temperature, $T_\infty$ can be written as:

$$q' = h \, (T_s - T_\infty) \tag{3.168}$$

where h is the *local heat transfer coefficient*.

This equation (3.168) is referred as *Newton's law of cooling*. Due to the variation of flow conditions from point to point, the values of $q'$ and h along the surface also vary. The *total heat transfer rate* may be obtained by integrating Eq. (3.168) over the entire surface, assuming a uniform surface temperature, $T_s$.

$$q = \int_A q' dA = (T_s - T_\infty) \int_A h \, dA \tag{3.169}$$

Equation (3.169) can be written as:

$$q = h^- A(T_s - T_\infty) \tag{3.170}$$

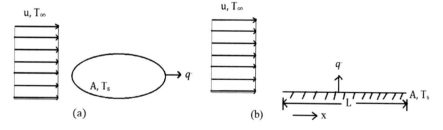

**Fig. 3.18**  Convection of heat on surface

where $h^-$ is the *average or total heat transfer coefficient* for entire surface.

Comparing Eqs. (3.169) and (3.170), the average and local convection coefficients are related by:

$$h_A^- = \frac{1}{A} \int_A h \, dA \tag{3.171}$$

In the case of a simple flat plate of unit width where h is a function of x alone (Fig. 3.18b).

$$h_L^- = \frac{1}{L} \int_0^L h \, dx \tag{3.172}$$

To determine the local heat flux or the total heat transfer rate, which may be determined from Eqs. (3.168) and (3.172) provided that the local and total heat transfer coefficients are known. The aim of any convection analysis is to get these coefficients first.

**Example 3.24** For forced convection on a heated horizontal plate the local heat transfer coefficient may be expressed as:

$$h_x(x) = C \, x^{-0.5}$$

where x is the distance from the leading edge of the plate and C is a coefficient independent of x.

Find out an expression for the ratio $h_L^-/h_x$ where $h_L^-$ is the average heat transfer coefficient between the leading edge and the location x.

**Solution**

From Eq. (3.171) The average heat transfer coefficient:

$$h_x^- = \frac{1}{x} \int_0^x C x^{-0.5} \, dx = \frac{C}{x} \int_0^x x^{-0.5} dx = \frac{C}{x} \left( \frac{x^{0.5}}{0.5} \right) = 2 \, C x^{-0.5} = 2 h_x.$$

Hence,

$$\frac{h_{\bar{x}}}{h_x} = 2.$$

Thus, the average heat transfer coefficient over a length x is twice the value of the local heat transfer coefficient at the Section x.

### 3.13.1 Natural Convective Heat Transfer

The fluid motion is produced due to change of density resulting from temperature different. The mechanism of heat transfer in this situation is called *free or natural convection*. Free convection is the principal mode of heat transfer from pipes, transmission lines, refrigerating coils, etc.

The movement of fluid in free convection occurs since the fluid particles in the of the hot object become warmer than the surrounding fluid due to local change of density. The warmer fluid is replaced by the colder fluid creating currents that are called convection currents. These currents originate when a body force (gravitational, centrifugal, electrostatic etc.) acts on a fluid in which there are density gradients. The force, which induces these convection currents, is called a *buoyancy force;* which is due to the presence of a density gradient within the fluid and a body force. Considering only those situations in which the body force is gravitational and the change in density is brought about by a temperature gradient.

#### 3.13.1.1   Free and Forced Convection Heat Transfer

Consider a heated vertical plate as shown in Fig. 3.19. The plate is maintained at a uniform temperature, $T_s$. A free convection boundary layer is formed when this heated plate is immersed in a quiescent fluid ($u_\infty = 0$) at a constant temperature $T_\infty$ (where $T_\infty < T_s$). Due to free convection currents, upward along the surface of the plate, velocity and thermal boundary layers are formed as shown in Fig. 3.19.

Assume a steady state, two-dimensional condition in which the gravity force acts in the negative x direction. Consider the flow to be incompressible except in the body force term resulting from the buoyancy. Assuming the boundary layer to be laminar, the appropriate equations of continuity, momentum and energy are obtained from Appendix II and III (Eqs. II.1, III.10), and Eq. (3.55).

(a)   Now from Appendix II Eq. (II.1), the equation of continuity:

**Fig. 3.19** Free convection boundary layer on a heated vertical plate

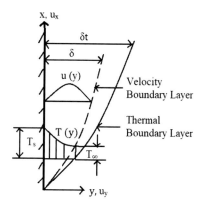

$$\frac{\delta\rho}{\delta t} + \frac{\delta(\rho\, u_x)}{\delta x} + \frac{\delta(\rho\, u_y)}{\delta y} + \frac{\delta(\rho\, u_z)}{\delta z} = 0$$

The steady state is defined as the state in which the density of the fluid does not change with time, i.e. $\left(\frac{\delta\rho}{\delta t}\right) = 0$

Again, for incompressible fluids (i.e. $\rho = $ constant), two-dimensional (i.e. $\frac{\delta u_z}{\delta z} = 0$);

continuity equation (II.1) reduces to:

$$\frac{\delta u_x}{\delta x} + \frac{\delta u_y}{\delta y} = 0 \tag{3.173}$$

(b)   Now from Appendix III Eq. III.10, Navier–Stokes equation, for x component of cartesian coordinates, can be written as:

$$\rho\left(\frac{\delta u_x}{\delta t} + u_x\frac{\delta u_x}{\delta x} + u_y\frac{\delta u_x}{\delta y} + u_z\frac{\delta u_x}{\delta z}\right) = -\frac{\delta p}{\delta x} + \mu\left[\frac{\delta^2 u_x}{\delta x^2} + \frac{\delta^2 u_x}{\delta y^2} + \frac{\delta^2 u_x}{\delta z^2}\right] + \rho\, g_x$$

Now for two dimensional flow: $u_z = 0$, so term $u_z\frac{\delta u_x}{\delta z} = 0$ and $\frac{\delta^2 u_x}{\delta z^2} = 0$

Since $\frac{\delta^2 u_x}{\delta x^2} = 0$, and steady-state condition: $\frac{\delta u_x}{\delta t} = 0$; hence the equation of momentum:

$$u_x\frac{\delta u_x}{\delta x} + u_y\frac{\delta u_x}{\delta y} = -\frac{1}{\rho}\frac{\delta p}{\delta x} + g + \mu\frac{\delta^2 u_x}{\delta y^2} \tag{3.174}$$

(c)   The equation of energy:

$$u_x\frac{\delta T}{\delta x} + u_y\frac{\delta T}{\delta y} = \frac{k}{\rho\, C_p}\frac{\delta^2 T}{\delta y^2} = \alpha\frac{\delta^2 T}{\delta y^2} \tag{3.175}$$

where

$$\alpha = \frac{k}{\rho\, C_p}$$

The pressure gradient term $\left(-\frac{\delta p}{\delta x}\right)$ can be easily evaluated by considering the edge of the boundary layer where $u_x \to 0$, $\rho \to \rho_\infty$ (density outside the boundary layer)

Thus

$$\frac{\delta p}{\delta x} = -\rho_\infty \cdot g \tag{3.176}$$

Or the change in pressure over a height dx is equal to the weight per unit area of the fluid element.

Substituting Eq. (3.176) into Eq. (3.174):

$$u_x \frac{\delta u_x}{\delta x} + u_y \frac{\delta u_x}{\delta y} = -\frac{1}{\rho}(-\rho_\infty \cdot g) - g + u_y \frac{\delta^2 u_x}{\delta y^2}$$

$$= \frac{g}{\rho}(\rho_\infty - \rho) + u_y \frac{\delta^2 u_x}{\delta y^2} \tag{3.177}$$

The change of density $(\rho_\infty - \rho)$ can be expressed in terms of the volumetric coefficient of thermal expansion, $\beta$:

$$-\frac{1}{\rho}\left(\frac{\delta \rho}{\delta T}\right)_p = \beta \quad \text{or} \quad \Delta\rho = -\beta\rho\Delta T \tag{3.178}$$

Thus

$$(\rho_\infty - \rho) = -\beta\,\rho(T_\infty - T) \tag{3.179}$$

Substituting Eq. (3.179) into Eq. (3.177) and x momentum equation becomes:

$$u_x \frac{\delta u_x}{\delta x} + u_y \frac{\delta u_x}{\delta y} = g\,\beta\,(T - T_\infty) + u_y \frac{\delta^2 u_x}{\delta y^2} \tag{3.180}$$

For a perfect gas,

$$\rho = \frac{p}{RT} \text{ and } \beta = -\frac{1}{\rho}\left(\frac{\delta \rho}{\delta T}\right)_p = -\frac{1}{\rho}\left(-\frac{p}{RT^2}\right) = \frac{1}{\rho}\left(\frac{\rho}{T}\right) = \frac{1}{T} \tag{3.181}$$

The set of governing equations for free convection can be written as:

$$\frac{\delta u_x}{\delta x} + \frac{\delta u_y}{\delta y} = 0 \tag{3.182}$$

$$u_x \frac{\delta u_x}{\delta x} + u_y \frac{\delta u_x}{\delta y} = g\,\beta\,(T - T_\infty) + u_y \frac{\delta^2 u_x}{\delta y^2} \tag{3.180}$$

$$u_x \frac{\delta T}{\delta x} + u_y \frac{\delta T}{\delta y} = \alpha \frac{\delta^2 T}{\delta y^2} \tag{3.175}$$

An inspection of Eqs. (3.173), (3.180) and (3.175) reveals that the momentum and energy equations of free convection are coupled and a solution for velocity profile demands a knowledge of the temperature distribution.

Introducing the following dimensionless variables:

$$x^* = \frac{x}{L}, \quad y^* = \frac{y}{L} \tag{3.183}$$

$$u_x^* = \frac{u_x}{u_o}, \quad u_y^* = \frac{u_y}{u_o}, \quad T^* = \frac{T - T_\infty}{T_s - T_\infty} \tag{3.184}$$

where L is a characteristic length and $u_o$ is a reference velocity. With these new variables, the Eqs. (3.173), (3.180), and (3.175) take the following dimensionless forms:

$$\frac{\delta u_x^*}{\delta x^*} + \frac{\delta u_y^*}{\delta y^*} = 0 \tag{3.185}$$

$$u_x^* \frac{\delta u_x^*}{\delta x^*} + u_y^* \frac{\delta u_x^*}{\delta y^*} = \frac{g\beta(T_s - T_\infty)L}{u_o^2} T^* + \frac{1}{Re} \frac{\delta^2 u_x^*}{\delta y^{*2}} \tag{3.186}$$

$$u_x^* \frac{\delta T^*}{\delta x^*} + u_y^* \frac{\delta T^*}{\delta y^*} = \frac{1}{Re\ L\ Pr} \frac{\delta^2 T^*}{\delta y^{*2}} \tag{3.187}$$

where

$$Re = \text{Reynolds number} = \left(\frac{u\ \rho\ d}{\mu}\right)$$

$$Pr = \text{Prandtl number} = \frac{\mu\ C_p}{k} = \frac{\mu}{\rho\ \alpha} \quad \left[\text{Since } \alpha = \frac{k}{\rho\ C_p}\right]$$

Again, the dimensionless group in Eq. (3.186) can be rearranged as:

$$\frac{g\beta(T_s - T_\infty)L}{u_o^2} = \frac{\left\{\frac{(\rho^2 L^3 g\beta\Delta T)}{\mu^2}\right\}}{\left(\frac{L\rho u_o}{\mu}\right)^2} = \frac{Gr}{Re^2} \tag{3.188}$$

where

$$Gr = \text{Grashof number} = \left\{\frac{(\rho^2 L^3 g\beta\Delta T)}{\mu^2}\right\} \tag{3.189}$$

The *Grashof number* (Gr) is the ratio of the buoyancy force to the viscous force in the fluid. This number is of great importance and plays a similar role in free convection as does the Reynolds number (Re) in forced convection. A critical value of the Grashof number is used to indicate transition from laminar to turbulent flow in free convection. Looking at Eqs. (3.183) and (3.184) may lead us to believe that the correlations of free convections would be of the form:

$$Nu = f(Re, Pr, Gr) \qquad (3.190)$$

where

$$Nu = \text{Nusselt number} \ = \frac{hL}{k_f}$$

But this is the case only when both free and forced convection occur simultaneously. In free convection, the only driving force is the buoyancy force due to which a flow field is set up and as such there is no external field present. Hence, the Reynolds number is not an important parameter in free convection, and the correlations for such a case are expected to be of the form:

$$Nu = f(Pr, Gr) \qquad (3.191)$$

## 3.13.2 Forced Convective Heat Transfer

Since the Grashof number is used to indicate transition from laminar to turbulent flow in free convection. For forced convection, the flow is turbulent, hence the Grashof number is not an important parameter in forced convection and the correlations [Eq. (3.187)] for such a case are expected to be of the form:

$$Nu = f(Re, Pr) \qquad (3.192)$$

or

$$Nu = C(Re)^e \cdot (Pr)^f \qquad (3.193)$$

The values of C, e, and f are independent of the nature of the fluid. Equation (3.193) is inferred from experimental measurements; it is termed *an empirical correlation*. However, the specific values of the coefficient C and the exponents e and f are varying with the nature of the surface geometry and the type of flow.

Nusselt number (Nu) is equal to the dimensionless temperature gradient at the surface and is a measure of the convection energy transfer occurring at the surface. Reynolds number (Re) is the ratio of inertial to viscous forces and it determines the nature of flow. Prandtl number (Pr) is a measure of relative effectiveness of momentum and energy diffusion in the velocity and thermal boundary layers.

Thus, the convection coefficient in forced convection heat transfer is a function of both Reynolds number and Prandtl number [Eq. (3.192)].

### 3.13.2.1  Dimensional Analysis Applied to Forced Convection Heat Transfer

Consider, a fluid is flowing across a heated tube of diameter, D. The mean velocity of fluid is u. The variables that affect the heat transfer coefficient, h are as follows:

(a)  Fluid velocity, u $(Lt^{-1})$
(b)  Tube diameter, D (L)
(c)  Thermal conductivity of fluid, k $(MLt^{-3}T^{-1})$
(d)  Fluid viscosity, $\mu (ML^{-1}t^{-1})$
(e)  Fluid density, $\rho (ML^{-3})$
(f)  Specific heat of fluid, $C_p (L^2t^{-2}T^{-1})$
(g)  Heat transfer coefficient, h $(Mt^{-3}T^{-1})$

Therefore, the equation for forced convection in the tube is assumed to have the following form:

$$h = C\, u^a\, D^b\, \mu^c\, k^d\, \rho^e\, C_p^f \qquad (3.194)$$

where C is a dimensionless constant, and a, b, c, ... are constants.

The variables in Eq. (3.194) may be expressed in terms of primary units.

$$Mt^{-3}T^{-1} = C(Lt^{-1})^a (L)^b (ML^{-1}t^{-1})^c (MLt^{-3}T^{-1})^d (ML^{-3})^e (L^2t^{-2}T^{-1})^f \quad (3.195)$$

Equation (3.195) to be dimensionally homogeneous the sum of the power exponents of a primary dimension on the left-hand side must be equal to the sum of the exponents of the same dimension on the right-hand side.

$$
\begin{aligned}
&\text{M: } 1 = c + d + e \quad \text{or } c = 1 - d - e & \text{(i)} \\
&\text{t: } -3 = -a - c - 3d - 2f \quad \text{or } a + c + 3d + 2f = 3 & \text{(ii)} \\
&\text{T: } -1 = -d - f \text{ or } d + f = 1 \text{ or } f = 1 - d & \text{(iii)} \quad (3.196) \\
&\text{L: } 0 = a + b - c + d - 3e + 2f & \text{(iv)}
\end{aligned}
$$

Now (ii) − (iv):  $a + c + 3d + 2f - (a + b - c + d - 3e + 2f) = 3$ or  $-b + 2c + 2d + 3e = 3$

Or  $-b + 2(1 - d - e) + 2d + 3e = 3$ or  $-b + e = 1$ or $b = e - 1$   (v)

Since $a + c + 3d + 2f = 3$ or $a + (1 - d - e) + 3d + 2(1 - d) = 3$ or $a - d - e + 3d - 2d = 3 - 3 = 0$

Or $a - e = 0$ or $a = e$

$$
\begin{aligned}
&\text{Since } c = 1 - d - e = 1 - (1 - f) - e = f - e \\
&\text{Hence } a = e,\ b = e - 1,\ c = f - e,\ d = 1 - f
\end{aligned}
\qquad (3.197)
$$

Substituting values of a, b, c, and d from Eq. (3.197) in Eq. (3.194):

$$h = C\, u^e\, D^{e-1}\, \mu^{f-e}\, k^{1-f}\, \rho^e\, C_p^f \qquad (3.198)$$

Therefore,

$$\frac{hD}{k} = C \left(\frac{Du\rho}{\mu}\right)^e \left(\frac{C_p\mu}{k}\right)^f \qquad (3.199)$$

These three ratios in Eq. (3.199) are the dimensionless groups such as Nusselt number (Nu), Reynolds number (Re), and Prandtl number (Pr):

$$Nu = C \, (Re)^e (Pr)^f \qquad (3.192)$$

In the original empirical correlation [Eq. (3.194)], $h$ is shown to be a function of six independent variables, Eq. (3.199) has reduced these to two variables appearing as dimensionless groups. If the values of constant, i.e. C, e, and f in Eq. (3.192) are found out by experiments for one set of Nu, Re, and Pr groups; these values would be valid for any other sets of dimensionless groups within the experimental range.

The main disadvantage of this technique is that some advance information is required on the mechanism of the process so that all significant variables can be included in the original equation. A discussion of dimensional analysis would not be completed without at least a brief mention of the Buckingham pi ($\pi$) theorem. This is a rule for deciding how many dimensionless groups to expect in the analysis of a given system.

$$r = (n - m). \qquad (3.131)$$

where r is number of dimensionless groups, n is number of variables and m is basic dimensions.

In Eq. (3.194): n = 6 and m = 4, so r = 6–4 = 2.

## 3.14  Thermal Radiation

Radiation is the third mode of heat transfer. All substances with body temperature above the absolute zero level continuously emit energy in the form of radiation. The emitted radiant energy is in direct proportion to the temperature of the substances. It may either be considered as being transported by electromagnetic waves in accordance with Maxwell's classical electromagnetic theory. Although there are many types of electromagnetic radiation, the nature of all radiation is same. It travels at the same speed as light ($2.998 \times 10^8$ m/s in vacuum). Wave lengths falling within the range of 0.1 to 100μ are called *thermal radiation.*

In contrast to heat transfer by conduction and convection, some medium (i.e. material or fluid) is required for transfer of energy from a higher temperature to a lower temperature; but no medium is needed for radiant interchange between two locations. The radiative energy will pass perfectly through a vacuum. Radiation becomes the only significant mode of heat transfer when no medium is present. Examples are (i) the heat dissipation from the filament of a vacuum tube or (ii) the heat loss through the evacuated walls of a thermos flask.

According to the quantum theory, the thermal radiation propagates in the form of discrete quanta, each quantum having an energy:

$$E = \hbar v \tag{3.200}$$

where ħ is Planck's constant ($6.625 \times 10^{-34}$ J.s) and $v$ is frequency of quantum.

Each quantum or *photon*, as it is often called, may be considered as a particle having energy, mass and momentum just like the molecule of a gas. Radiation can thus be considered as a *photon gas* flowing from one location to another with its particles governed by the following relations [3]:

$$\begin{aligned} E &= mC_o^2 \\ &= \hbar v \end{aligned} \tag{3.201}$$

Therefore,

$$m = \frac{\hbar v}{C_o^2}, \text{ and Momentum } = \frac{C_o \hbar v}{C_o^2} = \frac{\hbar v}{C_o} \tag{3.202}$$

where m is the mass, $C_o$ is the speed of light in vacuum $= \lambda$ (wavelength) $\times v$ (frequency).

The *emissive power*, $E_b$, of a black surface is defined as the energy emitted by the surface per unit area, per unit time, and is dependent upon a number of parameters among which are the surface material and roughness. As seen above, $E_b = q_b$.

At any given temperature, the amount of radiation emitted per unit wavelength varies at different wavelengths. For this purpose, the *monochromatic emissive power* ($W/m^2\mu$), $E_{b\lambda}$, of the surface is used. It is defined as the rate of energy radiated per unit area of the surface per unit wavelength.

$$E_b = \int_0^\infty E_{b\lambda} \cdot d\lambda \tag{3.203}$$

### 3.14.1  Absorption, Reflection, and Transmission

When incident radiation (also called irradiation) falls on a surface, (i) a part is reflected from the surface, (ii) a part is transmitted through the surface, and (iii) remainder is absorbed on the surface (as shown in Fig. 3.20).

**Fig. 3.20** Incident radiation on a surface

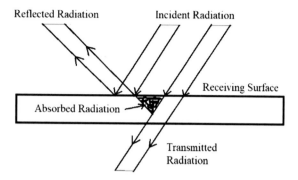

If q is the rate of heat received by the surface; out of this, the amount $q_\rho$ is reflected, $q_\tau$ is transmitted and $q_\alpha$ is absorbed. On the principle of conservation of energy, the total sum must be equal to the incident radiation, i.e. incident radiation = reflection + transmission + absorption.

$$q = q_\rho + q_\tau + q_\alpha \text{ or } \frac{q_\rho}{q} + \frac{q_\tau}{q} + \frac{q_\alpha}{q} = 1 \tag{3.204}$$

$$\rho + \tau + \alpha = 1 \tag{3.205}$$

where $\rho = \frac{q_\rho}{q}$, is the fraction of incident radiation reflected and is called the *reflectivity*.

$\tau = \frac{q_\tau}{q}$, is the fraction of incident radiation transmitted and is called *transmissivity*.

$\alpha = \frac{q_\alpha}{q}$, is the fraction of incident radiation absorbed and is called absorptivity.

Equation (3.204) holds for surfaces or layers of finite thickness. The following points should be noted about $\rho$, $\tau$, and $\alpha$:

- They are always positive values and their values lie between the limits 0 and 1, i.e. $0 \le \rho, \tau, \alpha \le 1$.
- $\rho = 0$ (i.e. $\tau + \alpha = 1$) represents a *non-reflecting* surface; $\rho = 1$ (i.e. $\tau = \alpha = 0$) represents a *perfect reflector*, i.e. it reflects all the incident radiation and does not absorb or transmit any part of it.
- $\tau = 0$ (i.e. $\rho + \alpha = 1$) represents an *opaque* surface. $\tau = 1$ (i.e. $\rho = \alpha = 0$) represents a *perfectly transparent* surface.
- $\alpha = 0$ ( i.e. $\rho + \tau = 1$) represents a *non-absorbing* surface (also called a *white* surface); $\alpha = 1$ ( i.e. $\rho = \tau = 0$) represents a *perfectly absorbing* surface (also called a *black* surface, if it is diffuse).

The definitions of $\rho$, $\tau$, and $\alpha$ given above are with respect to the total values of q, integrated with respect to the area of the surface, the solid angle in the hemispherical space above it and all the wavelengths of the spectrum.

Now define the monochromatic and directional values of $\rho$, $\tau$, and $\alpha$ by taking the corresponding values of $q_\rho$, $q_\tau$ and $q_\alpha$:

$$\rho_\lambda = \frac{q_{\rho\lambda}}{q_\lambda}, \quad \tau_\lambda = \frac{q_{\tau\lambda}}{q_\lambda} \quad \text{and} \quad \alpha_\lambda = \frac{q_{\alpha\lambda}}{q_\lambda} \tag{3.206}$$

where $q_\lambda$ represents the total heat flux per unit area received at the point at that wavelength, and $\rho_\lambda$ is the monochromatic reflectivity or the fraction of incident energy in the wavelength range $\lambda$ to $\lambda + d\lambda$, which is reflected.

Since

$$q_\rho = \int_0^\infty (q_{\rho\lambda})\, d\lambda = \int_0^\infty (\rho_\lambda \cdot q_\lambda)\, d\lambda \tag{3.207}$$

$$\rho = \frac{q_\rho}{q} = \frac{1}{q} \int_0^\infty (\rho_\lambda \cdot q_\lambda)\, d\lambda \tag{3.208}$$

Similar equations for monochromatic transmissivity, $\tau_\lambda$, and monochromatic absorptivity, $\alpha_\lambda$, can be derived.

Solids generally do not radiate heat unless the material is of very thin section. Metals absorb radiation within a fraction of a micrometer, and insulators within a fraction of a millimeter. Glasses and liquids absorb most of the radiation within a millimeter. Solids and liquids are therefore generally considered as opaque.

Gases such as hydrogen, oxygen, and nitrogen (and their mixtures such as air) have a transmissivity of practically unity. The radiation transfer through air is estimated using the relationships for radiation through a vacuum.

### 3.14.2  Concept of Black Body

A black body is an ideal body that absorbs all incident radiant energy and reflects or transmits none. This is true of radiation for all wavelengths and for all angles of incidence for a black body, therefore, $\rho = 0$, $\tau = 0$ and $\alpha = 1$. No actual body is perfectly black; the concept of a black body is an idealization with which the radiation characteristics of real bodies can be conveniently compared. A black body is a perfect absorber of incident radiation, as well as is a perfect emitter [3].

When describing the radiation characteristics of real surfaces, it is useful to introduce the concept of a *black body*. The black body is an ideal surface having the following properties [1]:

1. A black body absorbs all incident radiation, regardless of wavelength and direction.
2. For a prescribed temperature and wavelength, no surface can emit more energy than a black body.
3. Although the radiation emitted by a black body is a function of wavelength and temperature, it is independent of direction. That is, the black body is a diffuse emitter.

The total radiation emitted by a black body is a function of temperature only. If the temperature of the enclosure is now changed to a different uniform value, the black body will adjust its temperature until it is in thermal equilibrium with the enclosure.

### 3.14.2.1 Planck's Distribution Law

In 1900, Max Planck showed by quantum arguments that the spectral distribution of the radiation intensity of a black body is given by the equation [3]:

$$E_{b\lambda} = \pi I_{b\lambda} = \frac{2\pi C_1}{\lambda^5\{\exp(\frac{C_2}{\lambda T}) - 1\}} \tag{3.209}$$

where $I_{b\lambda}$ is the spectral intensity of the emitted radiation.
$C_1 = \hbar C_o^2 = 0.595 \times 10^{-8}$ Wm$^2$; ħ is Planck's constant $= 6.625 \times 10^{-34}$ J.s
$C_2 = \frac{\hbar C_o}{k} = 1.4387 \times 10^{-2}$ mk; k is Boltzmann constant $= 1.3805 \times 10^{-23}$ J/K.
Equation (3.209), known as the Planck distribution, is plotted in Fig. 3.21 where the hemispherical spectral emission power is given as a function of wavelength ($\lambda$) for selected temperature.

The curves show the following distinct characteristics of black body radiation:

1. The emitted radiation varies continuously with wavelength.
2. At any wavelength, the magnitude of the emitted radiation increases with increasing temperature.
3. The peak spectral emission power shifts towards a smaller wavelength at higher temperatures.

For a body at 1000 K, only a small amount of energy is in the visible region. When its temperature is raised further its color changes from red towards the violet end of the spectrum and at very high temperatures it becomes white, representing radiation

**Fig. 3.21** Special hemispherical emissive power for different temperatures

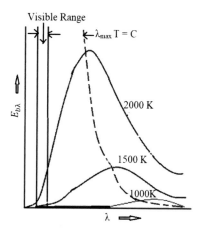

composed of a mixture of all wavelengths. Equation (3.209) is of great importance as it provides quantitative results for the radiation from a black body.

### 3.14.2.2 Wien's Displacement Law

From Fig. 3.21, it can be found that the black body spectral distribution has a maximum and that the corresponding wavelength $\lambda_{max}$ depends on temperature. The nature of this dependence may be obtained by differentiating Eq. (3.208) with respect to $\lambda$ and setting the result equal to zero.

$$\lambda_{max} \cdot T = C_3 \tag{3.210}$$

where $C_3$ is the third radiation constant $= 0.289 \times 10^{-2}$ m.K.

Equation (3.210) is known as Wien's displacement law, and the locus of points described by the law is plotted as the dashed line of Fig. 3.21. According to this result, the maximum spectral emissive power is displaced to shorter wavelengths with increasing temperature.

### 3.14.2.3 Stefan–Boltzmann Law

Substituting the Planck distribution, Eq. (3.209), into Eq. (3.203), the total emissive power of a black body, $E_b$ may be expressed as:

$$E_b = \int_0^\infty \frac{2\pi C_1}{\lambda^5 \left\{ \exp\left(\frac{C_2}{\lambda T}\right) - 1 \right\}} d\lambda \tag{3.211}$$

Performing the integration, the result obtained:

$$E_b = \sigma\, T^4 \tag{3.212}$$

where $E_b$ is the quantity of energy emitted per unit area and per unit time by the ideal radiator, i.e. black body (W/m$^2$), $\sigma$ is the Stefan–Boltzmann constant ($5.669 \times 10^{-8}$ W/m$^2$.K$^4$.s), and T is the absolute temperature in Kelvin.

Stefan–Boltzmann law said that the flux of heat energy emitted by radiation, from an ideal surface called *black body*, is proportional to its absolute temperature to the fourth power [3]; and this Eq. (3.212) is known as *Stefan–Boltzmann equation*, which is derived theoretically by Boltzmann and given empirically by Stefan.

### 3.14.3 Radiation from Non-black Surfaces

The concept of a black body is an idealization, which serves as a standard for real body performance. Most surfaces encountered in engineering applications do not behave like black bodies. The *emissivity* of a surface is a measure of how it emits radiant energy in comparison with a black surface at the same temperature [3]. The emissive power of an actual surface is expressed as a proportion of $E_b$ as:

$$E = \epsilon E_b$$

Or emissivity,

$$\epsilon = \frac{E}{E_b} \tag{3.213}$$

The emissivity of a surface is the ratio of the emissive power of the surface to the emissive power of a black body at the same temperature.

Equation (3.212) becomes:

$$E = \epsilon \sigma T^4 \tag{3.214}$$

The emissivity defined in Eq. (3.213) is also called the total emissivity because it represents the integrated behavior of the material over all wavelengths. The emissivity of a material varies with temperature and the wavelength of the radiation.

*Monochromatic emissivity* of a surface is the ratio of the monochromatic emissive power of the surface to the monochromatic emissive power of a black body at the same temperature and wavelength.

$$\epsilon_\lambda = \frac{E_\lambda}{E_{b\lambda}} \tag{3.215}$$

*Normal total emissivity*, $\epsilon_n$, is the ratio of the normal component of the total emissive of a surface, $E_n$ to the normal component of the total emissive power of a black body, $(E_b)_n$, at the same temperature.

$$\epsilon_n = \frac{E_n}{E_{bn}} \tag{3.216}$$

A *gray* body is defined such that the monochromatic emissivity, $\epsilon_\lambda$, of the body is independent of wavelength.

The total emissivity of the body is related to the monochromatic emissivity by:

$$E = \int_0^\infty \epsilon_\lambda E_{b\lambda} \, d\lambda \tag{3.217}$$

| Surface | Emissivity |
|---|---|
| Aluminum, highly polished | 0.04 |
| Brass, highly polished | 0.03 |
| Copper, highly polished | 0.03 |
| Copper, oxidized | 0.8 |
| Steel, mild polished | 0.07 |
| Steel, galvanized | 0.3 |
| Steel, rusted | 0.8 |
| Brick, fire clay | 0.7 |
| Brick, red | 0.9 |
| Glass, polished | 0.94 |
| Paper, white | 0.97 |

**Table 3.2** Normal total emissivity of various surfaces at ambient temperature

Again from Eq. (3.203):

$E_b = \int_0^\infty E_{b\lambda} \cdot d\lambda = \sigma T^4$ [from Eq. (3.212)]

Equation (3.213) becomes:

$$\epsilon = \frac{E}{E_b} = \frac{\int_0^\infty \epsilon_\lambda E_{b\lambda} \, d\lambda}{\sigma T^4} \qquad (3.218)$$

For a gray body, $\epsilon_\lambda$ = constant; Eq. (3.218) becomes:

$$\epsilon = \epsilon_\lambda = \text{constant}$$

The emissivity of a surface is a function of its nature and characteristics and is independent of the wavelength or the nature of the impinging radiation waves. Thus, it is essentially a surface property. The absorptivity of a surface is not a surface property because of its dependence on the nature of the incident radiation. Table 3.2 gives emissivity values for a few common surfaces.

## 3.15  Exercises

**Problem 3.1** A wall is 17 0mm thick and its area is 4.9 m$^2$. If its conductivity is 9.35 W/mK and surface temperatures are steady at 425 K and 316 K. Determine heat flow across the wall.

[Ans: 293.76 kW]

**Problem 3.2** Calculate the rate of heat loss for a brick wall of length 5 m, height 4 m, and thickness 0.25 m. The temperature of the inner surface is 110 °C and that of the outer surface is 40 °C. The thermal conductivity of brick is 0.7 W/mK. Also calculate the temperature at an interior point of the wall, 30 cm from the inner wall. [Ans: 3.92 kW, 26 °C]

**Problem 3.3** A hollow sphere used as a container of liquid chemical, has 10 cm inner diameter and 30 cm outer diameter, the thermal conductivity of sphere material is 50 W/mK. Its inner and outer surface temperatures are 300 °C and 100 °C, respectively. Calculate the heat flow rate through the sphere. [Ans: 9.42 kW]

**Problem 3.4** A wall of 0.6 m thick is constructed from material of thermal conductivity, 1.5 W/mK. The wall is insulated with a material having thermal conductivity, 0.3 W/mK; so that the heat loss per square meter will not exceed 1.4 kW. Calculate the thickness of insulation required. Given: the inner and outer surfaces temperatures are 1200 °C and 25 °C, respectively.

[Ans: 130 cm]

**Problem 3.5** (i) A gas fired furnace is made of a layer A (15 cm) of fireclay brick and a layer B (60 cm) of red brick (as shown in Figure). If the wall temperature inside the furnace is 1200 °C and that on the outside wall is 50 °C. Determine the amount of heat loss per square meter of the wall and temperature of the interface fireclay-red brick.

(ii) It is desired to replace the fireclay brick by dolomite brick and reduce the thickness of the red brick layer (50 cm) in this furnace. Calculate the thickness of the dolomite brick to ensure an identical loss of heat for the same inside and outside temperatures.

Given: k for fireclay $= 0.533$ W/mK, k for red brick $= 0.7$ W/mK and k for dolomite $= 0.113 + 2.3 \times 10^{-4}$ T W/mK. [Ans: (i) 1010.54 W/m², 916 °C; (ii) 14.4 cm]

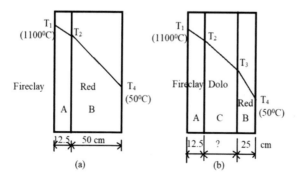

**Problem 3.6** An electrical conducting aluminum wire with a diameter of 1 mm is covered with a plastic insulation of thickness 1 mm. The temperature of its surroundings is 25 °C. Find the maximum current carried by the aluminum wire so that no part of the plastic is above 80 °C.

Given: k for aluminum 204 W/mK, k for plastic = 0.5 W/mK and h = 8 W/m²K. Specific electric resistance of aluminum, $\rho = 2.82 \times 10^{-8}$ Ωm.

[Ans: 3.02 A]

## 3.16  Questions

Q1. What do you understand by heat transfer? What are the modes of heat transfer?

Q2. Why heat transfer is important in most of the Metallurgical operations?

Q3. What do you understand by convection? What are the types of convection? Explain.

Q4. Discuss the Fourier's law and derive the equation. What do you understand by (−) ve sign of Fourier's equation?

Q5. What do you understand by heat flux? Derive the equation for heat flux in terms of heat conductivity.

Q6. Explain the following terms:

(a) Thermal resistance, (b) Thermal conductivity, and (c) Thermal diffusivity.

Q7. Derive the equation of thermal resistance for composite wall with the help of electrical circuits.

Q8. Derive the equation of heat conduction through a hollow cylinder from general heat conduction equation.

Q9. Derive the equation of heat conduction for composite cylinder.

Q10. Discuss about radiation of heat and derive the equation for total rate of heat transfer from the surface of hot body.

Q11. Write short notes:

(a) White body and Black body, (b) Contact resistance, (c) Radiation, (d) Wien's displacement law, (e) Stefan-Boltzmann law, (f) Planck distribution.

# References

1. Incropera FP, Dewitt DP (2006) Fundamentals of heat and mass transfer, 5th Edn, Wiley (Asia) Pvt Ltd, India
2. Bennett CO, Myers JE (1988) Momentum, heat, and mass transfer, 3rd Edn, McGraw Hill International Editions, Singapore
3. Sachdeva RC (1988) Fundamentals of engineering heat and mass transfer. Wiley Eastern Limited, New Delhi

# Chapter 4
# Mass Transfer

Basic concept, different modes of mass transfer, Fick's laws of diffusion, and mass flux in terms of velocity are discussed. General mass diffusion equation, equation of diffusion of gas through solid, and equation of motion of gas bubbles in liquid are derived. Mechanism of mass transfer, simultaneous heat and mass transfer, and heat and mass transfer to single particle are described.

## 4.1 Basic Concept

If there is a difference in the concentration of some chemical components in a mixture, then mass transfer must occur. By definition, the process of transfer of mass as a result of the concentration difference of chemical components in a system/mixture is known as *mass transfer*. A mixture consists of two or more chemical components.

Mass transfer is by far the most important aspect of transport phenomenon in the study of metallurgical processes. These processes invariably involve transfer of one or more chemical components from one phase to another phase. The rates of these processes are thus predominantly controlled by the mass transfer alone [1]. Nowadays, mass transfer along with momentum and heat transfers are also considered for the complete mathematical formulation of several extractive metallurgical processes. Such mathematical formulations in which all these aspects are considered simultaneous along with the thermodynamics and chemical kinetics of the involved reactions are known as *mathematical models*. These mathematical models find extensive applications in (i) dry process analysis for assessing the role of different variables on a process, or (ii) for the analysis of the performance of a reactor and its optimization.

The concentration of any chemical component in a mixture can be expressed in several ways:

1. *Mass concentration* (or *mass density*): This form of expression of concentration is denoted by the symbol $\rho$, which is equal to the ratio of mass ($m_i$) of the component to volume (V).

© The Author(s), under exclusive license to Springer Nature Singapore Pte Ltd. 2023
S. K. Dutta, *Fundamental of Transport Phenomena and Metallurgical Process Modeling*,
https://doi.org/10.1007/978-981-19-2156-8_4

$$\rho_i = \frac{m_i}{V}, \text{kg.m}^{-3} \tag{4.1}$$

This is also called *mass density*, $\rho_i$ of component i in a multi-component mixture. This is the mass of i per unit volume of the mixture.

Therefore, mixture mass density,

$$\rho = \sum_i \rho_i, \text{kg.m}^{-3} \tag{4.2}$$

2.  *Molar concentration* (or *molar density*): The molar concentration $C_i$ of component i is defined as the number of moles of component i per unit volume of the mixture.

$$C_i = \frac{m_i}{M_i V} = \frac{\rho_i}{M_i}, \text{kg.kmol}^{-1} \tag{4.3}$$

where $M_i$ is the molecular weight of the component i.

The total number of moles per unit volume of the mixture, that is, overall molar density,

$$C = \sum_i C_i \tag{4.4}$$

3.  *Mass fraction*: The mass fraction, $m_i$ is defined as the ratio of mass concentration of component i to the total mass density of the mixture.

$$m_i = \frac{\rho_i}{\rho} = \frac{\rho_i}{\sum_i \rho_i} = \frac{\left(\frac{m_i}{V}\right)}{\sum \left(\frac{m_i}{V}\right)} = \frac{m_i}{\sum m_i} \tag{4.5}$$

4.  *Mole fraction*: The mole fraction, $X_i$ in terms of the total mole concentration of the mixture is given by

$$X_i = \frac{C_i}{C} = \frac{C_i}{\sum C_i} \tag{4.6}$$

Again,

$$\sum m_i = 1 \quad \text{and} \quad \sum X_i = 1 \tag{4.7}$$

According to the ideal gas law:

$$p_i V = n_i RT \tag{4.8}$$

Again

$$C_i = \frac{n_i}{V} = \frac{p_i}{RT} \quad \left[\text{from Eq. (4.8)}\right] \tag{4.9}$$

From Eq. (4.3), $C_i = \frac{\rho_i}{M_i}$, therefore,

$$\rho_i = C_i M_i = \frac{\rho_i M_i}{RT} \tag{4.10}$$

## 4.2 Mass Transfer

Mass transfer in any system may take place in two different modes:

1. Mass transfer by diffusion.
2. Mass transfer by convection.

### 4.2.1 Mass Transfer by Diffusion

This is a microscopic mode of mass transfer like heat conduction and involves the movement of mass on an atomic scale. Mass diffusion from a region of high concentration to a region of low concentration in a system mixture of liquids or gases is called molecular diffusion. It occurs when a substance diffuses through a layer of stagnant fluid and may be due to concentration, temperature, or pressure gradients.

Mass diffusion occurs in liquid and solid as well as in gases. However, since mass transfer is strongly influenced by molecular spacing, diffusion occurs more readily in gases than in liquids, and more readily in liquids than in solids [2].

Considering a chamber in which two different gas components A and B, at the same temperature and pressure, are initially separated by a partition (Fig. 4.1), the left component has a high concentration (i.e. more molecules per unit volume) of gas A (open circle), whereas the right component is rich in gas B (dark circle). When the partition wall is removed, a driving potential comes into existence which tends to equalize the concentration difference. Mass transfer by diffusion occurred in the direction of decreasing concentration, and subsequently, there is a net transport of component A to the right and component B to the left. After a sufficiently long time, a uniform concentration of components A and B is achieved in two chambers and then further the mass diffusion is stopped.

#### 4.2.1.1 Fick's Laws of Diffusion

The rate equation for mass diffusion is known as *Fick's law* and the transfer of component A in a binary mixture of A and B may be expressed in the form as

**Fig. 4.1** Diffusion of gases
in two chambers

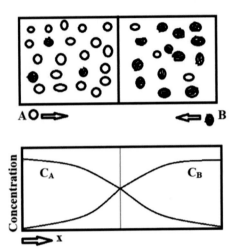

$$J'_{Ax} = -\rho D_{AB}\left(\frac{\delta m_A}{\delta x}\right) \tag{4.11}$$

where $J'_{Ax}$ is mass flux of component A in the x direction (kg.m$^{-2}$.s$^{-1}$). It is the amount of component A that is transferred per unit time and per unit area perpendicular to the direction of transfer, and it is proportional to the mixture mass density $\left[\rho = (\rho_A + \rho_B),\ \text{kg.m}^{-3}\right]$, and to the gradient in the component mass fraction, $m_A = \frac{\rho_A}{\rho}$.

$D_{AB}$ is the *diffusion coefficient* or *mass diffusivity* for a binary mixture of components A and B (m$^2$.s$^{-1}$). It is a transport property of the material. It does not depend upon the unit of mass.

Fick's law of diffusion is analogous to

(i)    Fourier law of heat conduction:

$$\frac{q}{A} = -k\frac{dT}{dx} \tag{3.7}$$

(ii)   Newton's law of viscosity:

$$\tau = -\mu\frac{du_x}{dy} \tag{1.5}$$

By comparison of the above equations, it can be observed that Fourier Eq. (3.7) describes the transfer of heat energy due to temperature gradient and Newton Eq. (1.5) describes the transfer of momentum due to velocity gradient, while Fick's law of diffusion describes the transfer of mass due to mass concentration gradient.

Fick's law of diffusion can be written in the following forms:

(1) In terms of *mass fraction*:

$$J_i' = -\rho D_i \left( \frac{\delta m_i}{\delta x} \right) \tag{4.12}$$

(2) In terms of *mass concentration*:

$$J_i' = -D_i \left( \frac{\delta \rho_i}{\delta x} \right) \tag{4.13}$$

(3) In terms of *molar concentration*:

$$J_i'^* = -D_i \left( \frac{\delta C_i}{\delta x} \right) \tag{4.14}$$

(4) In terms of *mole fraction*:

$$J_i'^* = -C D_i \left( \frac{\delta X_i}{\delta x} \right) \tag{4.15}$$

where $J_i'^*$ is the *molar flux* of component i ($kmol.m^{-2}.s^{-1}$).

### 4.2.1.2 Mass Flux in Terms of Velocity

Although mass diffusion may result from a temperature gradient, a pressure gradient, or an external force, as well as from a concentration gradient, assume that these additional effects are not present or are negligible. The most problematic and the dominant driving force is the component concentration gradient. This condition is referred to as *ordinary diffusion*.

The second restrictive condition is that the fluxes are measured relative to coordinates that are more with the average velocity of the mixture. If the mass or molar flux of a component is expressed relative to a fixed set of coordinates, Eqs. (4.12)–(4.15) are not generally valid. To obtain an expression for the mass flux relative to a fixed coordinate system, consider component A in a binary mixture of A and B. The mass flux $\left( n_A'' \right)$ relative to a fixed coordinate system is related to the component absolute velocity ($u_A$) by

$$n_A'' = \rho_A . u_A \tag{4.16}$$

and

$$n_B'' = \rho_B . u_B \tag{4.17}$$

Total mass flux for mixture,

$$n'' = \rho.u = n''_A + n''_B \tag{4.18}$$

Therefore,

$$\rho.u = \rho_A.u_A + \rho_B.u_B \quad \text{or} \quad u = m_A.u_A + m_B.u_B \tag{4.19}$$

From Eq. (4.5), $m_i = \frac{\rho_i}{\rho}$.

The mass average velocity (u) is a useful parameter of the binary mixture since it needs to be only multiplied by the total mass density to obtain the total mass flux with respect to fixed axes [2].

(1)  Now the mass flux of component A relative to the mass average velocity of the mixture:

$$J'_A = \rho_A(u_A - u) \tag{4.20}$$

where $J'_A$ is relative or diffusive mass flux, $(u_A - u)$ is the relative velocity with respect to the fluid.

From Eq. (4.20),

$$\rho_A.u_A = n''_A = J'_A + \rho_A.u \quad \text{[from Eq. (4.16)]} \tag{4.21}$$

This Eq. (4.21) indicates that there are two contributions to the absolute mass flux of component $A(n''_A)$: (i) a contribution due to diffusion (i.e. due to the motion of component A relative to the mass average motion of the mixture), and (ii) a contribution due to the motion of component A with the mass average velocity of the mixture.

Now substituting Eqs. (4.11) and (4.18) to Eq. (4.21) we get

$$n''_A = -\rho \, D_{AB}\left(\frac{\delta m_A}{\delta x}\right) + \frac{\rho_A}{\rho}\left(n''_A + n''_B\right) = -\rho \, D_{AB}\left(\frac{\delta m_A}{\delta x}\right) + m_A\left(n''_A + n''_B\right) \tag{4.22}$$

[from Eq. (4.5): $m_A = \frac{\rho_A}{\rho}$]

Equation (4.22) is also known as Fick's law with respect to stationary coordinates.

(2)  Now the mass flux of component B relates to the mass average velocity of the mixture (i.e. diffusion flux):

$$J'_B = \rho_B(u_B - u) \quad \text{[similar to Eq. (4.20)]} \tag{4.23}$$

Now adding Eqs. (4.20) and (4.23)

$$J_A' + J_B' = \rho_A(u_A - u) + \rho_B(u_B - u)$$
$$= (\rho_A\, u_A + \rho_B\, u_B) - u(\rho_A + \rho_B) = \rho u - u\rho = 0 \qquad (4.24)$$

[since $(\rho_A u_A + \rho_B u_B) = \rho u$ and $\rho = (\rho_A + \rho_B)$]
Now,

$$J_B' = -\rho\, D_{BA}\left(\frac{\delta m_B}{\delta x}\right) \quad \text{[similar to Eq. (3.11)]} \qquad (4.25)$$

Now adding Eqs. (4.11) and (4.25)

$$J_A' + J_B' = -\rho\, D_{AB}\left(\frac{\delta m_A}{\delta x}\right) - \rho\, D_{BA}\left(\frac{\delta m_B}{\delta x}\right) = 0 \qquad (4.26)$$

[Since $m_A + m_B = 1$, therefore $\left(\frac{\delta m_A}{\delta x}\right) + \left(\frac{\delta m_B}{\delta x}\right) = 0$ or $\left(\frac{\delta m_A}{\delta x}\right) = -\left(\frac{\delta m_B}{\delta x}\right)$
Therefore, $-\rho D_{AB}\left(\frac{\delta m_A}{\delta x}\right) = -\rho D_{BA}\left(\frac{\delta m_B}{\delta x}\right) = \rho\, D_{BA}\left(\frac{\delta m_A}{\delta x}\right)$ so $D_{AB} = D_{BA}$.]
The absolute mass flux of component B may be expressed, similar to Eq. (4.22),
as

$$n_B'' = -\rho\, D_{AB}\left(\frac{\delta m_B}{\delta x}\right) + m_B\left(n_A'' + n_B''\right) \qquad (4.27)$$

Equations (4.22) and (4.27) are for mass fluxes, and the same procedures can be used to obtain results on a molar basis. The absolute molar fluxes of components A and B may be expressed as

$$N_A'' = C_A\, u_A \quad \text{and} \quad N_B'' = C_B\, u_B \qquad (4.28)$$

Now

$$N'' = \left(N_A'' + N_B''\right) = (C_A\, u_A + C_A\, u_B) = C\, u^* \qquad (4.29)$$

The molar average velocity for mixture ($u^*$) is calculated from Eq. (4.29):

$$u^* = \frac{C_A}{C}u_A + \frac{C_B}{C}u_B = X_A u_A + X_B u_B \qquad (4.30)$$

The significance of the molar average velocity ($u^*$) is that when multiplied by the total molar concentration (C), it provides the total molar flux $\left(N''\right)$ with respect to a fixed coordinate system.

Diffusion flux ($J_A^*$) may be related to the mixture molar average velocity ($u^*$):

$$J_A^* = C_A\left(u_A - u^*\right) \qquad (4.31)$$

Equation (4.31) is similar to Eq. (4.20) and

$$N''_A = J^*_A + C_A.u^*$$ (4.32)

Equation (4.32) is similar to Eq. (4.21).
Therefore,

$$N''_A = -CD_i\left(\frac{\delta X_i}{\delta x}\right) + X_A(N''_A + N''_B)$$ (4.33)

[From Eq. (4.15): $J'^*_A = -CD_A\left(\frac{\delta X_A}{\delta x}\right)$, $Cu^* = (N''_A + N''_B)$, and $\frac{C_A}{C} = X_A$]
This Eq. (4.33) is again similar to Eq. (4.22).
For the binary mixture,

$$J^*_A + J^*_B = 0$$ (4.34)

Equation (4.34) is similar to Eq. (4.24).

## 4.3 General Mass Diffusion Equation

The general equation for mass transfer can be derived from a similar analogous heat transfer equation. Consider a homogeneous medium consisting of a binary mixture of components A and B. Let the medium is stationary, i.e. the mass average or molar average velocity of the mixture is zero everywhere, and mass transfer may occur only by diffusion.

There are concentration gradients in each of the x, y, and z coordinate directions; first define a differential control volume ($\Delta x$, $\Delta y$, and $\Delta z$) within the medium and consider the processes that influence the distribution of component A (Fig. 4.2). With the concentration gradients, diffusion must result in the transfer of component A through the control surfaces.

The mass balance of component A diffusing through the control volume in the stationary medium B is given by

**Fig. 4.2** Differential control volume

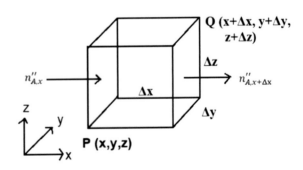

$$\left[\{(\text{Input of A}) - (\text{Output of A})\} + (\text{Generation of A})\right] = (\text{Accumulation of A})$$
(4.35)

1. Input of A from three faces of control volume at point P:

$$M'_{A,in} = n''_{A,x}.\Delta y \Delta z + n''_{A,y}.\Delta x \Delta z + n''_{A,z}.\Delta x \Delta y$$
(4.36)

where $n''_A$ is the absolute mass flux.

2. Output of A from the other three faces of control volume at point Q:

$$M'_{A,out} = n''_{A,x+\Delta x}.\Delta y \Delta z + n''_{A,y+\Delta y}.\Delta x \Delta z + n''_{A,z+\Delta z}.\Delta x \Delta y$$
(4.37)

3. There may be volumetric (i.e. homogeneous) chemical reaction occurring throughout the medium. The rate at which component A is generated within the control volume due to such reaction may be expressed as

$$M'_{A,g} = n'_A (\Delta x \Delta y \Delta z)$$
(4.38)

where $n'_A$ is the rate of increase of the mass of component A per unit volume of the mixture.

4. Accumulation of mass of component A in the control volume due to its mass diffusion in the control volume per unit time:

$$M'_{A,ac} = \frac{\delta \rho_A}{\delta t}.(\Delta x \Delta y \Delta z)$$
(4.39)

Now substituting Eqs. (4.36) to (4.39) into Eq. (4.35):

$$\left\{(M'_{A,in} - M'_{A,out}) + M'_{A,g}\right\} = M'_{A,ac}$$
(4.40)

$$\rightarrow \left[\left\{\begin{matrix}(n''_{A,x} - n''_{A,x+\Delta x}).\Delta y \Delta z + (n''_{A,y} - n''_{A,y+\Delta y}).\Delta x \Delta z \\ +(n''_{A,z} - n''_{A,z+\Delta z}).\Delta x \Delta y\end{matrix}\right\} + n'_A.(\Delta x \Delta y \Delta z)\right]$$
$$= \left\{\frac{\delta \rho_A}{\delta t}.(\Delta x \Delta y \Delta z)\right\}$$
(4.41)

$$\rightarrow \left\{\frac{(n''_{A,x} - n''_{A,x+\Delta x})}{\Delta x}\right\} + \left\{\frac{(n''_{A,y} - n''_{A,y+\Delta y})}{\Delta y}\right\} + \left\{\frac{(n''_{A,z} - n''_{A,z+\Delta z})}{\Delta z}\right\} + n'_A$$
$$= \frac{\delta \rho_A}{\delta t}$$
(4.42)

$$\rightarrow \quad \frac{\delta \rho_A}{\delta t} = -\left(\frac{\delta n''_{A,x}}{\delta x} + \frac{\delta n''_{A,y}}{\delta y} + \frac{\delta n''_{A,z}}{\delta z}\right) + n'_A \tag{4.43}$$

[since $\lim\limits_{\Delta x \to 0}\left\{\frac{(n''_{A,x+\Delta x}-n''_{A,x})}{\Delta x}\right\} = \frac{\delta n''_{A,x}}{\delta x}$]

For a stationary medium, the mass average velocity (u) is zero, and from Eq. (4.21),

$$n''_A = J'_A + \rho_A.u = J'_A + \rho_A.0 = J'_A$$

Now Eq. (4.43) becomes as follows:

$$\frac{\delta \rho_A}{\delta t} = -\left(\frac{\delta J'_{A,x}}{\delta x} + \frac{\delta J'_{A,y}}{\delta y} + \frac{\delta J'_{A,z}}{\delta z}\right) + n'_A \tag{4.44}$$

Since

$$J'_{Ax} = -\rho \, D_{AB}\left(\frac{\delta m_A}{\delta x}\right) \tag{4.11}$$

Equation (4.44) further modifies to

$$\frac{\delta \rho_A}{\delta t} = \left[\frac{\delta}{\delta x}\left\{\rho D_{AB}\left(\frac{\delta m_A}{\delta x}\right)\right\} + \frac{\delta}{\delta y}\left\{\rho D_{AB}\left(\frac{\delta m_A}{\delta y}\right)\right\} + \frac{\delta}{\delta z}\left\{\rho D_{AB}\left(\frac{\delta m_A}{\delta z}\right)\right\}\right]$$
$$+ n'_A \tag{4.45}$$

Since $\rho$ and $D_{AB}$ are constants

$$\rho \, D_{AB}\left[\frac{\delta^2 m_A}{\delta x^2} + \frac{\delta^2 m_A}{\delta y^2} + \frac{\delta^2 m_A}{\delta z^2}\right] + n'_A = \frac{\delta \rho_A}{\delta t}$$

or

$$\left[\frac{\delta^2 \rho_A}{\delta x^2} + \frac{\delta^2 \rho_A}{\delta y^2} + \frac{\delta^2 \rho_A}{\delta z^2}\right] + \frac{n'_A}{D_{AB}} = \frac{1}{D_{AB}}.\frac{\delta \rho_A}{\delta t} \tag{4.46}$$

[since $m_A = \frac{\rho_A}{\rho}$]

Equation (4.46) is similar to heat equation:

$$\left[\left(\frac{\delta^2 T}{\delta x^2} + \frac{\delta^2 T}{\delta y^2} + \frac{\delta^2 T}{\delta z^2}\right) + \frac{q}{k}\right] = \frac{1}{\alpha}\frac{\delta T}{\delta t} \tag{3.51}$$

In terms of molar concentration, Eq. (4.15): $J'^*_i = -C D_i\left(\frac{\delta X_i}{\delta x}\right)$; similar equation can be obtained from Eq. (4.44):

$$\left[\frac{\delta}{\delta x}\left\{CD_{AB}\left(\frac{\delta X_A}{\delta x}\right)\right\} + \frac{\delta}{\delta y}\left\{CD_{AB}\left(\frac{\delta X_A}{\delta y}\right)\right\} + \frac{\delta}{\delta z}\left\{CD_{AB}\left(\frac{\delta X_A}{\delta z}\right)\right\}\right] + n'_A$$

$$= \frac{\delta C_A}{\delta t} \qquad (4.47)$$

If C and $D_{AB}$ are constants, and $C_i = C.X_i$, then Eq. (4.47) becomes:

$$\left[\frac{\delta^2 C_A}{\delta x^2} + \frac{\delta^2 C_A}{\delta y^2} + \frac{\delta^2 C_A}{\delta z^2}\right] + \frac{n'_A}{D_{AB}} = \frac{1}{D_{AB}} \cdot \frac{\delta C_A}{\delta t} \qquad (4.48)$$

Similar equations can be derived for cylindrical and spherical coordinates (shown in *Appendix V*).

## 4.3.1 Steady-State Diffusion Through Plain Membrane

Considering mass diffusion of fluid A through a plain membrane (means sheet-like connective tissue or lining in animal or vegetable), whose thickness (L) is very small in comparison with other dimensions (Fig. 4.3). The mass concentrations of the fluid at the opposite wall are $C_{A1}$ and $C_{A2}$, respectively.
    Assuming:

(i)   steady-state condition, i.e. $\frac{\delta C_A}{\delta t} = 0$.
(ii)  no chemical reaction occurred, i.e. no generation of component A, i.e. $n'_A = 0$.
(iii) mass transfer takes place only in one direction, i.e. x direction, so

$$\frac{\delta^2 C_A}{\delta y^2} = 0 \text{ and } \frac{\delta^2 C_A}{\delta z^2} = 0.$$

Hence, Eq. (4.48) becomes

$$\frac{\delta^2 C_A}{\delta x^2} = 0 \qquad (4.49)$$

**Fig. 4.3** Mass diffusion of fluid A through a plain membrane

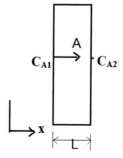

By integration

$$\frac{\delta C_A}{\delta x} = C_1 \tag{4.50}$$

where $C_1$ is an integration constant.
On further integration

$$C_A = C_1 x + C_2 \tag{4.51}$$

where $C_2$ is another integration constant.
Boundary condition at $x = 0, C_A = C_{A1}$; from Eq. (4.51),

$$C_{A1} = C_2 \tag{a}$$

and at $x = L, C_A = C_{A2}$; from Eq. (4.51), $C_{A2} = C_1 L + C_2 = C_1 L + C_{A1}$, so

$$C_1 = \left( \frac{C_{A2} - C_{A1}}{L} \right) \tag{b}$$

Putting the values of $C_1$ and $C_2$ from (a) and (b) in Eq. (4.51)

$$C_A = \left( \frac{C_{A2} - C_{A1}}{L} \right) x + C_{A1} \tag{4.52}$$

Mass transfer rate for molar concentration [from Eq. (4.14)] is

$$J_A^{\prime *} = -D_A \left( \frac{\delta C_A}{\delta x} \right) = -D_A \frac{\delta}{\delta x} \left\{ \left( \frac{C_{A2} - C_{A1}}{L} \right) x + C_{A1} \right\} \text{ [from Eq. (4.52)]}$$

$$= -D_A \left( \frac{C_{A2} - C_{A1}}{L} \right) = \left\{ \frac{(C_{A1} - C_{A2})}{\left( \frac{L}{D_A} \right)} \right\} \tag{4.53}$$

where $\left( \frac{L}{D_A} \right)$ is known as *diffusional resistance*.
Equation (4.53) is similar to heat equation,

$$q_x = \left( \frac{T_1 - T_2}{\left( \frac{L}{kA} \right)} \right) \tag{3.21}$$

where $\left( \frac{L}{kA} \right)$ is the *thermal resistance*.

## 4.3.2 Steady-State Equimolar Counter Diffusion

Equimolar counter diffusion between components A and B of a binary gas mixture is defined as an isothermal diffusion process in which each molecule of component A is replaced by each molecule of component B and vice-versa.

Consider two large chambers A and B, connected by a passage in such a way that each molecule of gas A is replaced by a molecule of gas B and vice-versa (Fig. 4.4). The total pressure, $p = p_A + p_B$ is uniform throughout, and the concentration of components is maintained constant in each of the chambers.

For ideal gas:

$$C_i = \frac{p_i}{RT}, \quad \text{and} \quad \rho_i = C_i M_i = \frac{p_i M_i}{RT} \tag{4.54}$$

Now, Fick's law of diffusion in terms of mass concentration [from Eq. (4.13)] is given by

$$J_i' = -D_i \left( \frac{\delta \rho_i}{\delta x} \right) = -D_i \frac{\delta}{\delta x} \left( \frac{p_i M_i}{RT} \right) = -D_i \left( \frac{M_i}{RT} \right) \left( \frac{\delta p_i}{\delta x} \right) \tag{4.55}$$

For components A and B, Eq. (4.55) becomes

$$J_A' = -D_{AB} \left( \frac{M_A}{RT} \right) \left( \frac{\delta p_A}{\delta x} \right) \tag{4.56}$$

$$J_B' = -D_{BA} \left( \frac{M_B}{RT} \right) \left( \frac{\delta p_B}{\delta x} \right) \tag{4.57}$$

Components A and B are diffusing toward their lower concentration gradient. Since

$$p = p_A + p_B \tag{4.58}$$

**Fig. 4.4** Counter diffusion of two gases

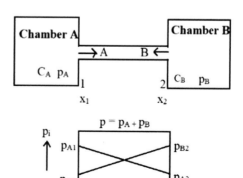

Differentiating with respect to x

$$\frac{dp}{dx} = \frac{dp_A}{dx} + \frac{dp_B}{dx} \tag{4.59}$$

The total pressure of the system remains constant under steady-state condition, i.e. $\frac{dp}{dx} = 0$.

Therefore,

$$\frac{dp}{dx} = 0 = \frac{dp_A}{dx} + \frac{dp_B}{dx} \quad \text{or} \quad \frac{dp_A}{dx} = -\frac{dp_B}{dx} \tag{4.60}$$

Further, under steady-state conditions, the total mass flux, relative to stationary conditions, must be zero. From Eq. (4.26):

$$J'_A + J'_B = 0 \text{ or } J'_A = -J'_B \tag{4.61}$$

$$-D_{AB}\left(\frac{M_A}{RT}\right)\left(\frac{\delta p_A}{\delta x}\right) = -\left[-D_{BA}\left(\frac{M_B}{RT}\right)\left(\frac{\delta p_B}{\delta x}\right)\right] \quad \{\text{from eqs. (4.56) and (4.57)}\}$$

$$= -D_{BA}\left(\frac{M_A}{RT}\right)\left(\frac{\delta p_A}{\delta x}\right) \tag{4.62}$$

{from Eq. (4.62) and for equimolar components: $M_A = M_B = M$}.

Therefore,

$$D_{AB} = D_{BA} = D \tag{4.63}$$

Equation (4.63) means that, for equimolar diffusion, the diffusion coefficient for the diffusion of gas A into gas B is equal to the diffusion coefficient for the diffusion of gas B into gas A.

Assuming the diffusion coefficient is constant, Eq. (4.56) for component A becomes

$$J'_A = -D\left(\frac{M}{RT}\right)\left(\frac{\delta p_A}{\delta x}\right)$$

Integrating between two points: $J'_A \int_1^2 dx = -D\left(\frac{M}{RT}\right)\int_1^2 dp_A$

or

$$J'_A(x_1 - x_2) = -D\left(\frac{M}{RT}\right)(p_{A1} - p_{A2}) \quad \text{or} \quad J'_A = D\left(\frac{M}{RT}\right)\left(\frac{p_{A1} - p_{A2}}{x_2 - x_1}\right) \tag{4.64}$$

Similarly, Eq. (4.57) for component B becomes

$$J'_B = D\left(\frac{M}{RT}\right)\left(\frac{p_{B1} - p_{B2}}{x_2 - x_1}\right) \tag{4.65}$$

The value of the mass diffusion coefficient for a binary mixture of gases can be calculated by using the following equation:

$$D_{AB} = 0.0043\left\{\frac{T^{3/2}}{p\left(V_A^{1/3} + V_B^{1/3}\right)^2}\right\}\left[\frac{1}{M_A} + \frac{1}{M_B}\right]^{1/2} \tag{4.66}$$

where

| | |
|---|---|
| $p =$ | total pressure in atmosphere $= p_A + p_B$. |
| $T =$ | absolute temperature, K |
| $V_A, V_B =$ | molecular volume of components A and B at normal boiling point $(m^3.k\ mol^{-1})$. |
| $M_A, M_B =$ | molecular weight of components A and B. |

### 4.3.3 Mass Diffusion Through Stagnant Fluid

Consider the case of vaporization of metal A at the bottom of a narrow tube (Fig. 4.5). The metal vapor along with gas B is flowing over the upper portion of the tube. The system is isothermal and the convection in the tube is negligible; therefore, all mass transfer from the surface of the metal to the gas stream must take place by diffusion.

Assume:

(1)   steady-state condition,
(2)   no chemical reaction occurs, i.e. no generation of any component,

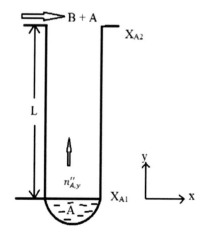

**Fig. 4.5** Mass diffusion via stagnant fluid

(3)    one direction mass diffusion process.

From general Eq. (4.43) in terms of mass flux,

$$\frac{\delta \rho_A}{\delta t} = -\left( \frac{\delta n''_{A,x}}{\delta x} + \frac{\delta n''_{A,y}}{\delta y} + \frac{\delta n''_{A,z}}{\delta z} \right) + n'_A \qquad (4.43)$$

For steady-state condition:

$$\frac{\delta \rho_A}{\delta t} = 0 \qquad (a)$$

For no generation of component:

$$n'_A = 0 \qquad (b)$$

For one direction (i.e. y direction only) mass diffusion process:

$$\frac{\delta n''_{A,x}}{\delta x} = 0 \text{ and } \frac{\delta n''_{A,z}}{\delta z} = 0 \qquad (c)$$

Equation (4.43) becomes

$$\frac{\delta n''_{A,y}}{\delta y} = 0 \qquad (4.67)$$

where $n''_{A,y}$ is the molar flux of the metal vapor A in the y direction.

To correlate the molar flux to the concentration gradient in the y direction, from Eq. (4.33)

$$n''_{A,y} = -C D_{AB} \left( \frac{dX_A}{dy} \right) + X_A \left( n''_{A,y} + n''_{B,y} \right) \qquad (4.68)$$

At steady-state condition, molecules of A move away from the evaporating surface while component B remains stationary. Therefore, $n''_{B,y} = 0$.

Equation (4.68) becomes

$$n''_{A,y} = -C D_{AB} \left( \frac{dX_A}{dy} \right) + X_A . n''_{A,y} \quad \text{or} \quad n''_{A,y} = -\left\{ \frac{C D_{AB}}{(1 - X_A)} \right\} \left( \frac{dX_A}{dy} \right) \qquad (4.69)$$

Putting the value of $n''_{A,y}$ [from Eqs. (4.67) to (4.69)]

$$\frac{d}{dy} \left[ -\left\{ \frac{C D_{AB}}{(1 - X_A)} \right\} \left( \frac{dX_A}{dy} \right) \right] = 0 \qquad (4.70)$$

If the molar density (C) and the diffusivity ($D_{AB}$) are constants, then Eq. (4.70) simplifies to

$$\frac{d}{dy}\left[-\left\{\frac{1}{(1-X_A)}\right\}\left(\frac{dX_A}{dy}\right)\right] = 0 \tag{4.71}$$

1st integration of Eq. (4.71):

$$\left[-\left\{\frac{1}{(1-X_A)}\right\}\left(\frac{dX_A}{dy}\right)\right] = C_1 \tag{4.72}$$

2nd integration of Eq. (4.72):

$$\ln(1-X_A) = C_1\,y + C_2 \tag{4.73}$$

where $C_1$ and $C_2$ are integration constants.

Putting boundary condition: at $y = 0$, $X_A = X_{A1}$, Eq. (4.73) becomes

$$\ln(1-X_{A1}) = C_2 \tag{4.74}$$

Again, putting boundary condition: at $y = L$, $X_A = X_{A2}$, Eq. (4.73) becomes
$\ln(1-X_{A2}) = C_1\,L + C_2 = C_1\,L + \ln(1-X_{A1})$ [from Eq. (4.74)]
Therefore,

$$\ln\left[\frac{(1-X_{A2})}{(1-X_{A1})}\right] = C_1\,L \quad\text{or}\quad C_1 = \left(\frac{1}{L}\right)\ln\left[\frac{(1-X_{A2})}{(1-X_{A1})}\right] \tag{4.75}$$

Putting the values of $C_1$ and $C_2$ in Eq. (4.73)

$$\ln(1-X_A) = \left(\frac{y}{L}\right)\ln\left[\frac{(1-X_{A2})}{(1-X_{A1})}\right] + \ln(1-X_{A1})$$

or

$$\ln\left[\frac{(1-X_A)}{(1-X_{A1})}\right] = \ln\left[\frac{(1-X_{A2})}{(1-X_{A1})}\right]^{\left(\frac{y}{L}\right)} \quad\text{or}\quad \frac{(1-X_A)}{(1-X_{A1})} = \left[\frac{(1-X_{A2})}{(1-X_{A1})}\right]^{\left(\frac{y}{L}\right)} \tag{4.76}$$

**Example 4.1** The molecular weights of the components A and B of a gas mixture are 24 and 48, respectively. The molecular weight of the gas mixture is found to be 30. If the mass concentration of the mixture is 1.2 kg.m$^{-3}$. Determine: (i) mole fraction, (ii) mass fraction and (iii) total pressure if the temperature of the mixture is 290 K.

**Solution**

Given: $M_A = 24$, $M_B = 48$, $M = 30$, $\rho = 1.2$ kg.m$^{-3}$, $T = 290$ K.

Molar concentration of the mixture, $C = \frac{\rho}{M} = \frac{1.2}{30} = 0.04$

Since $C_i = \frac{\rho_i}{M_i}$, or $\rho_i = C_i M_i$

so $\rho_A = C_A M_A = 24 C_A$ and $\rho_B = C_B M_B = 48 C_B$

Therefore,

$$\rho_A + \rho_B = 24 C_A + 48 C_B = \rho = 1.2 \tag{a}$$

Again

$$C_A + C_B = C = 0.04 \tag{b}$$

Equation (b) is multiplied by 24:

$$24 C_A + 24 C_B = 0.96 \tag{c}$$

Substitute from Eq. (a) in Eq. (c): $24 C_B = 0.24$; therefore, $C_B = 0.01$
Now from Eq. (b): $C_A + C_B = 0.04$ or $C_A = 0.04 - C_B = 0.04 - 0.01 = 0.03$

(i)  Now mole fraction: $X_i = \frac{C_i}{C}$

So $X_A = \frac{C_A}{C} = \frac{0.03}{0.04} = 0.75$ and $X_B = \frac{C_B}{C} = \frac{0.01}{0.04} = 0.25$

(ii)  Mass fraction: $m_i = \frac{\rho_i}{\rho}$

So

$$m_A = \frac{\rho_A}{\rho} = \frac{0.72}{1.2} = 0.6 \quad (\text{since } \rho_A = 24 C_A = 24 \times 0.03 = 0.72)$$

$$m_B = \frac{\rho_B}{\rho} = \frac{0.48}{1.2} = 0.4 \quad (\text{since } \rho_B = 48 C_B = 48 \times 0.01 = 0.48)$$

(iii)  Total pressure at the temperature, 290 K
Since $\rho_i = \frac{p_i M_i}{RT}$

therefore, $p = \frac{\rho RT}{M} = \frac{1.2 \times 8.314 \times 290}{30} = 96.44$ kPa.

**Example 4.2** A vessel contains a binary mixture of oxygen and nitrogen with partial pressure in the ratio of 0.21 and 0.79 at 15 °C. The total pressure of the mixture is $1.1 \times 10^5$ N.m$^{-2}$. Calculate: (i) molar concentration, (ii) mass densities, (iii) mass fraction and (iv) mole fraction.

**Solution**
$T = 15 + 273 = 288$ K. $p = 1.1 \times 10^5$ N.m$^{-2}$.
$p_{O2} = 0.21 \times 1.1 \times 10^5$ N.m$^{-2} = 0.231 \times 10^5$ N.m$^{-2}$, and $p_{N2} = 0.79 \times 1.1 \times 10^5$ N.m$^{-2} = 0.869 \times 10^5$ N.m$^{-2}$.

(i)  Molar concentration: $C_i = \frac{p_i}{RT}$

$$C_{O2} = \frac{p_{O2}}{RT} = \frac{0.231 \times 10^5}{8.314 \times 288} = 9.647 \text{ mol.m}^{-3} = 9.647 \times 10^{-3} \text{ kmol.m}^{-3}.$$

$$C_{N2} = \frac{p_{N2}}{RT} = \frac{0.869 \times 10^5}{8.314 \times 288} = 36.292 \text{ mol.m}^{-3} = 36.292 \times 10^{-3} \text{ kmol.m}^{-3}.$$

(ii)  Mass densities: $\rho_i = C_i M_i$

$$\rho_{O2} = C_{O2} M_{O2} = 9.647 \times 10^{-3} \times 32 = 0.309 \text{ kg.m}^{-3}$$
$$\rho_{N2} = C_{N2} M_{N2} = 36.292 \times 10^{-3} \times 28 = 1.016 \text{ kg.m}^{-3}$$

(iii)  Mass fraction: $m_i = \frac{\rho_i}{\rho}$

Now $\rho = \rho_{O2} + \rho_{N2} = 0.309 + 1.016 = 1.325 \text{ kg.m}^{-3}.$

Therefore,

$$m_{O2} = \frac{\rho_{O2}}{\rho} = \frac{0.309}{1.325} = 0.233$$

$$m_{N2} = \frac{\rho_{N2}}{\rho} = \frac{1.016}{1.325} = 0.767$$

(iv)  Mole fraction: $X_i = \frac{C_i}{C}$

Now $C = C_{O2} + C_{N2} = 0.00965 + 0.0363 = 0.04595 \text{ kmol.m}^{-3}.$

Therefore,

$$X_{O2} = \frac{C_{O2}}{C} = \frac{0.00965}{0.046} = 0.21$$

$$X_{N2} = \frac{C_{N2}}{C} = \frac{0.0363}{0.046} = 0.79$$

→ These are equal to the partial pressure fractions.

**Example 4.3**  From the following data, calculate the diffusion coefficient for $NH_3$ in air at 27 °C temperature and one atm. pressure:

$NH_3$ (gas A) : molecular weight $= 17$, molecular volume $= 0.0264 \text{ m}^3.\text{kmol}$
Air (gas B) : molecular weight $= 29$, molecular volume $= 0.0306 \text{ m}^3.\text{kmol}$

**Solution**

Given: $M_A = 17$, $V_A = 0.0264 \text{ m}^3. \text{kmol}$; $M_B = 29$, $V_B = 0.0306 \text{ m}^3. \text{kmol}.$
$T = 27 + 273 = 300 \text{ K, p} = 1 \text{ atm.}$
The mass diffusion coefficient for a binary mixture of gases:

$$D_{AB} = 0.0043 \left\{ \frac{T^{3/2}}{p\left(V_A^{1/3} + V_B^{1/3}\right)^2} \right\} \left[ \frac{1}{M_A} + \frac{1}{M_B} \right]^{1/2}$$

$$= 0.0043 \left\{ \frac{(300)^{3/2}}{1.(0.0264)^{1/3} + (0.0306)^{1/3})^2} \right\} \left[ \frac{1}{17} + \frac{1}{29} \right]^{1/2} = 18.3 \text{ m}^3.\text{s}^{-1}.$$

**Example 4.4**  A steel rectangular container having walls 16 mm thick is used to store gaseous hydrogen at elevated pressure. The molar concentration of hydrogen in the steel at the inside and outside surfaces are $1.2 \text{ kmol.m}^{-3}$ and 0, respectively. Assume the diffusion coefficient for hydrogen in steel as $0.248 \times 10^{-12} \text{ m}^2.\text{s}^{-1}$. Calculate the molar diffusion flux for hydrogen through the steel surface.

**Solution**

Given: $L = 16 \text{ mm} = 0.016 \text{ m}$, $C_{A1} = 1.2 \text{ kmol.m}^{-3}$, $C_{A2} = 0$; $D_A = 0.248 \times 10^{-12}$ $\text{m}^2.\text{s}^{-1}$.

Therefore, mass transfer rate in terms of molar concentration, i.e. molar diffusion flux $J_A^{'*}$:

$$J_A^{'*} = -D_A \left( \frac{C_{A2} - C_{A1}}{L} \right) = D_A \left( \frac{C_{A1} - C_{A2}}{L} \right)$$

$$= (0.248 \times 10^{-12}) \left( \frac{1.2 - 0}{0.016} \right) = 18.6 \times 10^{-12} \text{ kmol.s.m}^2.$$

## 4.4   Diffusion of Gas Through Solid

Gaseous reduction of dense spherical pellets is the best example of diffusion of gas through solid. Overall reduction of dense hematite pellets with hydrogen gas is as follows:

$$Fe_2O_3 \text{ (s)} + 3H_2 \text{ (g)} = 2 Fe \text{ (s)} + 3H_2O \text{ (g)} \tag{4.77}$$

If the gas velocity in the bulk phase is high, diffusion of gases through the gas film will be negligible. Hence the rate of reduction will be controlled by diffusion of the gas through the porous reduced shell (Fig. 4.6). If the total pressure of the gas is the same at every point in the reduced shell, then the amount of hydrogen gas diffusing into the shell will be equal to that of the water vapor diffusing out.

Assume:

(i)    considering the diffusion of only one gaseous component,
(ii)   concentration of gas at the surface of the pellet is equal to the concentration of gas at the bulk,
(iii)  concentration of gas at the interface will be equilibrium gas composition,
(iv)   under steady-state condition and one-directional diffusion process.

Now considering the diffusion equation in the spherical coordinate system (from Appendix V), the pellet shows a spherical symmetry (Fig. 4.6).

**Fig. 4.6** Diffusion of gas through solid

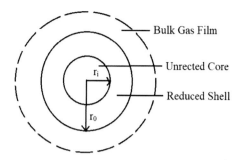

Bulk Gas Film

Unrected Core

Reduced Shell

$$\frac{1}{r^2}\frac{\delta}{\delta r}\left(CD_{AB}r^2\frac{\delta X_A}{\delta r}\right) + \frac{1}{r^2 Sin\theta^2}\frac{\delta}{\delta \varphi}\left(CD_{AB}\frac{\delta X_A}{\delta \varphi}\right)$$
$$+ \frac{1}{r^2 Sin\theta}\frac{\delta}{\delta \theta}\left(CD_{AB}Sin\theta\frac{\delta X_A}{\delta \theta}\right) + N_A' = \frac{\delta C_A}{\delta t} \tag{4.78}$$

Equation (4.78) can be simplified by considering C and $D_{AB}$ as constants, and multiplying by ($\frac{r^2}{CD_{AB}}$) we get

$$\frac{\delta}{\delta r}\left(r^2\frac{\delta C_A}{\delta r}\right) + \frac{1}{Sin\theta^2}\frac{\delta}{\delta \varphi}\left(\frac{\delta C_A}{\delta \varphi}\right) + \frac{1}{Sin\theta}\frac{\delta}{\delta \theta}\left(Sin\theta\frac{\delta C_A}{\delta \theta}\right) + N_A'\left(\frac{r^2}{CD_{AB}}\right)$$
$$= \left(\frac{r^2}{CD_{AB}}\right)\frac{\delta C_A}{\delta t} \tag{4.79}$$

For steady-state: $\frac{\delta C_A}{\delta t} = 0$, and no extra chemical spices are generated:

$$N_A' = 0 \tag{a}$$

One-dimensional reaction front (i.e. in r direction):

$$\frac{\delta C_A}{\delta \varphi} = 0 \text{ and } \frac{\delta C_A}{\delta \theta} = 0 \tag{b}$$

Equation (4.79) becomes

$$\frac{\delta}{\delta r}\left(r^2\frac{\delta C}{\delta r}\right) = 0 \tag{4.80}$$

By first integration of Eq. (4.80)

$$r^2\frac{\delta C}{\delta r} = C_1 \text{ or } \delta C = C_1\left(\frac{\delta r}{r^2}\right) \tag{4.81}$$

Again, further second integration of Eq. (4.81)

$$C = -\left(\frac{C_1}{r}\right) + C_2 \tag{4.82}$$

where $C_1$ and $C_2$ are integration constants.

Now at boundary conditions:

(i) at $r = r_0$, $C = C_b$      (c)                                          (c)

(ii) at $r = r_i$, $C = C_e$      (d)                                         (D)

where $C_b$ is the bulk gas concentration at the surface of pellet of radius $r_0$, and $C_e$ is the equilibrium gas concentration for the reaction at the interface situated at a radius $r_i$.

Equation (4.82) becomes

$$C_b = -\left(\frac{C_1}{r_0}\right) + C_2 \quad \text{and} \quad C_e = -\left(\frac{C_1}{r_i}\right) + C_2 \tag{4.83}$$

Therefore,

$$C_b - C_e = C_1\left(\frac{1}{r_i} - \frac{1}{r_0}\right) = C_1\left(\frac{r_0 - r_i}{r_i r_0}\right)$$

Hence,

$$C_1 = (C_b - C_e)\left(\frac{r_i r_0}{r_0 - r_i}\right) \tag{4.84}$$

By putting value of $C_1$ in Eq. (4.83) we get

$$C_b = -\left(\frac{C_1}{r_0}\right) + C_2 = -\left(\frac{C_b - C_e}{r_0}\right)\left(\frac{r_i r_0}{r_0 - r_i}\right) + C_2$$

Therefore,

$$C_2 = C_b + \left\{\frac{(C_b - C_e)r_i}{r_0 - r_i}\right\} = C_b + \left(\frac{C_b r_i - C_e r_i}{r_0 - r_i}\right) \tag{4.85}$$

Now putting values of $C_1$ and $C_2$ in Eq. (4.82) we get

$$C = -\left(\frac{C_1}{r}\right) + C_2$$

Therefore,

$$C = -\left(\frac{C_b - C_e}{r}\right)\left(\frac{r_i r_0}{r_0 - r_i}\right) + \left\{C_b + \left(\frac{C_b r_i - C_e r_i}{r_0 - r_i}\right)\right\}$$

or

$$C_b - C = \left(\frac{C_b - C_e}{r}\right)\left(\frac{r_i r_0}{r_0 - r_i}\right) - \left(\frac{C_b r_i - C_e r_i}{r_0 - r_i}\right) = \left(\frac{C_b - C_e}{r_0 - r_i}\right)\left[\left(\frac{r_i r_0}{r}\right) - r_i\right]$$

or

$$\left(\frac{C_b - C}{C_b - C_e}\right) = \left(\frac{r_0 - r}{r_0 - r_i}\right)\left(\frac{r_i}{r}\right) \tag{4.86}$$

If the overall reaction is controlled by the diffusion in the reacted shell and if $\alpha$ moles of the gas react with $\beta$ moles of the hematite pellet lending to the formation of the product, then an equation can be written based on mass balance. According to this, the amount of gas, $m_g$ diffusing in from the surface of the pellet is equal to the amount of gas reacted.

Thus,

$$m_g = -4\pi r_0^2\left(D\frac{\delta C}{\delta r}\right)_{r=r_0} = \left(\frac{\alpha}{\beta}\right)\left(\frac{\rho}{M}\right)(4\pi r_i^2)\left(\frac{dr_i}{dt}\right) \tag{4.87}$$

where $\rho$ is the density of solid and M is its molecular weight.

From Eq. (4.81):

$$\frac{\delta C}{\delta r} = \left(\frac{C_1}{r^2}\right) = \left\{\frac{(C_b - C_e)}{r^2}\right\}\left(\frac{r_i r_0}{r_0 - r_i}\right) \quad \text{[from Eq. (4.84)]} \tag{4.88}$$

Now, in Eq. (4.88) substitute from Eq. (4.87):

$$D\left(\frac{r_i r_0}{r_0 - r_i}\right)(C_b - C_e) = \left(\frac{\alpha}{\beta}\right)\left(\frac{\rho}{M}\right)(r_i^2)\left(\frac{dr_i}{dt}\right)$$

or

$$\left(\frac{dr_i}{dt}\right) = \left\{\frac{Dr_0(C_b - C_e)}{r_i(r_0 - r_i)}\right\}\left(\frac{M}{\rho}\right)\left(\frac{\beta}{\alpha}\right) \tag{4.89}$$

By integrating Eq. (4.89) we get

$$\int\left(r_i - \frac{r_i^2}{r_0}\right)dr_i = \left\{\frac{\beta M(C_b - C_e)D}{\alpha\rho}\right\}\int dt$$

or

$$\left(\frac{r_i^2}{2} - \frac{r_i^3}{3r_0}\right) = \left\{\frac{\beta M(C_b - C_e)D}{\alpha\rho}\right\} \cdot t + I \tag{4.90}$$

where I is an integration constant.

Now putting the boundary condition $r_i = r_0$ at $t = 0$ in Eq. (4.90),

$$\left(\frac{r_0^2}{2} - \frac{r_0^3}{3r_0}\right) = \left\{\frac{\beta M(C_b - C_e)D}{\alpha\rho}\right\} \times 0 + I \text{ or } I = \frac{r_0^2}{6} \tag{4.91}$$

Putting the value of I in Eq. (4.90) we get

$$\left[\frac{r_i^2}{2} - \frac{r_i^3}{3r_0} - \frac{r_0^2}{6}\right] = \left\{\frac{\beta M(C_b - C_e)Dt}{\alpha\rho}\right\} \tag{4.92}$$

If $\rho$ is the density of the pellet containing $W_O$ as the initial amount of oxygen present in it, then

$$W_O = \left(\frac{48}{160}\right)\rho\left(\frac{4\pi r_0^3}{3}\right) \tag{4.93}$$

If $r_i$ is the radius of the reaction interface at any instant, then $W_i$ is the amount of oxygen present in the pellet at that instant and will be given by the equation:

$$W_i = \left(\frac{48}{160}\right)\rho\left(\frac{4\pi r_i^3}{3}\right) \tag{4.94}$$

The fraction of reduction (f) is defined as the amount of oxygen removed at any instant as a fraction of the total oxygen present in the pellet initially.

Therefore,

$$f = \frac{W_O - W_i}{W_O} = \left[\frac{0.3 \times \rho\left\{\frac{4\pi(r_0^3 - r_i^3)}{3}\right\}}{0.3 \times \rho\left(\frac{4\pi r_0^3}{3}\right)}\right] = \left(\frac{r_0^3 - r_i^3}{r_0^3}\right) = 1 - \left(\frac{r_i}{r_0}\right)^3 \tag{4.95}$$

Hence,

$$r_i = r_0(1 - f)^{1/3} \tag{4.96}$$

Substitute the value of $r_i$ in Eq. (4.92):

$$[(2f - 3) + 3(1 - f)^{2/3}] = \left\{\frac{6\beta M(C_b - C_e)}{\alpha\rho r_0^2}\right\}Dt \tag{4.97}$$

Hence, $[(2f-3)+3(1-f)^{2/3}]$ is directly proportional to time (t), the rate controlled by diffusion through a porous shell.

**Example 4.5**  A dense hematite pellet ($1.5 \times 10^{-2}$ m in diameter) is reduced at 700 °C by hydrogen gas under one atm. pressure. The diffusion coefficient of hydrogen in a porous reduced shell is found to be $1.7 \times 10^{-5}$ m$^2$.s$^{-1}$. Derive an equation for the rate of movement of the reaction interface. Also, calculate the fractional reduction and the amount of oxygen removed from the pellet in 3 h. The density of the pellet is 5300 kg.m$^{-3}$, and the atomic weights of Fe and O are 56 and 16, respectively.
Given: $Fe_2O_3$ (s) + $3H_2$ (g) = 2 Fe (s) + $3H_2O$ (g), $\Delta G^0 = 6.7$ kJ at 700 °C.

**Solution**

$$\text{Weight of pellet} = \left(\frac{4\pi r_0^3}{3}\right)\rho = \frac{4}{3} \times \frac{22}{7} \times \left(\frac{1.5 \times 10^{-2}}{2}\right)^3 \times 5300$$

$$= 9.37 \times 10^{-3} \text{ kg.}$$

Amount of oxygen initially present in the pellet,

$$W_O = \left(\frac{48}{160}\right)\rho\left(\frac{4\pi r_i^3}{3}\right) = 0.3 \times 9.37 \times 10^{-3}$$

$$= 2.81 \times 10^{-3} \text{ kg.}$$

Since $\Delta G^0 = -RT \ln k$
or

$$\ln k = \frac{-\Delta G^0}{RT} = \frac{-6700}{8.314 \times 973} = -0.828 \text{ or } k = 0.437$$

Again

$$k = \left(\frac{p_{H2O}}{p_{H2}}\right)^3 = \left(\frac{1 - p_{H2}}{p_{H2}}\right)^3 \quad \left[\text{Since } p_{H2} + p_{H2O} = 1 \text{ atm, } p_{H2O} = 1 - p_{H2}\right]$$

or

$$\left(\frac{1 - p_{H2}}{p_{H2}}\right) = k^{1/3} = (0.437)^{1/3} = 0.759, \quad \text{or } p_{H2} = 0.569 \text{ atm}$$

Concentration of hydrogen at bulk phase,

$$C_b = \left(\frac{p_{H2}}{RT}\right) = \left(\frac{1}{0.082 \times 973}\right) = 12.5 \times 10^{-3} \text{ kmol.m}^{-3}$$

Concentration of hydrogen at reaction interface,

$$C_e = \left(\frac{p_{H2}}{RT}\right) = \left(\frac{0.569}{0.082 \times 973}\right) = 7.13 \times 10^{-3}\, kmol.m^{-3}$$

Therefore, $(C_b - C_e) = (12.5 - 7.13) \times 10^{-3} = 5.37 \times 10^{-3}\, kmol.m^{-3}$
According to Eq. (4.89):

$$\left(\frac{dr_i}{dt}\right) = \left\{\frac{Dr_0(Cb - Ce)}{r_i(r_0 - r_i)}\right\}\left(\frac{M}{\rho}\right)\left(\frac{\beta}{\alpha}\right)$$

$$= \frac{(1.7 \times 10^{-5})(7.5 \times 10^{-3})(5.37 \times 10^{-3})}{r_i(7.5 \times 10^{-3} - r_i)} \times \left(\frac{160}{5300}\right)\left(\frac{1}{3}\right)$$

$$= \left(\frac{0.69 \times 10^{-11}}{r_i(7.5 \times 10^{-3} - r_i)}\right)$$

Therefore, $[7.5 \times 10^{-3}r_i - r_i^2]dr_i = (0.69 \times 10^{-11})dt$
By integration: $\left[\left(\frac{7.5 \times 10^{-3}r_i^2}{2}\right) - \left(\frac{r_i^3}{3}\right)\right] = (0.69 \times 10^{-11})t$
For $t = 3\, h = 10{,}800\, s$, since $r_i$ is very small, so $(r_i)^3 \to 0$.
Therefore, $r_i = 4.46 \times 10^{-3}\, m$
For fractional reduction,

$$f = 1 - \left(\frac{r_i}{r_0}\right)^3 = 1 - \left(\frac{4.46 \times 10^{-3}}{7.5 \times 10^{-3}}\right)^3 = 1 - (0.595)^3 = \mathbf{0.79}$$

Since $f = \frac{W_O - W_i}{W_O}$
therefore, the amount of oxygen removed

$$(W_O - W_i) = f \times W_O = 0.79 \times 2.81 \times 10^{-3}$$
$$= \mathbf{2.22 \times 10^{-3}\, kg = 2.22\, g}$$

## 4.5    Motion of Gas Bubbles in Liquid

Many metallurgical processes related to the extraction and refining of metals involve the interaction of a gas and a liquid phase, e.g. steelmaking. The gaseous phase is introduced in reactive systems as bubbles rise through the liquid. The shape and velocity of a bubble depend upon its volume, which in turn also affects its velocity.

The shape and velocity of a bubble are expressed as a function of the bubble Reynolds number ($Re_b$):

$$Re_b = \frac{d_b u_b \rho_l}{\mu_l} \tag{4.98}$$

where $d_b$ is the diameter of a spherical bubble, $u_b$ is the velocity of the bubble, $\rho_l$ is the density of the liquid, and $\mu_l$ is the viscosity of the liquid.

(i)   For very small bubbles with Reynolds number equal to or less than 2 they behave like rigid solid spheres. They follow the Stokes law and thus their terminal velocity $u_\infty$ is given by

$$u_\infty = \frac{d_b^2 g(\rho_l - \rho_g)}{18\mu_l} = \frac{2r_b^2 g(\rho_l - \rho_g)}{9\mu_l} \tag{4.99}$$

(ii)  Bubbles with Reynolds number between 2 and 400 are still spherical in shape but their rising velocity is far greater than that predicted by Stokes law [Eq. (4.99)].
(iii) Bubbles having Reynolds number between 400 and 5000 are either spherical or ellipsoidal and rise in a spiral path through the liquid.
(iv)  Bubbles having Reynolds number above 5000 are of a spherical-cup shape and rise at a terminal velocity given by the equation:

$$u_\infty = 1.02 \times \left(\frac{gd_b}{2}\right)^{1/2} \tag{4.100}$$

**Example 4.6**  Argon gas is purged through liquid steel from the bottom of the ladle at 1600 °C. Assuming the bubbles to be a spherical-cup shape of equivalent spherical bubble diameter 10 mm, calculate the residence time of the gas in a bath of depth 0.5 m. The density and viscosity of liquid steel at 1600 °C are given as 7000 kg.m$^{-3}$ and $2 \times 10^{-2}$ Poise, respectively.

**Solution**

Since the Reynolds number of the bubble is not known and assuming the bubbles are spherical-cup shape, then Eq. (4.100) can be applied:
  Terminal velocity,

$$u_\infty = 1.02 \times \left(\frac{gd_b}{2}\right)^{1/2} = 1.02 \times \left(\frac{9.81 \times 0.01}{2}\right)^{1/2} = 0.226 \, \text{m.s}^{-1}$$

Now bubbles Reynolds number,

$$Re_b = \frac{d_b u_b \rho_l}{\mu_l} = \left(\frac{0.01 \times 0.226 \times 7000}{0.2 \times 10^{-2}}\right) = 7910$$

(since 10 Poise = 1 kg.m$^{-1}$.s$^{-1}$, so $2 \times 10^{-2}$ Poise $= 0.2 \times 10^{-2}$ kg.m$^{-1}$.s$^{-1}$)
So, residence time $= \frac{distance}{velocity} = \frac{0.5}{0.226} = \mathbf{2.21 \, s}$.

## 4.6  Mechanism of Mass Transfer

Considering a piece of mild steel plate whose surface is to be carburized by the flow of CO gas over it. There are two basic modes of mass transfer: (i) diffusion mass transfer and (ii) convective mass transport [3].

CO gas will be transported to the steel surface–gas interface as a result of convective mass transport, where it will dissociate according to the Boudouard reaction:

$$2CO \rightleftharpoons C + CO_2 \tag{4.101}$$

for which

$$k_e = \frac{[wt\%C]p_{CO_2}}{p_{CO}^2} \tag{4.102}$$

where $k_e$ is the equilibrium constant for reaction (4.101).

The newly produced carbon molecule enters the steel surface as a result of diffusion mass transfer and the gaseous product $CO_2$ leaves the interface and returns to the gaseous atmosphere as a result of convective mass transport. Profiles of carbon concentration, during the carburization process, are shown in Fig. 4.7. It is important to note that carbon will continue to diffuse into the surface of steel until its chemical potential in the steel is equal to its chemical potential in the gaseous atmosphere of the surrounding.

(i)   Diffusion mass transfer: Carbon will be transported inwards to the center of the steel plate slowly as a result of molecular diffusion mechanisms. Fick state that the rate at which a solute (i.e. carbon) diffuses through a stationary solvent (i.e. steel section) should be proportional to the cross-sectional area perpendicular to the mass flow, the concentration gradient in the solid at the location.

$$N'_{A,x,t} \propto A_{cs} \tag{4.103}$$

$$\propto \left(\frac{dc}{dx}\right)_{x,t} \tag{4.104}$$

**Fig. 4.7**  Carbon concentration vs distance from surface

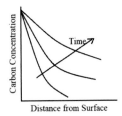

Therefore,

$$N'_{A,x,t} = -D_{AB}A_{CS}\left(\frac{dc}{dx}\right)_{x,t} \tag{4.105}$$

where

$N'_{A,x,t}$     is the number of mol of solute A diffusing per unit time in the x direction at time t across a plane located at x (mol/s),

$A_{CS}$     is the cross-sectional area perpendicular to the x direction ($m^2$),

$\left(\frac{dc}{dx}\right)_{x,t}$     is the concentration gradient at location x and time t ($mol/m^3$.m, i.e. $mol/m^4$),

$D_{AB}$     is the binary diffusion coefficient of solute A in B ($m^2/s$).

(ii)    Convective mass transfer: The mechanism of $CO_2$ transport, and $CO_2$ transport to the plate surface (interface), is an example of forced convective mass transfer. As a result of the normal no-slip condition between the gas and solid phase molecules at the interface, molecular diffusion processes will predominate in the immediate vicinity of the interface. However, a little way out (say 0.1 mm), this mechanism of mass transfer will give way to one of convective mass transfer, since the gas mixture begins to move in a direction and with a speed approaching those of the bulk flow. The amount of $CO_2$ removed will depend upon the difference in $CO_2$ concentration between the interface and the bulk.

$$N'_A = kA_{cs}\left(c'_A - c_A^\infty\right) \tag{4.106}$$

where $N'_A$ is the rate of mass transfer of A (mol/s), k is the convective mass transfer coefficient (m/s), $c'_A$ is the interfacial concentration of species A (i.e. $CO_2$) ($mol/m^3$), and $c_A^\infty$ is the bulk concentration of species A (i.e. $CO_2$) in the gas phase ($mol/m^3$).

## 4.7  Simultaneous Heat and Mass Transfer

### 4.7.1  Change of Phase Due to Melting

The solidification and melting of metal alloys present several interesting problems on simultaneous heat and mass transfer. Consider melting of a large steel scrap, which is immersed in a liquid bath of an oxygen steelmaking furnace. Steel scrap contains very low carbon, while the concentration of carbon in the liquid bath is a function of the oxygen lancing time.

The rate of melting at a surface of the scrap in contact with the molten liquid bath may be written as

$$\left(\frac{dM}{dt}\right)_h = \frac{h(T_b - T_{mp})}{\Delta H_T \rho_s} \qquad (4.107)$$

where $\left(\frac{dM}{dt}\right)_h$ is the rate of melting to heat transfer, expressed as the linear rate of advance of liquid–solid interface. h is the heat transfer coefficient between surface and melt. $T_b$ and $T_{mp}$ are the bulk temperature and melting point of scrap, respectively. $\Delta H_T$ is the effective latent heat of melting, including sensible heat required to bring solid to melting point. $\rho_s$ is the density of solid.

Equation (4.107) would be applied to one component liquid–solid system, where the melting temperature would be constant. However, in the Fe–C system under consideration, $T_{mp}$ is not fixed but may be decreased because of the diffusion of carbon to the solid surface.

If the bath temperature $T_b$ is lower than the melting point of the scrap, the melting process is controlled entirely by carbon diffusion to the surface, and its rate is expressed by

$$\left(\frac{dM}{dt}\right)_m = \left\{\frac{K_m(C_b - C_0)}{f_C}\right\} \qquad (4.108)$$

where $\left(\frac{dM}{dt}\right)_m$ is the rate of melting due to mass transfer, expressed as a linear rate of advance of the interface. $K_m$ is the mass transfer coefficient between surface and melt. $C_b$ and $C_0$ are the concentration of carbon in bulk and scrap, respectively. $f_C$ is the weight of carbon that must be transferred per unit weight of metal melted.

While $f_C$ will vary during the lancing, assuming that as a first approximate, $f_C \cong (C_b - C_0)$.

Therefore, Eq. (4.108) becomes

$$\left(\frac{dM}{dt}\right)_m \cong K_m \qquad (4.109)$$

The heat and mass transfer coefficients will depend on the flow conditions in the bath. As an illustration of the relative contributions of heat and mass transfer to melting, let us assume that the appropriate heat transfer correlation is for natural convection from a vertical plate, in the laminar region.

The heat transfer coefficient is, in general, expressed in terms of a dimensionless number called Nusselt's number, defined as

$$Nu = \frac{hL}{k} \qquad (4.110)$$

where L is the characteristic length of the system, and k is the thermal conductivity of the fluid.

$$\text{Grashof number, Gr} = \left\{ \frac{(\rho^2 L^3 g \beta \Delta T)}{\mu^2} \right\} \tag{4.111}$$

where $\beta$ is the thermal coefficient of isobaric volume expansion $= -\left\{ \frac{1}{\rho} \left( \frac{\delta \rho}{\delta t} \right)_p \right\}$

$$\text{Prandtl number, } Pr = \frac{C_p \mu}{k} \tag{4.112}$$

where $C_p$ is specific heat.

Now Nusselt's number can be written as

$$Nu = \frac{hL}{k} = \left\{ \frac{(0.508 Pr^{1/2})}{(0.95 + Pr^{1/4})} \right\} . Gr^{1/4} \tag{4.113}$$

Therefore,

$$Gr^{1/4} = \left\{ \frac{hL(0.95 + Pr^{1/4})}{k(0.508 Pr^{1/2})} \right\} \tag{4.114}$$

Again Sherwood number,

$$Sh = \frac{K_m L}{D_{A-B}} \tag{4.115}$$

Schmidt's number,

$$Sc = \frac{\mu}{\rho D_{A-B}} \tag{4.116}$$

Therefore,

$$Sh = \frac{K_m L}{D_{A-B}} = \left\{ \frac{(0.902 Sc^{1/2})}{(0.861 + Sc)^{1/4}} \right\} \left( \frac{Gr'}{4} \right)^{1/4} \tag{4.117}$$

where $Gr'$ is Grashof's number for mass transfer.

Therefore,

$$(Gr')^{1/4} = \left[ \frac{\{4^{1/4} K_m L (0.861 + Sc)^{1/4}\}}{D_{A-B}(0.902 Sc^{1/2})} \right] \tag{4.118}$$

To determine the relative values of h and $K_m$, it has been assumed that the Grashof number based on thermal and concentration driven natural convection are approximately equal, i.e. $Gr \cong Gr'$.

By combining Eqs. (4.114) and (4.118),

$$\left\{\frac{hL(0.95 + Pr^{1/4})}{k(0.508\,Pr^{1/2})}\right\} = \left[\frac{\{4^{1/4}K_m L(0.861 + Sc)^{1/4}\}}{D_{A-B}(0.902\,Sc^{1/2})}\right]$$

or

$$\left(\frac{K_m}{h}\right)\left(\frac{k}{D_{A-B}}\right) = \left[\left\{\frac{(0.95 + Pr^{1/4})}{(0.508\,Pr^{1/2})}\right\}\left\{\frac{(0.902\,Sc^{1/2})}{(4^{1/4})(0.861 + Sc)^{1/4}}\right\}\right]$$

or

$$\left(\frac{K_m}{h}\right) = \left[\left\{\frac{(0.95 + Pr^{1/4})}{(0.508\,Pr^{1/2})}\right\}\left\{\frac{(0.902\,Sc^{1/2})}{(4^{1/4})(0.861 + Sc)^{1/4}}\right\}\right]\left(\frac{D_{A-B}}{k}\right) \quad (4.119)$$

Now combining Eqs. (4.107) and (4.109),

$$\left\{\frac{\left(\frac{dM}{dt}\right)_m}{\left(\frac{dM}{dt}\right)_h}\right\} = K_m\cdot\left[\frac{(\Delta H_T \rho_S)}{\{h(T_b - T_{mp})\}}\right] = \left(\frac{K_m}{h}\right)\cdot\left[\frac{(\Delta H_T \rho_S)}{(T_b - T_{mp})}\right] \quad (4.120)$$

$$= \left[\left\{\frac{(0.95 + Pr^{1/4})}{(0.508\,Pr^{1/2})}\right\}\left\{\frac{(0.902\,Sc^{1/2})}{(4^{1/4})(0.861 + Sc)^{1/4}}\right\}\right]\left(\frac{D_{A-B}}{k}\right)\cdot\left[\frac{(\Delta H_T \rho_S)}{(T_b - T_{mp})}\right]$$

$$(4.121)$$

In practice, $(T_b - T_{mp})$ may vary from 0 to 150 K and $(\Delta H_T)$ from 232.64 to 698.0 kJ/kg depending on the time of the blowing. Therefore, the contribution of mass transfer to the melting process may vary from 100 to 10% of the total effect. The diffusion of carbon plays the predominant role in the initial stages of the blowing when the bath temperature is below the melting point of the scrap.

**Example 4.7** Find out the ratio of rate of melting due to mass transfer and rate of melting to heat transfer, if melting point of mild steel (0.14% C) scrap is 1523 °C and the bath temperature is 1650 °C. The following data are given: $D_{C\text{-}Fe} = 5 \times 10^{-9}$ m²/s, $\rho_{Fe} = 7100$ kg/m³, $\mu_{Fe} = 6.7 \times 10^{-3}$ kg/m.s, k = 29.43 J/s.m.K, $C_{p,\,Fe} = 255.9$ J/kg.K, $\Delta H_T = 271.95 \times 10^3$ J/kg.

**Solution**

From Eq. (4.116):

$$\text{Schmidt's number, } Sc = \frac{\mu}{\rho D_{A-B}} = \left(\frac{6.7 \times 10^{-3}}{7100 \times (5 \times 10^{-9})}\right) = 188.73$$

From Eq. (4.112):

$$\text{Prandtl number, } Pr = \frac{c_p \mu}{k} = \left(\frac{255.9 \times 6.7 \times 10^{-3}}{29.43}\right) = 58.26 \times 10^{-3} = 0.058$$

From Eq. (4.119):

$$\left(\frac{K_m}{h}\right) = \left[\left\{\frac{\left(0.95 + Pr^{1/4}\right)}{\left(0.508 Pr^{1/2}\right)}\right\}\left\{\frac{\left(0.902 Sc^{1/2}\right)}{(4^{1/4})(0.861 + Sc)^{1/4}}\right\}\right]\left(\frac{D_{A-B}}{k}\right)$$

$$= \left[\left\{\frac{\left(0.95 + \left(0.058^{0.25}\right)\right)}{0.508 \times \left(0.058^{0.5}\right)}\right\}\left\{\frac{\left(0.902 \times 188.73^{0.5}\right)}{(4^{1/4})(0.861 + 188.73)^{1/4}}\right\}\right]\left(\frac{5 \times 10^{-9}}{29.43}\right)$$

$$= 4.86 \times 10^{-9}$$

From Eq. (4.120):

$$\left\{\frac{\left(\frac{dM}{dt}\right)_m}{\left(\frac{dM}{dt}\right)_h}\right\} = \left(\frac{K_m}{h}\right) \cdot \rho_S \cdot \left[\frac{(\Delta H_T)}{(T_b - T_{mp})}\right] = \left(4.86 \times 10^{-9} \times 7100\right)\left[\frac{(\Delta H_T)}{(T_b - T_{mp})}\right]$$

$$= \left[\frac{\left(0.345 \times 10^{-4}\right)(\Delta H_T)}{(T_b - T_{mp})}\right] = \left[\frac{\left(0.345 \times 10^{-4}\right)\left(271.95 \times 10^3\right)}{(1923 - 1796)}\right]$$

$$= \mathbf{0.074}$$

(Here $\Delta H_T = 271.95 \times 10^3$ J/kg, $T_b = 1650\ °C = 1923$ K and $T_{mp} = 1523\ °C = 1796$ K.)

## 4.7.2 Heat and Mass Transfer to Single Particle

### 4.7.2.1 Forced Convection

The functional dependence of the heat and mass transfer coefficients may be expressed in terms of dimensionless numbers.

Thus, for heat transfer

$$Nu \propto Re_p^n . Pr^m \qquad (4.122)$$

where Nu is Nusselt number $= \frac{hd_p}{k}$, $Re_p$ is Reynolds number of particle $= \frac{d_p u_b \rho}{\mu}$, Pr is Prandtl number $= \frac{C_p \mu}{k}$, h is heat transfer coefficient, $d_p$ is diameter of the particle, k is thermal conductivity of the fluid, $u_b$ is bulk velocity of the fluid, $\rho$ is density of the fluid, $\mu$ is viscosity of fluid, and n, m are exponential constants.

Similarly, the mass transfer coefficient can be expressed as

$$Sh \propto Re_p^{n'} . Sc^{m'} \tag{4.123}$$

where Sh is Sherwood number $= \frac{k_d d_p}{D_{A-B}}$, Sc is Schmidt number $= \frac{\mu}{\rho D_{A-B}}$, $k_d$ is mass transfer coefficient, $D_{A-B}$ is diffusivity coefficient, and $n'$, $m'$ are exponential constants.

Now consider the value of the heat and mass transfer coefficients in the limiting case where there is no relative motion between fluid and particle, at steady-state conditions.

The rate of heat conduction across a gas shell of differential thickness dr is expressed as (Fig. 4.8)

$$q_c = 4\pi r^2 k \left( \frac{dT}{dr} \right) \tag{4.124}$$

where $q_c$ is the rate of heat transfer by conduction, k is thermal conductivity of gas, T is temperature of gas at radius, r.

→ Integrating Eq. (4.124) for the boundary conditions:

T = $T_b$ at r = $r_b$ (outer radius of boundary layer) and T = $T_p$ at r = $r_p$ (particle radius).

$$q_c \int_{r=r_p}^{r=r_b} \left( \frac{dr}{r^2} \right) = 4\pi k \int_{T=T_p}^{T=T_b} dT$$

$$\rightarrow \quad q_c \left[ \frac{1}{r_p} - \frac{1}{r_b} \right] = 4\pi k (T_b - T_p) \quad \text{or} \quad q_c = \frac{4\pi k (T_b - T_p)}{\left[ \frac{1}{r_p} - \frac{1}{r_b} \right]} \tag{4.125}$$

**Fig. 4.8** Single particle surrounded by bulk gas

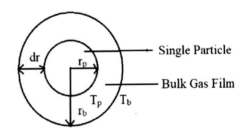

Single Particle

Bulk Gas Film

However, by definition, in a stagnant fluid the thickness of the boundary layer thickness is infinite, and consequently, $(1/r_b) \cong 0$.

Therefore,

$$q_c = 4\pi k\, r_p (T_b - T_p) \qquad (4.126)$$

Assuming the existence of an equivalent heat transfer coefficient, then $q_c$ can be expressed as

$$q_c = h\left(4\pi r_p^2\right)(T_b - T_p) \qquad (4.127)$$

where h is heat transfer coefficient between gas and particle.

Now comparing Eqs. (4.126) and (4.127):

$$h\, r_p \cong k \qquad (4.128)$$

Therefore, Nusselt number

$$Nu = \frac{h d_p}{k} = \frac{2 h r_p}{k} = \frac{2k}{k} = 2 \qquad (4.129)$$

Therefore, in the limiting case of a spherical particle in a motionless fluid, the Nusselt number at steady-state conditions has the value of 2.

Similar conditions apply for mass transfer. Thus, at zero Reynolds number, and in the absence of natural convection, it can be written as

$$Nu = Sh = 2 \qquad (4.130)$$

On the basis of Eq. (4.130), Eqs. (4.122) and (4.123) can be completed as

$$Nu = 2 + B\, Re_p^n \cdot Pr^m \qquad (4.131)$$

$$Sh = 2 + B'\, Re_p^{n'} \cdot Sc^{m'} \qquad (4.132)$$

where B and $B'$ are constants.

## 4.8 Exercises

**Problem 4.1**: The molecular weight of the components A and B of a gas mixture are 20 and 40, respectively. The molecular weight of gas mixture is found to be 25. If the mass concentration of the mixture is $1.1$ kg.m$^{-3}$. Determine: (i) mole fraction, (ii) mass fraction and (iii) total pressure if the temperature of the mixture is 300 K.

[Ans: (i) 0.75, 0.25; (ii) 0.6, 0.4; (iii) 109.74 kPa]

**Problem 4.2**: A vessel contains a binary mixture of CO and $CO_2$ with partial pressure in the ratio 3 and 1 at 25 °C. The total pressure of the mixture is $1.0 \times 10^5$ N.m$^{-2}$. Calculate: (i) molar concentration, (ii) mass densities, (iii) mass fraction and iv) mole fraction.

[Ans: (i) $30.272 \times 10^{-3}$, $10.091 \times 10^{-3}$ kmol.m$^{-3}$; (ii) 0.848, 0.444 kg.m$^{-3}$; (iii) 0.656, 0.344 ad iv) 0.75, 0.25]

**Problem 4.3**: Argon gas is purged through liquid steel from the bottom of the ladle at 1600 °C. Assuming the bubbles to be spherical-cup shape of equivalent spherical bubble diameter 8 mm, calculate the residence time of the gas in a bath of depth 60 cm. Also find out Reynolds number. The density and viscosity of liquid steel at 1600 °C are given as 7000 kg.m$^{-3}$ and $2 \times 10^{-2}$ Poise, respectively.

[Ans: 2.97 s and Re = 5656]

**Problem 4.4**: A dense hematite pellet (12 mm in diameter) is reduced at 1100 °C by hydrogen gas under one atm. pressure. The diffusion coefficient of hydrogen in porous reduced shell is found to be $1.7 \times 10^{-5}$ m$^2$.s$^{-1}$. Derive an equation for the rate of movement of the reaction interface. Also calculate the fractional reduction and the amount of oxygen removed from the pellet in 120 min. The density of pellet is 5300 kg.m$^{-3}$, and atomic weights of Fe and O are 56 and 16, respectively.
Given: $1/3$ $Fe_2O_3$ (s) + $H_2$ (g) = $2/3$ Fe (s) + $H_2O$ (g), $\Delta G^0 = 27{,}251.77 - 32.19$ T J.

[Ans: 0.96, 1.38 g]

## 4.9   Questions

Q1. Explain the following terms:

(a) Mass concentration, (b) Molar concentration, (c) Mass fraction, (d) Mole fraction.

Q2. Discuss Fick's law of diffusion. What is the analogous to other laws ? Explain.
Q3. What do you understand by mass flux? Derive equation for mass flux in terms of velocity.
Q4. Derive the equation for diffusion of gas through a spherical solid by using general diffusion equation.
Q5. Derive the equation for steady-state diffusion through a plain membrane.
Q6. Derive the equation for steady-state equimolar counter diffusion of gases.
Q7. Derive the equation for simultaneous heat and mass transfer to single particle.

Q8. What do you understand by Sherwood number, Schmidt's number, Greshof number for mass transfer? What is the relation between them?

# References

1. Mohantry AK (2012) Rate processes in metallurgy. PHI Learning Pvt Ltd, New Delhi
2. Incropera FP, Dewitt DP (2006) Fundamentals of heat and mass transfer, 5th ed. John Wiley & Sons
3. Guthrie RIL (1992) Engineering in process metallurgy. Oxford University Press, USA

# Chapter 5
# Basic Concept of Models

Basic concept and types of models, types of similarity and techniques for ensuring similarity are discussed. Development of semi-rigorous physical model, preliminary ad hoc measurements, classification of mathematical modeling, development of a mathematical model, the role of the mathematical model in process analysis, solution of the equations, and concept of scaling are described. Finite-difference approximations, ordinary and partial differential equations, solutions of ordinary differential equations, and numerical solutions of partial differential equations are also discussed in this chapter.

## 5.1 Basic Concept

Process modeling means a miniature representation of the process. Now the question arises: Why do people go for modeling?

Modeling can be done due to:

(i)    Process improvement,
(ii)   Process optimization,
(iii)  Development of new technology, and
(iv)   Trouble shooting.

The basic aim of modeling is:

(i)    Better understanding,
(ii)   Process optimization, and
(iii)  Process control.

Models are classified into four types [1]:

(i)    Physical models,
(ii)   Mathematical models,

© The Author(s), under exclusive license to Springer Nature Singapore Pte Ltd. 2023     221
S. K. Dutta, *Fundamental of Transport Phenomena and Metallurgical Process Modeling*,
https://doi.org/10.1007/978-981-19-2156-8_5

(iii)   Empirical models, and
(iv)    Computer models.

(I)     Physical models: Physical models are based on the principle of similarity. They
        have helped to understand the processes, e.g. modeling of liquid metal flows
        and their mass transfer processes in converters, ladles, tundishes, and molds.

Physical models for a given situation are constructed by respecting four different
states of similarity, namely [2]:

(a)     Geometrical (i.e. same shape),
(b)     Mechanical or dynamical (i.e. same ratio of forces),
(c)     Thermal (i.e. same ratio of heat transfer rates), and
(d)     Chemical (i.e. same ratio of chemical reaction rates).

        Based on these similarities, a full-scale system is scaled down (i.e. photo-type →
model) or a laboratory-scale model is scaled up (i.e. model → photo-type), as shown
in Fig. 5.1.

(II)    Mathematical models: Mathematical models are based on physico-chemical
        laws. Mathematical models with proper tuning to actual process conditions
        have been very successful in the design, development, and control of processes.
        Basically, a mathematical model consists of a set of algebraic or differential
        equations, which may be used to simulate a process.

To develop a realistic mathematical model based on the fundamentals of transport
phenomena (and kinetics), it is necessary to have a good physical view of the system.
Then the mathematical model may be developed by combining existing known
relationships in order to represent the system [3].
        In many cases, these relationships exist (e.g. the rising velocity of gas bubbles in
melts, radiation heat transfer between surfaces, etc.). One may readily proceed with
the development of an appropriate mathematical model. However, in many other

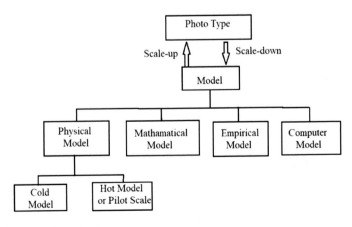

**Fig. 5.1** Classification of models

instances, one just does not have an adequate physical basis for the construction of an approximate mathematical model. Furthermore, the very nature of the real industrial system may preclude direct observations from which such a physical picture may be deduced (e.g. it is difficult to observe the hearth region of the blast furnace, the interior of an argon-stirred ladle, etc.). In other cases, while one could represent the system in terms of appropriate differential equations, the system of equations that would ultimately result is too complex. Under such conditions, another useful tool, i.e. physical model, can be developed. Quite often physical model studies are essential for the development of good mathematical models.

(III)    Empirical models: Empirical models are based on data analysis of the process. The availability of a large and reliable database that contains data of all relevant variables is essential for empirical modeling. Depending on the technique of analysis, these models are classified [1] as (i) regression analysis, (ii) time series, and iii) neural network. Regression analysis is mostly used if a linear relationship is expected between the dependent variable and other variables. When the dependent variable at any time is dependent on its previous value, time series is used. The neural network is used when a nonlinear functional relationship is expected between the dependent variable and other variables. Neural network is the most versatile of the three methods, but it requires a very large database.

(IV)    Computer models: Computer models are the combination of mathematical or empirical models with computer graphics. Hence, these models miniature the actual process on the screen and are extremely useful as a simulator.

## 5.2  Physical Model

The fluid flow phenomena in the iron and steelmaking processes cannot be experimentally investigated because of the very high temperature, size of the reactors, and systems not being transparent [1]. So, these are studied using cold models. Cold model techniques have been widely used to understand and analyze the fluid flow in a blast furnace, converter, tundish, ladle furnace, continuous castings, etc. Cold models are also used for stress analysis.

In constructing a physical model of a process [3], the following requirements are necessary:

- By changing the materials that are easily handled and frequently the scale of the operations.
- The main objective of modeling is to achieve a true representation of the system using materials and equipment such that measurements may be made more conveniently, and in a cost-effective manner.

Typical examples of a physical model include the modeling of an argon-stirred ladle holding molten steel by using the air–water system; the use of water in a glass container to represent the behavior of a ladle-tundish system in continuous casting, etc. In these cases, molten steel or gases are modeled using water or air at room temperature so that the velocity measurements can be easily made.

Physical models may be classified into three groups:

1. Rigorous physical models,
2. Semi-rigorous models, and
3. Preliminary ad hoc measurements.

(1) Rigorous physical models: They are constructed by certain rules or similarity criteria. Under these conditions the quantitative measurements made during model experiments (e.g. velocity fields, temperature profiles, and reaction rates) may be translated directly through appropriate scaling, to describe the behavior of the real systems.

(2) Semi-rigorous models: They are the alternative approach to study the relevant physical phenomena with the objective of providing information for the construction of a mathematical model. The behavior of the real industrial system would then be predicted from the physical model.

(3) Preliminary ad hoc measurements: They are the third approach to conduct ad hoc experiments to acquire a feel for the system as a preliminary to mathematical modeling studies.

## 5.2.1 Development of Rigorous Physical Model

The main requirement for the model to represent the real system is the observation of the quantitative criteria for similarity. First, let us discuss the principles of similarity. There exist many states of similarity. However, here discussion can be confined to (i) geometric similarity, (ii) kinematic similarity, (iii) dynamic similarity, (iv) thermal similarity, and (v) chemical similarity.

### 5.2.1.1  Geometric Similarity

Geometric similarity is the similarity of shape. Systems are geometrically similar when the ratio of any length in one system to the corresponding length in the other system is the same everywhere. This ratio is termed the *scale factor*. While geometric similarity is one of the most obvious requirements in modeling, often it may not be possible to attain perfect geometric similarity.

If the model and the prototype are geometrically similar, then the Froude number (Fr) must be the same for the model and the prototype, i.e.

$$\mathrm{Fr}_{(m)} = \mathrm{Fr}_{(p)}; \ \text{or} \ \frac{u_m^2}{gl_m} = \frac{u_p^2}{gl_p} \tag{5.1}$$

where $u_m$ and $u_p$ are the velocity for model and prototype;

$l_m$ and $l_p$ are the length of the model and prototype, respectively.
g is gravity due to acceleration.

**Example 5.1** Liquid steel is poured into a billet mold (0.15 m × 0.15 m) through a 0.05 m diameter submerged nozzle at a mean linear velocity of 0.5 m/s. Now, a water model for this process is developed, which is twice the size of the prototype. Find out the mean velocity for the model and the nature of the flow. $\rho$ for liquid steel is 7100 kg/m$^3$, $\mu$ for liquid steel is 5 × 10$^{-3}$ kg/m.s and $\rho$ for water is 1000 kg/m$^3$, $\mu$ for liquid steel is 10$^{-3}$ kg/m.s.

**Solution**

Since the liquid steel flow is gravity driven, the only criteria to be considered are as follows:

1. The model and the prototype must be geometrically similar.
2. The Froude number (Fr) must be the same for the model and the prototype.
3. The conditions must be such that the flow is turbulent in the nozzle for both the model and the prototype.

(1) The geometric similarity is readily observed. There are many occasions when the model is larger than the prototype, especially when the prototype itself is not very big and when one wishes to make precise measurements.

(2) The condition (2) can be written from Eq. (5.1):

$$\frac{u_m^2}{g l_m} = \frac{u_p^2}{g l_p}$$

or

$$\frac{u_m^2}{u_p^2} = \frac{g l_m}{g l_p} = \frac{l_m}{l_p}$$

or

$$\frac{u_m}{u_p} = \sqrt{\frac{l_m}{l_p}} = \sqrt{2}$$

Therefore, $u_m = \sqrt{2} u_p = \sqrt{2} \times 0.5 = 0.71$ m/s.

Reynolds number, $\text{Re} = \frac{\rho l u}{\mu}$

$\text{Re}_{(p)} = \frac{\rho l u}{\mu} = \frac{7100 \times 0.05 \times 0.5}{5 \times 10^{-3}} = 3.55 \times 10^4$ (i.e. turbulent flow)

Again, $\text{Re}_{(m)} = \frac{\rho l u}{\mu} = \frac{1000 \times 0.1 \times 0.71}{10^{-3}} = 7.1 \times 10^4$ (i.e. turbulent flow).

### 5.2.1.2  Kinematic Similarity

Kinematic similarity represents the similarity of motion. It is observed between two systems if, in addition to being geometrically similar, the velocities at corresponding locations in the two systems are in the same fixed ratio.

### 5.2.1.3  Dynamic Similarity

Dynamic similarity represents the similarity of forces. It is observed between two systems when the magnitude of forces at corresponding locations in each system is in a fixed ratio.

### 5.2.1.4  Thermal Similarity

Non-isothermal systems must satisfy the criteria for thermal similarity. In other words, the rate of heat transfer by the various mechanisms, such as conduction, convection, and radiation must be of a fixed ratio for the model and the photo-type. Under such conditions, when the systems are geometrically and kinematically similar, then the temperature profiles in the model and in the photo-type are in a fixed ratio throughout.

### 5.2.1.5  Chemical Similarity

Chemical similarity concerns the establishment of the necessary conditions in the model so that the rate of chemical reaction at any location is proportional to the rate of the same reaction at the corresponding time and location in the photo-type. It is, therefore, necessary for all conditions that contribute to the overall rate of the chemical reaction to be simulated in the model.

The concentration of a chemical component at a specific location in the chemical reactor depends on: (a) the initial concentration, (b) the rate at which the particular component is produced or reacted by chemical reaction, (c) the rate at which it diffuses (by molecular or eddy diffusion), and (d) the rate at which it is transported by bulk movement of material to and from the location of the reaction.

In turn, these phenomena depend on the temperature, the concentration gradient, and the flow pattern in the reactor. It can be concluded that chemical similarity imposes inevitably thermal and kinematic similarity, and in addition, it requires a concentration profile within the model that corresponds to the photo-type. In practice, the chemical similarity of a process presents several problems because it requires the proportionality of time, temperature, and concentration between model and photo-type. The rate of chemical reaction is very sensitive to temperature.

In many pyrometallurgical systems involving the interaction of melts (slag-metal kinetics) or gases and melts, the rate of chemical reactions is often not rate-controlling, so the attainment of chemical similarity is not an important consideration; the similarity of the heat and mass transfer rates must be accomplished. For certain gas–solid reactions, such as iron oxide reduction or coke gasification, at least partial control by chemical kinetics cannot be ruled out, and here chemical similarity is important.

In many cases, it is simply impossible to satisfy all the similarity criteria simultaneously. Under such conditions, certain compromises must be made in modeling only one key aspect of the process.

## 5.2.2 Techniques for Ensuring Similarity

(i) Geometric similarity: Geometric similarity of the model and photo-type is the first requirement that must be satisfied. This is readily done by many systems by simply keeping the ratio of any length in a system to a corresponding length in the other system the same. In other words, the same scale factor uses for each linear dimension. If the photo-type is a cylindrical vessel of height h and of diameter d and is filled with melt to a height $h_0$, then the corresponding dimensions for the model are $\beta h$, $\beta d$, and $\beta h_0$, where $\beta$ is the scale factor.

It should be noted that in most cases the quantity $\beta$ cannot be selected arbitrarily. In some cases, one cannot scale every dimension exactly, for instance, using a porous plug to introduce inert gas into molten metal. It may not be possible to scale the size of the pores in the porous plug in the photo-type and in the model.

(ii) Dynamic similarity: Since most of the modeling work involving metallurgical systems concerns fluid (gas/liquid), the need to observe dynamic similarity is perhaps one of the most important criteria that must be fulfilled. Dynamic similarity represents the similarity of forces.

The principal forces (for dynamic similarity) are as follows:

(a) Inertial force,
(b) Pressure force,
(c) Viscous force,
(d) Gravitational force,
(e) Surface tension,
(f) Elastic forces, and
(g) Electromagnetic force.

(a)   Inertial force:

Inertial force α mass × acceleration

$$\alpha \, \rho L^3 \times \left(u^2/L\right)$$

$$\alpha \left(\rho L^2 u^2\right) \tag{5.2}$$

where ρ is the density, L is the length, and u is the velocity.

(b)   Pressure force:

   The average pressure is calculated by dividing the normal force, pushing against a plane area, by area, i.e.

$$\text{Pressure} = \frac{Force}{Area} = \left(\frac{F}{L^2}\right) = \left(\rho u^2\right) \tag{5.3}$$

where F is the force [i.e. mass × acceleration $= (\rho L^2 \, u^2)$].

(c)   Viscous force:

$$\text{Viscous force acting on the fluid} = \text{shear stress} \times \text{area}$$

$$= \mu \left(\frac{du_x}{dy}\right) L^2 \tag{5.4}$$

where μ is viscosity, $u_x$ is fluid velocity in x direction (which is α u), and $\left(\frac{du_x}{dy}\right)$ is the change in velocity per length (y α L).
   Hence,

$$\text{Viscous force } \alpha \, (\mu uL) \tag{5.5}$$

(d)   Gravitational force:

$$\text{Gravitational force } \alpha \left(\rho L^3 g\right) \tag{5.6}$$

(e)   Surface tension:

$$\text{Surface tension force } \alpha \, (\sigma L) \tag{5.7}$$

where σ is surface tension.

(f)    Elastic forces:

The solid has moved from its initial position because of the applied stresses and when imposed stresses are removed, the solid returns to its initial position due to elastic force.

(g)    Electromagnetic force:

$$\text{Electromagnetic force} \propto \left(L^2 B^2 \sigma\right) \tag{5.8}$$

where B is the magnetic flux intensity.

From the combination of the ratios of these and other forces, the dimensionless groups are formed, which are independent of process parameters.

The general comments are as follows:

1.    In modeling a system consisting of one continuous phase (e.g. a gas jet in a gas that results from forced flow), geometric similarity must be observed, and the nozzle Reynolds numbers (for fluid flow) must be the same.
2.    In modeling a system consisting of one continuous phase resulting from gravity-driven flow (e.g. a teemed stream, using a submerged nozzle in continuous casting), then in addition to geometric similarity, the Froude numbers (surface behavior) must be the same.
3.    In thermal natural convection-driven flows, the geometric similarity must be observed; in addition, the Grashof–Prandtl product and the Froude numbers (ratio of inertial to viscous forces) should be the same.
4.    While in principle, it should be possible to model dispersed systems, such as atomization processes, the breakup of gas jets in liquids, or slag-metal emulsions through the simultaneous adjustment of the Froude and the Weber numbers (bubble formation), there are many practical difficulties posed by the very high surface tension of liquid metals. For this reason, most of the successful modeling efforts have addressed essentially one-phase systems.

For example, Fig. 5.2 shows a sketch of an electroslag refining (ESR) process. It is seen that this consists of a consumable electrode immersed in a molten slag phase through which a current is being passed. The passage of the current causes heat generation in the slag phase, which in turn causes the electrode to melt. The molten metal droplets fall through the slag phase and some of the chemical impurities are removed during that stage [3]. The metal droplets accumulate in the form of a molten metal pool at the bottom, which then solidifies as an ingot.

This is clearly a complex process; it can be modeled mathematically. In seeking to use a physical model to test the appropriateness of the mathematical model, we need to restrict some specific aspects of the system.

Molten metal droplets flow in the slag phase is driven by the combination of buoyancy and electromagnetic forces, and restrict the attention to the electromagnetically

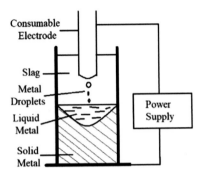

Fig. 5.2  Electroslag refining process

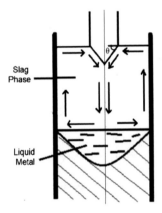

**Fig. 5.3**  Electromagnetically driven flow in slag phase of ESR

driven component of the flow. The electromagnetically driven flow in the slag phase is shown in Fig. 5.3.

To attain geometric and dynamic similarities at the plane corresponding to the free surface of the physical model, it will correspond to a vertical central section of the slag phase in the real system. Geometric similarity is readily attained. In seeking dynamic similarity with respect to the electromagnetically driven flow component, we need to consider the Reynolds number and the magnetic Reynolds number.

In this case, it is not necessary that the values of the magnetic Reynolds numbers be the same, but it is sufficient that the magnetic Reynolds number should be much less than unity for both the model and the prototype. A small value of the magnetic Reynolds number ensures that the rate at which the magnetic field propagates is much faster than the convective processes within the system. Thus, the effect of fluid convection on the magnetic field may be neglected.

(iii)  Thermal similarity: Physical models would not develop for purely heat conduction problems because transient one, two, or even three-dimensional heat conduction is now readily tackled using numerical methods in an essentially routine manner. Physical modeling may be usefully undertaken in cases of thermal buoyancy-driven flows, such as convection in ladles or molds. In addition to establishing geometric similarity in the study of these systems, it is also necessary that the Grashof number (molecular diffusivity/thermal diffusivity) and Prandtl number (momentum diffusivity/thermal diffusivity) be the same.

Two complicating factors should be noted here:

1. The Prandtl number for metals will be significantly less than unity, which means that the thermal boundary layer will be thicker than the velocity boundary layer. Another important consequence of this finding is that for transient systems (i.e. in heating, cooling, or solidifying melts held in containers) thermal equilibrium will be attained much faster than equilibrium with respect to fluid flow; in other words, the fluid flow field will lag behind the temperature field.
2. In most thermal problems of practical importance, one cannot specify the inner wall temperatures of a container, but rather more complex boundary conditions have to be introduced, involving transient conduction through walls of finite thickness followed by an additional thermal interchange with a heater or the environment.

There may be serious practical difficulties in attaining true similarity for these two sets of circumstances. However, useful physical modeling of convection in glass tanks by using glycerol or other viscous fluid can be done.

(iv) Chemical similarity: The attainment of the chemical similarity between the model and a photo-type would be extremely difficult. In studies involving reacting systems where the rates of chemical reaction are chemically controlled, the best way to proceed is to obtain the actual rate expressions by separate experimentation and then incorporate these into a suitably developed mathematical model.

(v) Scaling: One of the important considerations in the construction of a physical model is to decide on a proper scale. Under certain conditions, the requirements of similarity are specified that may be uniquely fixed by these considerations. Under other conditions (e.g. in many flow situations such as in modeling fluid flow in tundishes, argon-stirred ladles, and molds of continuous casting units) the requirements must satisfy geometric similarity and dynamic similarity such that the Froude numbers are equal.

## 5.2.3  Development of Semi-Rigorous Physical Model

The development of a rigorous model is often not possible, because one simply cannot satisfy all the similarity criteria. For modeling the behavior of a gas bubble injection into the molten metal pool, it should satisfy simultaneously the equality of the Froude and Weber numbers in the model and the photo-type, which would not be possible. Similarly, it may not be practically possible to model the atomization of molten metals using room temperature liquids because the high surface tension of the molten metal could not be reproduced.

The difficulties are inherent in satisfying all the similarity criteria, an alternative approach to the development of an experimental analytical study, and an objective to obtain a quantitative relationship between the key process variables, e.g. the rising velocity of large (d > 2 cm) spherical gas bubbles in liquids is given by the relation:

$$u_b = 1.02(gd_B/2)^{1/2} \tag{5.9}$$

where $u_b$ is the rising velocity, g is the acceleration due to gravity, and $d_B$ is the bubble diameter.

It had been seen that this relationship [Eq. (5.9)] applied equally well to molten metals.

### 5.2.4  Preliminary Ad Hoc Measurements

The development of semi-rigorous models or the detailed study of key physical phenomena might require quite an extensive commitment of time and resources. In some cases, it may be helpful to conduct a preliminary investigation by performing some simple ad hoc measurements. Such measurements may be particularly helpful as a preliminary to mathematical model development when the investigator does not have a good physical feel for the system and the governing equations would have to be set up arbitrarily.

In the design of ad hoc experiments, an investigator must start with a clear-cut definition of the objectives. There are no clear guidelines for planning preliminary ad hoc measurements, except for the fact that they are less rigorous than the experimental programs. Careful planning, a clear definition of the objectives, and a good level of awareness of the limitations are essential components.

## 5.3  Mathematical Modeling

Mathematical modeling has evolved over the years into a critical research tool for quantification of various phenomena. Basically, a mathematical model consists of a set of algebraic or differential equations, which may be used to simulate a process. Since the relationships considered may not be exact and the resulting predictions may be close approximations, the term *model* is used as opposed to *law* [4].

In combination with careful experimental measurements, a mathematical model can be a powerful tool in the development of new or existing metallurgical processes.

Depending on the objectives of a study, a mathematical model offers many advantages as follows:

1. It can increase the understanding of a process in fundamental terms.
2. It can be used offline to study the influence of different variables on the operation of a process at a significantly lower cost than on a full-scale experimental investigation.
3. It can assist in the evaluation of results from in-plant trials in a very short time.
4. For a computer program, there is very little difficulty in having a very large or small temperature, vessel dimensions, etc. Thus, a full-scale system with liquid steel as the fluid can be modeled very conveniently.
5. A model may be invaluable and in scale-up and design.
6. A model can be used for the purpose of process control and optimization.

Mathematical models have wide applications. In iron and steelmaking, mathematical modeling has been applied for the following [5]:

(a) Process control,
(b) Optimization,
(c) Process analysis,
(d) Artificial intelligence, and so on.

## 5.3.1 Classification of Mathematical Models

Mathematical models can be classified broadly into four categories as follows:

1. Mechanistic or fundamental models,
2. Population-balance models,
3. Empirical models, and
4. Artificial intelligence (AI)-based models.

### 5.3.1.1 Mechanistic or Fundamental Models

Mechanistic or fundamental models are based on mechanisms involving transport phenomena and chemical reaction rates in a process. Mechanistic models are based on fundamental scientific principles such as conservation of mass, energy, and so on.

Chemical and physical laws are applied to obtain mathematical equations relating to independent and dependent variables. Examples of the laws involved are conservation of mass, energy, and momentum; Fourier's law (for conduction heat transfer), and Fick's law (for molecular diffusion).

These are based on physical or chemical laws such as thermodynamics equilibria, chemical kinetics, heat/fluid flow, mass transfer, deformation processing, etc. Such models can be used in situations where an adequate mathematical representation of the system being modeled is available [4]. A critical factor in the success of a mechanistic model is the selection of appropriate boundary conditions, assumptions, etc.

Typical areas in which these models have been very successfully used include solidification, thermal stresses and crack formation during continuous casting, fluid flow/heat transfer in continuous casting tundishes, flash smelting processes, blast furnace process, furnace design, and rolling processes.

### 5.3.1.2 Population-Balance Models

Population-balance model is also a specific kind of mechanistic model that is based on the principle of conservation. This model has been used extensively in the mineral processing field. It has been developed to describe processes having distributed properties, such as particle size in a ball mill. Typical examples of these models include the determination of particle size distribution in grinding circuits and the study of inclusion removal in steel processing (e.g. in ladle/tundish/degassing units).

### 5.3.1.3 Empirical Models

Empirical models are based on the determination of mathematical equations linking dependent and independent variables using measurements of the variables in an operating plant. The fitting of relationships is usually based on statistical analysis, but a fundamental knowledge of the process may influence the form of the equations. These models are based on actual measurements in a process system. Adequate precautions must be taken before generalizing the results from these models since they are rigorously valid for the conditions prevailing during the measurement of various parameters in the system.

### 5.3.1.4 Artificial Intelligence-Based Models

Artificial intelligence (AI)-based models are input–output type models. In the development of such models, empirical information, as well as mechanistic models, play crucial roles. These models satisfy overall relationships and are used for scheduling, cost analysis, etc. Sometimes, they can be effectively compiled with a mechanistic modeling approach.

## 5.4  Development of Mathematical Model

During the mathematical modeling, researchers tended to begin from the first principle, using elementary control volume. However, the task of present-day modelers has been considerably simplified due to the availability of the following [4]:

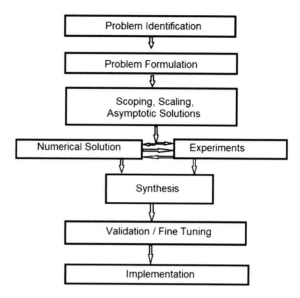

**Fig. 5.4** Flowchart for a mathematical model

a. Powerful and inexpensive hardware,
b. A wide range of software packages,
c. A wealth of literature on the subjects, precedents, analogous problems, etc.

These developments have resulted in the identification of effective strategies for developing mathematical models. A flowchart summarizing an effective approach to developing a mathematical model is presented in Fig. 5.4.

Perhaps the most difficult and critical step is of identifying the problem. What is required? Requirement is a logical analysis of the key parameters which affect the system, followed by a mathematical description of the relationships between the parameters, using appropriate differential equations and boundary conditions. A thorough understanding of the process is a prerequisite, which may call for some preliminary investigations using a scaled-down physical model.

Based on the above analysis, the problem is formulated and followed by scaling and scoping to arrive at nearer solutions. Scaling refers to the estimation of the order of magnitude of key parameters like velocity and time, which will provide some physical insight without computer calculations. This exercise will help to throw some light on the precise answer sought from the model and will be useful for calibrating the computed results. Subsequently, the numerical solutions and experimental program should be planned for execution through a complimentary and interactive approach. Measurements will be required both for some parameters which cannot be calculated, e.g. rate constants, viscosity, thermal conductivity, etc., and for validation of the model predictions. The latter is a critical factor in the successful implementation of mathematical models and must be carried out with extreme care. At this stage, the model can also be fine-tuned to improve the accuracy of prediction.

**Table 5.1** Commonly used mathematical laws/equations and their applications [4, 5][a]

| S. no | Law/Equation | Application |
|-------|--------------|-------------|
| 1 | Navier–Stokes eqs | Fluid flow (BOF, ladle, tundish, etc.) |
| 2 | Fourier's law | Heat conduction (solid as well as liquid state processing) |
| 3 | Convective transport | Heat and mass transfer phenomena in flow system (melting, temperature loss, dissolution, mixing, etc.) |
| 4 | Fick's law | Mass diffusion (solid as well as liquid state processing) |
| 5 | Thermodynamics | Phase equilibria |
| 6 | Kinetics | Prediction of reaction rates |
| 7 | Constitutive relationships | Deformation processing |
| 8 | Maxwell's equations | Electrodynamics (steel processing in induction furnace, EMF stirring in continuous casting) |
| 9 | Turbulence transport | Turbulent flow (liquid steel processing operations such as gas stirring, ladle filling, etc.) |
| 10 | Volume advection equations | Multi-phase flows (gas-slag-metal interactions in ladle, tundish, etc.) |

[a]Reproduced with permission from *IIM Metal News, IIM Kolkata, India*

### 5.4.1  Components of Mathematical Model

Two basic components of any mathematical model are (i) physical/chemical laws that represent the system and (ii) appropriate boundary conditions. Some of the commonly used laws/equations are presented in Table 5.1.

Standard boundary conditions include slag-metal equilibria, different modes of heat or mass transfer, gas-metal equilibria, solid-melt equilibria, and symmetry.

### 5.4.2  Role of Mathematical Model in Process Analysis

As already mentioned, a mathematical model can be a powerful tool in process development. However, it is not the only tool that the process engineer will need to employ in process analysis because a measurement program is normally necessary as well. The types of measurements the process engineer may need to make in the study of a process are:

1. Measurements of concentration, temperature, and fluid flow or mixing in an existing process.
2. Measurements on pilot plants that are designed according to four states of similarity with actual plants geometric, mechanical, thermal, and chemical.
3. Measurements of physical models that are designed to obey only one or two similarity criteria.

These types of measurements are not only complementary to mathematical modeling but are usually vital to determining empirical constants in the model and are needed to check the adequacy of model pre-conditions. Thus, an integrated approach to process analysis involves both mathematical modeling and measurements in plant or in the laboratory on a smaller scale. In many respects, the mathematical model acts as a vital link between the research laboratory and the operating process.

One example of this integrated approach is the mathematical modeling of the direct reduction process, in which iron ore pellets are reduced with coal in a rotary kiln. The model required the measurement in the laboratory of empirical constants to characterize coal reactivity, iron ore reducibility, and the rate of coal devolatilization.

## 5.4.3 Developing of Mathematical Model

Whether a mathematical model is inherently complex or simple, its development should proceed through five stages:

(I)     Preparation,
(II)    Mathematical formulation,
(III)   Solution of the equations,
(IV)    Validation of the model, and
(V)     Application of the model.

### 5.4.3.1 Preparation

The development of a mathematical model should begin with a set of objectives. Is a model of a casting for metal required simply to predict the temperature distribution in an ingot or is the model required to provide a solution to a cracking problem in the ingot? This is a very important state of model development, and in the industrial setting it requires that all parties involved: operating, quality control, research, or industrial engineering, and so on, and are consulted before proceeding. Input from these groups, coupled with a thorough discussion of the problem, is essential to ensure that it is well defined.

During this stage of problem formulation, a clear idea of the complexity of the mathematical model should emerge, and the model should be made as simple as possible while maintaining a stronghold on reality. Does the model have to predict several dependent variables such as temperature, concentration, velocity, and pressure or is only one of these important? Does the model need to be three-dimensional or can it be simplified to one or two dimensions? Does the model need to be formulated to predict unsteady state or steady-state behavior? Are adequate data available to characterize the phenomena of interest, or can they be estimated with accuracy? These questions need to be answered to estimate the time required to complete the modeling work, and the cost of the model in meeting the original objectives.

### 5.4.3.2  Mathematical Formulation

In many respects, a mathematical formulation is the easiest phase of model development. At this point, many decisions on the model should already have been made, such as steady or unsteady state, the coordinate system, number of dimensions, and the need to employ variable properties as follows:

(a)  Assumptions,
(b)  Balance,
(c)  Boundary and initial conditions,
(d)  Fluid flow,
(e)  Heat transfer,
(f)  Mass transfer, and
(g)  Electromagnetic phenomena.

(a)  Assumptions: In further setting out a mathematical description of the process, it is usually necessary to make several assumptions. These may relate to the state of mixedness (back mixed or plug flow) of a fluid, the initial temperature of a solid, or the radiative properties of a surface (black or gray body). Depending on the complexity of the process and the objectives of the modeling study, the number of assumptions made may be very large.

(b)  Balance: A mathematical model must conform to the fundamental laws of conservation, and this is the basis of the mathematical formulation. The quantity to be conserved may be mass, energy, or momentum, depending on the nature of the model. To apply conservation, a part of the system must be isolated, and thus a volume element must be defined. Four types of volume elements are possible, as shown in Fig. 5.5. The choice of an element depends on the number of independent space variables that are required to solve the problem [3].

As shown in Fig. 5.5, the use of an infinitesimal volume will lead to equations expressed in terms of three space variables, an infinitesimal area to two space variables, an infinitesimal section to one space variable, and an overall microvolume will yield equations that contain no space variables. Of course, more information may be obtained from the three-dimensional case, but it is gained at the expense of increased computational cost and sometimes computational difficulties. Therefore, it is preferable to employ the minimum number of dimensions consistent with an adequate mathematical description of the process and to choose the volume element accordingly.

Once the volume element has been defined, a balance is performed on the quantity to be conserved (e.g. heat, mass, or momentum). The balance can be written generally as

(Input through surfaces of volume element) − (Output through surfaces of volume element)
+(Generation inside of volume element) − (Consumption inside volume element)
= (Net accumulation in volume element)

$$(5.10)$$

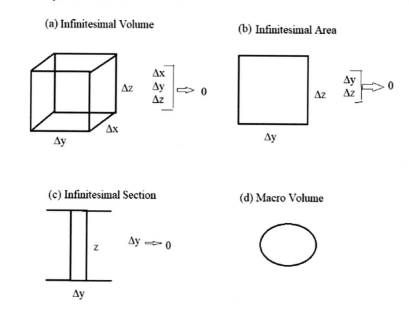

**Fig. 5.5**  Different types of volume elements

The first two terms in Eq. (5.10) include all important transport mechanisms by which the conserved quantity moves into or out of volume element. These mechanisms are bulk motion (convection) and diffusion for heat, mass, and momentum; and in the case of heat transfer, a third mechanism radiation may also be operative. Usually, area of the normal face of the volume element yields a rate.

The second two terms in Eq. (5.10) account for changes to the conserved quantity within the volume element. Thus, for instance, heat may be generated or consumed by a homogeneous chemical reaction occurring within the confines of the volume element. Phase transformations in solid material are another example of a process that generates or consumes heat throughout the volume.

The final term in the balance expresses the buildup or depletion of the conserved quantity within the volume element and gives the change of the dependent variable with time. If this term is zero, the process, by definition, is at a steady state.

The balances yield the governing equations that relate the dependent variables, concentration from mass balances, temperature from heat balances, and velocity and pressure from momentum balances to the independent variable: space and time. The equation may be a partial or an ordinary differential equation, depending on the number of independent variables, or the equation may be a single algebraic equation if the dependent variable does not vary with space and time.

(c)  Boundary and initial conditions: Mathematical formulation of a process is not complete within the specification of initial conditions (for unsteady-state problems) and boundary conditions, subject to which the governing equation is solved. A set of n boundary conditions is required for each nth-order derivative in the differential equation and an initial condition is needed for each time derivative, which usually is first order.

For example, in the case of two-dimensional, unsteady-state heat conduction in a slab in a reheating furnace for which the governing equation is

$$\left[ \frac{\delta}{\delta x}\left\{ k\left(\frac{\delta T}{\delta x}\right)\right\} + \frac{\delta}{\delta y}\left\{ k\left(\frac{\delta T}{\delta y}\right)\right\}\right] = \rho\, C_p\left(\frac{\delta T}{\delta t}\right) \tag{5.11}$$

with four boundary conditions (two for each second order derivative) and one initial condition need to be specified.

There are three types of boundary conditions:

(i)     Specification of the dependent variable at the boundary,
(ii)    Expression of equilibrium at the boundary between two phases, and
(iii)   Continuity of heat, mass, or momentum at a boundary surface.

(i)     Specification of the dependent variable at the boundary: A good example is a no-slip condition at a stationary solid boundary past which a fluid is flowing.

$$\text{At } x = 0, \qquad\qquad u_0 = 0$$
$$\text{(at solid surface) (fluid velocity is zero)} \tag{5.12}$$

Another example is the specification of temperature at the surface of a slab:

$$\text{At } x = 0, \quad T = T_0 \text{ or } T = T_0(t) \tag{5.13}$$

However, in most heating and cooling problems this boundary condition is not likely to be as useful or realistic.

(ii)    Expression of equilibrium at the boundary between two phases: For heat transfer problems in which there is no contact resistance at the boundary between two phases, the temperature of the phases immediately on either side of the boundary must be equal (i.e. there is thermal equilibrium locally at the boundary). Thus, for phases 1 and 2 in contact:

$$\text{At } x = 0, T_{0,1} = T_{0,2} \tag{5.14}$$

Another case, in which local equilibrium may be specified, is that of mass transfer accompanied by an instantaneous chemical reaction at the boundary between two phases, such as gas and metal or metal and slag. This is important to the extraction and refining operations. Under these conditions, chemical equilibrium prevails locally at the boundary, so the interfacial concentrations of reactants and products are related by the equilibrium constant. The general reaction can be written as

$$aA + bB = dD + eE \tag{5.15}$$

where A, B, D, and E are the chemical species; and a, b, d, and e are mol of chemical species.

The equilibrium constant, in terms of molar concentrations at the interface, $c^*$, of the above reaction can be written as

$$k = \frac{(c_D^*)^d (c_E^*)^e}{(c_A^*)^a (c_B^*)^b} \tag{5.16}$$

(iii)   Continuity of heat, mass, or momentum at a boundary surface: In heat flow problems involving conduction, it is common to express as a boundary condition the continuity between conduction internally at the boundary and external heat transfer from and to the boundary. In mathematical terms, this boundary condition is written as follows:

$$\text{At } x = 0, \quad -k \left( \frac{\delta T}{\delta x} \right)_{x=0} = q_0' \tag{5.17}$$

where $q_0'$ maybe a surface heat flux.

The surface boundary condition for a slab losing heat to the environment predominantly by radiation would be as follows:

$$\text{At } x = 0, \quad -k \left( \frac{\delta T}{\delta x} \right)_{x=0} = \sigma \varepsilon \left( T_0^4 - T_a^4 \right) \tag{5.18}$$

where $\sigma$ is the Stefan–Boltzmann constant, $\varepsilon$ is the emissivity of the slab surface, $T_0$ is the temperature (K) of the surface, and $T_a$ is the ambient temperature (K).

(d)   Fluid flow: Laminar fluid flow problems are quite rarely encountered in the metallurgy process (with the exception of entrainment, the flotation of inclusions, lubrication, and the movement of viscous slags). This may be represented by writing down the equation of continuity, which is in essence a mass balance:

$$\nabla \rho u = 0 \tag{5.19}$$

and the equation of motion, often termed the Navier–Stokes equation, is a differential momentum balance:

$$\frac{\delta (\rho u)}{\delta t} = -\nabla \rho u u + \nabla \mu \nabla u + \rho F_b - \nabla p \tag{5.20}$$

where $\frac{\delta(\rho u)}{\delta t}$ is the accumulation of momentum, $\nabla \rho u u$ is the convective transport of momentum, $\nabla \mu \nabla u$ is the diffusive transport of momentum, $\rho F_b$ is the body force, and $\nabla p$ is the force due to pressure.

Again, $\rho$ is the fluid density, u is the velocity vector (which has three components:, $u_x, u_y$ and $u_z$), $\mu$ is the molecular viscosity, $F_b$ is the body force vector (which has three components: $F_x$, $F_y$, and $F_z$) and may be due to gravity and/or electromagnetic forces, and p is the pressure.

The equations of continuity and motion must be solved together because in a physical sense both the conservation of matter and the conservation of momentum must be satisfied.

(e)  Heat transfer: Conduction of heat in a solid may be represented by Fourier's law, which may be written as

$$\rho C_p \left( \frac{\delta T}{\delta t} \right) = \nabla k \nabla T + q^{\cdot} \tag{5.21}$$

where $C_p$ is the specific heat, T is temperature, k is the thermal conductivity, and is $q^{\cdot}$ the rate of heat generation (e.g. due to induction, chemical reactions, or a phase change).

In general, when thermal energy is being transferred in fluids, in addition to conduction, convection, caused by the bulk motion of the fluid, may also play an important role. In a fluid undergoing bulk motion, the conservation of thermal energy may be expressed as

$$\rho C_p \left( \frac{\delta T}{\delta t} \right) = -\rho C_p u. \nabla T + \nabla k_{eff}. \nabla T + q^{\cdot} \tag{5.22}$$

where $\rho C_p \left( \frac{\delta T}{\delta t} \right)$ is accumulation, $\rho C_p u \nabla T$ is convective transport, $\nabla k_{eff}. \nabla T$ is conductive transport may include turbulent transport, and $q^{\cdot}$ is heat generation.

The principal difference between Eqs. (5.21) and (5.22) is the presence of the convective heat flow term ($\rho C_p u. \nabla T$) together with the fact that for a turbulent system the effective thermal conductivity ($k_{eff}$) is no longer a material property, but a function of the fluid flow field as well.

(f)  Mass transfer:
(i)  Diffusion: For low concentration of the diffusing species and in the absence of convection, the conservation of the transferred species is given by Fick's law:

$$\frac{\delta C_A}{\delta t} = D_{A-B} \nabla^2 C_A + R_A \tag{5.23}$$

where $D_{A-B}$ is the binary diffusion coefficient, $C_A$ is the concentration of the diffusing species, and $R_A$ is the rate at which species A is being generated due to a homogeneous chemical reaction.

It should be noted that Eq. (5.23) is applicable only when the diffusing species is present at low concentration levels and/or when there is equimolar counter-diffusion.

(ii)  Convective mass transfer: In the presence of fluid motion, mass transfer takes place both by convection and by diffusion; and under these conditions, the conservation of the diffusing species in a binary system may be written as

$$\frac{\delta C_A}{\delta t} = \mathbf{u} \cdot \nabla C_A + \nabla D_{eff} \cdot \nabla C_A + R_A \tag{5.24}$$

where $\frac{\delta C_A}{\delta t}$ is accumulative, $\mathbf{u} \cdot \nabla C_A$ is convective transport, $\nabla D_{eff} \cdot \nabla C_A$ is diffusion transport including turbulent component, and $R_A$ is generation of A due to chemical reaction.

(g)  Electromagnetic phenomena: Some of the equipment in process metallurgy in the flow of electric current is used as control of the process. Examples of such equipment are induction furnace, submerged arc furnaces, and the electrolytic cells used to produce aluminum or magnesium. The currents are significant in that they may (i) generate electromagnetic forces that drive the circulation of melts, (ii) provide electrical heating, and (iii) bring about electrochemical reactions.

The electromagnetic force appears as a body force in the Navier–Stokes Eq. (5.20). The electromagnetic force vector is due to the interaction of the current and magnetic fields and is given by

$$F_b = \mathbf{J} \cdot \mathbf{B}. \tag{5.25}$$

where J is the current density vector and B is the magnetic induction vector.

In direct current (DC) systems, Eq. (5.25) is used directly since J and B do not vary with time. In alternate current (AC) systems, however, J and B are varying.

### 5.4.3.3  Solution of the Equations

To solve the equations that have been obtained, one of two paths may be followed. An analytical solution may be used to achieve an equation giving values of the dependent variable continuously with space and/or time. Alternatively, i.e. numerical solution in which the differential equation is transformed into algebraic equations, and values of the dependent variable are determined at discrete intervals of space and/or time may be undertaken.

#### 5.4.3.4  Validation of the Model

The fourth stage in the development of a mathematical model, its validation, is the most difficult stage. At the same time, the importance of model validation cannot be overemphasized. It should be apparent by now that mold building is based on approximations and assumptions that necessarily cause the model to be removed somewhat from reality. Therefore, the model must be checked against the real world to determine its adequacy in making predictions that are accurate enough to be useful. In other words, measurements must be made of temperature, composition, velocity, pressure, and so on in the process for comparison to the model prediction.

This is not to say that if model predictions do not agree with measurements, the model is invalid. Particularly under industrial conditions, measurements also can be in error.

#### 5.4.3.5  Application of the Model

The final stage in model development is applying the model to meet the project objectives. Even though the model has been validated, caution should still be exercised in interpreting and using the model predictions.

### 5.4.4  Solution of the Equations

To solve the equations that have been obtained, one of two paths may be followed [3]:

1.  An analytical solution may be sought to achieve an equation giving values of the dependent variable continuously with space and/or time.
2.  A numerical solution in which the differential equation is transformed into algebraic equations.

An analytical solution is preferable, but for most practical problems it is not achievable owing to the complexity of the mathematical equations and associated boundary or initial conditions. For simple problems, an analytical solution may be achievable, and even for more complicated situations, it may be employed to check a numerical solution under special or limiting conditions.

The significant difference between an analytical solution and a numerical one is that the former presents a functional relationship (usually an algebraic equation) between the dependent and independent variables. The numerical solution is precisely the numerical value of the dependent variable at particular values of the independent variables. The analytical solution is thus available continuously, whereas the numerical solution is available only at a finite number of *nodes*, usually arranged in a *mesh*.

### 5.4.4.1 Concept of Scaling

Before proceeding with the analytical or numerical solutions, a useful intermediate step is available through the concept of scaling. When the governing equations are made dimensionless in the proper way, the dimensionless groups that emerge may serve as an excellent semiquantitative indication of the overall system behavior that can be expected. By proper choice of these dimensionless parameters, useful order of magnitude estimates regarding relationships among the key process parameters may be obtained.

### 5.4.4.2 Finite-Difference Approximations

The mathematical basis for numerical methods can be seen by considering the definition of a first derivation:

$$\frac{dy}{dx}\Big|_x = \lim_{\Delta x \to 0}\left[\frac{y_{x+\Delta x} - y_x}{\Delta x}\right] \qquad (5.26)$$

The dependent variable is only available at nodes separated by finite intervals of space or time.

$$\frac{dy}{dx}\Big|_{xi} \cong \left[\frac{y_{i+1} - y_i}{\Delta x}\right] \qquad (5.27)$$

where $x_i$ and $y_i$ are the independent and dependent variables, respectively, at the node designated i, and $\Delta x = x_{i+1} - x_i$. This finite-difference approximation is frequently called the forward difference. Other common finite-difference approximations to derivatives are

$$\frac{dy}{dx}\Big|_{xi} \cong \left[\frac{y_{i+1} - y_{i-1}}{2\Delta x}\right] \quad \text{(central difference)} \qquad (5.28)$$

$$\frac{dy}{dx}\Big|_{xi} \cong \left[\frac{y_i - y_{i-1}}{\Delta x}\right] \quad \text{(backward difference)} \qquad (5.29)$$

Similarly, the approximations to a second derivative can be written as

$$\frac{d^2y}{dx^2}\Big|_{xi} \cong \left[\frac{y_{i+1} - 2y_i + y_{i-1}}{(\Delta x)^2}\right] \qquad (5.30)$$

All these are derivable by simple manipulation of Taylor series expansions of $y_i$ and $x_i$.

### 5.4.4.3   Ordinary Differential Equations

In some process models, the dependent variable is sought only as a function of a single independent variable. For example, a model of a rotary kiln may have been formulated to predict concentration and temperature profiles in the axial direction only at a steady state. Another example is a model of a thin steel strip being cooled by water jets on a run-out table in which the axial temperature profile of the strip is predicted at a steady state. In these cases, one or more first-order ordinary differential equations will be formulated from the balances.

Considering the first-order ordinary differential equation:

$$\frac{dy}{dx} = f(x, y) \tag{5.31}$$

with a boundary condition $y = y_0$ at $x = x_0$. To solve this equation, from the discussion above, the first requirement is to divide the range of the independent variable into discrete intervals of size $\Delta x$. Now, $\Delta x$ will not vary with x. The solution, then, will be to obtain the values of $y(y_1, y_2, y_3, \ldots, y_n)$ at the specified discretely separated values of x ($x_1, x_2, x_3, \ldots, x_n$). This is done by starting at $x_0$ (where $y = y_0$) and marching forward with a step size of $\Delta x$.

This can be accomplished most simply by approximating the derivative with the forward finite-difference approximation as follows:

$$\frac{y_1 - y_0}{\Delta x} = \frac{dy}{dx}\Big|_{x0} = f(x_0, y_0) \tag{5.32}$$

Since $x_0$, $y_0$ and $f(x_0, y_0)$ are known, $y_1$ can be calculated immediately by rearranging the Eq. (5.32) as

$$y_1 = y_0 + \Delta x \cdot f(x_0, y_0) \tag{5.33}$$

Similarly,

$$y_2 = y_1 + \Delta x \cdot f(x_1, y_1) \tag{5.34}$$

and so on

$$y_n = y_{n-1} + \Delta x \cdot f(x_{n-1}, y_{n-1}) \tag{5.35}$$

This is Euler's method, which has the advantage of simplicity but is inaccurate unless $\Delta x$ is kept so small that the pace with which the solution marches forward is very slow.

Equations higher than the first order can be handled by simple substitutions:

$$\frac{d^2y}{dx^2} + \frac{dy}{dx} = f(x, y) \tag{5.36}$$

can be reduced to two simultaneous ordinary differential equations:

$$\frac{dy}{dx} = z \quad \text{or} \quad \frac{dz}{dx} = \frac{d^2y}{dx^2} \tag{5.37}$$

Therefore, from Eq. (5.36):

$$\frac{dz}{dx} = f(x, y) - z \tag{5.38}$$

which can be marched forward simultaneous starting with an initial condition:

$$\frac{dy}{dx} = z = z_0 \text{ and } y = y_0 \text{ at } x = x_0. \tag{5.39}$$

### 5.4.4.4  Partial Differential Equations

The equations contain two or more independent variables, such as time and spatial position. A common example is the cooling of a slab during rolling, in which the unsteady state conduction of heat within the slab describing by a one-dimensional second-order partial differential equation:

$$k\left(\frac{\delta^2 T}{\delta z^2}\right) = \rho C_p\left(\frac{\delta T}{\delta t}\right) \tag{5.40}$$

A model of this equation will be discussed in some detail in Sect. 5.4.3.2. Equation (5.40) is a parabolic type of partial differential equation that can be written more generally as

$$\frac{\delta^2 u}{\delta x^2} = \frac{\delta u}{\delta t} \tag{5.41}$$

Other types of partial differential equation are elliptic:

$$\frac{\delta^2 u}{\delta x^2} + \frac{\delta^2 u}{\delta y^2} = 0 \tag{5.42}$$

and hyperbolic:

$$\frac{\delta^2 u}{\delta x^2} + \frac{\delta^2 u}{\delta y^2} = \frac{\delta^2 u}{\delta t^2} \tag{5.43}$$

The elliptic partial differential equation arises in steady-state problems such as heat flow through the copper mold of a continuous casting machine operating at steady state.

The second-order partial differential equations can be solved by finite-difference methods, which is another numerical technique that has traditionally been used in stress analysis.

## 5.5  Solution of Ordinary Differential Equations

Considering the general first-order differential equation:

$$\frac{dy}{dx} = f(x, y) \tag{5.44}$$

with the initial condition $y = y_0$ at $x = x_0$; therefore,

$$y(x_0) = y_0 \tag{5.45}$$

The methods so developed can be applied to the solution of system of first-order equations, and will yield the solution in one of the two forms [6]:

(1)  A series for y in terms of powers of x, from which the value of y can be obtained by direct substitution. The methods of Taylor and Picard belong to this class.
(2)  A set of tabulated values of x and y are produced. Methods of Euler, Runge–Kutta, etc. belong to this class. These methods are also called step-by-step methods or marching methods.

### 5.5.1  Taylor's Series

Considering first-order differential equation:

$$\frac{dy}{dx} = y' = f(x, y) \tag{5.44}$$

with the initial condition $y = y_0$ at $x = x_0$; therefore,

$$y(x_0) = y_0 \tag{5.45}$$

If $y(x)$ is the exact solution of Eq. (5.45), then the Taylor's series for $y(x)$ around $x = x_0$ is given by

$$y(x) = y_0 + (x - x_0)y'_0 + \frac{(x - x_0)^2}{2!}y''_0 + \cdots \qquad (5.46)$$

If the values of $y'_0, y''_0, \ldots$ are known, then Eq. (5.46) gives a power series for y. Using the formula for total derivatives, it can be written as

$$y'' = f' = f_x + y'f_y = f_x + ff_y \qquad (5.47)$$

where the suffixes denote partial derivatives with respect to the variable concerned. Similarly,

$$y''' = f'' = f_{xx} + f_{xy}\,f + f\left[f_{yx} + f_{yy}\,f\right] + f_y\left[f_x + f_y\,f\right]$$

$$= f_{xx} + 2ff_{xy} + f^2 f_{yy} + f_x f_y + ff_y^2 \qquad (5.48)$$

This method is best understood by Example 5.2.

**Example 5.2** From the Taylor series for $y(x)$, find $y(0.1)$ correct to four decimal places if $y(x)$ satisfies: $y' = x - y^2$ and $y_0 = 1$.

**Solution**

Taylor series for $y(x)$ is given by

$$y(x) = 1 + xy'_0 + \frac{x^2}{2!}y''_0 + \frac{x^3}{3!}y'''_0 + \frac{x^4}{4!}y_0^{'v} + \frac{x^5}{5!}y_0^{v} + \ldots\ldots$$

The derivatives $y'_0, y''_0, \ldots$, etc. are obtained as follows:

$$y'(x) = x - y^2, \qquad \therefore y'_0 = 0 - (y_0)^2 = 0 - 1^2 = -1$$

$$y''(x) = 1 - 2y\,y', \qquad \therefore y''_0 = 1 - 2y_0\,(-1) = 1 + 2 = 3$$

$$y'''(x) = -2y\,y'' - 2(y')^2, \therefore y'''_0 = -2y_0(3) - 2(-1)^2 = -8$$

$$y^{iv}(x) = -2y\,y''' - 2y'\,y'' - 4y'\,y'' = -2y\,y''' - 6y'\,y'', \therefore y_0^{v}$$
$$= -2y_0(-8) - 6(-1)3 = 34$$

$$y^{v}(x) = -2y\,y^{iv} - 8y'y''' - 6(y'')^2, \therefore y_0^{v} = -2y_0\,(34) - 8(-1)(-8) - 6(3)^2 = -186$$

Hence Taylor series becomes

$$y(x) = 1 - x + \frac{3}{2}x^2 - \frac{4}{3}x^3 + \frac{17}{12}x^4 - \frac{31}{20}x^5 + \ldots$$

To obtain the value of y(0.1) correct to four decimal places, and it is found that terms up to $x^4$ should be considered.

Therefore,

$$y(0.1) = 1 - 0.1 + 1.5(0.1)^2 - 1.333\,(0.1)^3 + 1.417\,(0.1)^4$$

$$= 1 - 0.1 + 0.015 - 1.333 \times 10^{-3} + 1.417 \times 10^{-4} = \mathbf{0.9138}$$

### 5.5.2 Picard's Method

$$\frac{dy}{dx} = y' = f(x, y) \tag{5.44}$$

Integrating the differential equation in Eq. (5.44):

$$y = y_0 + \int_{x_0}^{x} f(x, y)dx \tag{5.49}$$

Equation (5.49) in which the unknown function y appears under the integral sign is called an integral equation. Such an equation can be solved by the method of successive approximations in which the first approximation to y is obtained by putting $y_0$ for y on the right side of Eq. (5.49):

$$y^{(1)} = y_0 + \int_{x_0}^{x} f(x, y_0)dx \tag{5.50}$$

The integral on the right can now be solved and the resulting $y^{(1)}$ is substituted for y in the integral of Eq. (5.49) to obtain the second approximation $y^{(2)}$.

$$y^{(2)} = y_0 + \int_{x_0}^{x} f\left(x, y^{(1)}\right)dx \tag{5.51}$$

Doing in this way, getting $y^{(3)}$, $y^{(4)}$, ..., $y^{(n-1)}$ and $y^{(n)}$ where

$$y^{(n)} = y_0 + \int_{x_0}^{x} f\left(x, y^{(n-1)}\right)dx \tag{5.52}$$

with $y^{(0)} = y_0$.

**Example 5.3** Using Picard's method, solve the equation: $y' = x + y^2$, if $y = 1$ when $x = 0$.

**Solution**

Therefore, first approximation is: $y^{(1)} = 1 + \int_0^x (x + 1)dx = 1 + x + \frac{x^2}{2}$.

At $x = 0$ therefore, $y^{(1)} = 1$.

Then, second approximation is:

$$y^{(2)} = 1 + \int_0^x \left( x + \left( 1 + x + \frac{x^2}{2} \right)^2 \right)$$

$$dx = 1 + \int_0^x \left( x + (1 + x)^2 + 2(1 + x)\frac{x^2}{2} + \left( \frac{x^4}{4} \right) \right)dx$$

$$= 1 + \int_0^x \left( 1 + 3x + 2x^2 + x^3 + \left( \frac{x^4}{4} \right) \right)dx = 1 + x + \frac{3x^2}{2} + \frac{2x^3}{3} + \frac{x^4}{4} + \frac{x^5}{20}$$

Now at $x = 0$, $y^{(2)} = 1$.

## 5.5.3 Euler's Method

This method gives the solution in the form of a set of tabulated values.

To solve the equation.

$$y' = f(x, y), \quad y(x_0) = y_0 \tag{5.44}$$

for values of y at $x = x_r = x_0 + rh$ (where $r = 1, 2, 3, \ldots$).

Integrating Eq. (5.44)

$$y_1 = y_0 + \int_{x_0}^{x_1} f(x, y)dx \tag{5.53}$$

Assuming that $f(x, y) = f(x_0, y_0)$ in $x_0 \leq x \leq x_1$, this gives Euler's formula:

$$y_1 = y_0 + hf(x_0, y_0) \tag{5.54}$$

Similarly, for the range $x_1 \leq x \leq x_2$,

$$y_2 = y_1 + \int_{x_0}^{x_1} f(x, y)dx \tag{5.55}$$

Substituting $f(x_1, y_1)$ for $f(x, y)$ in the range $x_1 \leq x \leq x_2$

$$y_2 = y_1 + h f(x_1, y_1) \tag{5.56}$$

Doing in this way, the general formula can be obtained:

$$y_{n+1} = y_n + h f(x_n, y_n) \quad \text{(where } n = 0, 1, 2, \ldots) \tag{5.57}$$

The process is very slow and to obtain reasonable accuracy with Euler's method, take a smaller value for h. Because of this restriction on h, the method is unsuitable for practical use and a modification of it, known as the modified Euler method, gives more accurate results.

**Example 5.4** Solve by Euler's method: $y' = -y$ with the condition $y(0) = 1$.

**Solution**

Equation (5.54) with $h = 0.01$ gives

$$y_1 = y_0 + h f(x_0, y_0) = 1 + 0.01(-1) = 0.99$$

$$y_2 = y_1 + h f(x_1, y_1) = 0.99 + 0.01(-0.99) = 0.9801$$

$$y_4 = y_3 + h f(x_3, y_3) = 0.9703 + 0.01(-0.9703) = 0.9606$$

The exact solution is $y = e^{-x}$ and from this value at $x = 0.04$ is $0.9608$.

## 5.5.4  Runge–Kutta Methods

Euler's method is less efficient in practical problems since it requires h to be small for obtaining reasonable accuracy. The Runge–Kutta methods are designed to give greater accuracy and they possess the advantage of requiring only the function values at some selected points on the subinterval [6].

Instead of approximating $f(x, y)$ by $f(x_0, y_0)$, the integral of Eq. (5.4) by means of the trapezoidal rule obtains

$$y_1 = y_0 + \frac{h}{2}[f(x_0, y_0) + f(x_1, y_1)] \tag{5.58}$$

Thus, by substituting $y_1 = y_0 + hf(x_0, y_0)$ on the right-hand side of Eq. (5.58) we obtain

$$y_1 = y_0 + \frac{h}{2}[f_0 + f(x_0 + h, y_0 + hf_0)] \tag{5.59}$$

where $f_0 = f(x_0, y_0)$.

Now consider $k_1 = hf_0$ and $k_2 = hf((x_0 + h, y_0 + k_1)$.

then Eq. (5.59) becomes

$$y_1 = y_0 + \frac{1}{2}[k_1 + k_2] \tag{5.60}$$

which is the second-order Runge–Kutta formula. The error in this formula can be shown to be of order $h^3$ by expanding both sides by Taylor's series.

## 5.6 Numerical Solution of Partial Differential Equations

The second-order linear partial differential equation is in the form:

$$A\frac{\delta^2 u}{\delta x^2} + B\frac{\delta^2 u}{\delta x \delta y} + C\frac{\delta^2 u}{\delta y^2} + D\frac{\delta u}{\delta x} + E\frac{\delta u}{\delta y} + Fu = G \tag{5.61}$$

which can be written as

$$A\,u_{xx} + B\,u_{xy} + C\,u_{yy} + D\,u_x + E\,u_y + Fu = G \tag{5.62}$$

Laplace's equation:

$$u_{xx} + u_{yy} = 0 \tag{5.63}$$

Laplace's Eq. (5.63) can be solved by the following formulas [6]:

(a)  Standard five-point formula: The value of u at any point is the mean of its values at the four neighboring points. This is called the *standard five-point formula* (as shown in Fig. 5.6a).

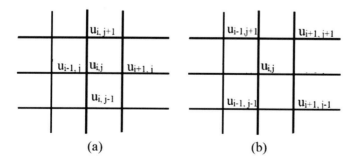

(a)                              (b)

**Fig. 5.6**  **a** Standard five-point formula and **b** diagonal five-point formula

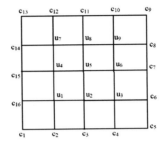

**Fig. 5.7** Interior mesh points

$$u_{i,j} = \frac{1}{4}\left(u_{i+1,j} + u_{i-1,j} + u_{i,j+1} + u_{i,j-1}\right) \tag{5.64}$$

(b)   Diagonal five-point formula: By using the function values at the diagonal points and is called the *diagonal five-point formula* (as shown in Fig. 5.6b).

$$u_{i,j} = \frac{1}{4}\left(u_{i-1,j-1} + u_{i+1,j-1} + u_{i-1,j+1} + u_{i+1,j+1}\right) \tag{5.65}$$

Now let R be a square region so that it can be divided into a network of small squares of side h. Let the values of $u(x, y)$ on the boundary c be given by $c_i$ and let the interior mesh points be (as shown in Fig. 5.7).

By using the diagonal five-point formula and compute $u_5$, $u_7$, $u_9$, $u_1$, and $u_3$. Therefore,

$$u_5 = \frac{1}{4}(c_1 + c_5 + c_9 + c_{13})$$

$$u_7 = \frac{1}{4}(c_{15} + u_5 + c_{11} + c_{13})$$

$$u_9 = \frac{1}{4}(u_5 + c_7 + c_9 + c_{11})$$

$$u_1 = \frac{1}{4}(c_1 + c_3 + u_5 + c_{15})$$

$$u_3 = \frac{1}{4}(c_3 + c_5 + c_7 + u_5) \tag{5.66}$$

Now compute the remaining quantities, i.e. $u_8$, $u_4$, $u_6$, and $u_2$ by the standard five-point formula:

$$u_8 = \frac{1}{4}(u_5 + c_{11} + u_7 + u_9)$$

$$u_4 = \frac{1}{4}(u_1 + u_7 + c_{15} + u_5)$$

$$u_6 = \frac{1}{4}(u_3 + u_9 + u_5 + c_7) \tag{5.67}$$

$$u_2 = \frac{1}{4}(c_3 + u_5 + u_1 + u_3)$$

When once all the $u_i (i = 1, 2, 3, \ldots, 9)$ are computed, their accuracy can be improved by any of the iterative methods.

(i)  Jacobi's method: Let $u_{i,j}^{(n)}$ denote the nth iterative value of $u_{i,j}$. An iterative procedure to solve is for the interior mesh points.

$$u_{i,j}^{(n+1)} = \frac{1}{4}\left[u_{i-1,j}^{(n)} + u_{i+1,j}^{(n)} + u_{i,j-1}^{(n)} + u_{i,j+1}^{(n)}\right] \tag{5.68}$$

This is called the point Jacobi method.

(iii)  Gauss–Seidel method: The method uses the latest iterative values available and scans the mesh points systematically from left to right along successive rows. The iterative formulae are as follows:

$$u_{i,j}^{(n+1)} = \frac{1}{4}\left[u_{i-1,j}^{(n+1)} + u_{i+1,j}^{(n)} + u_{i,j-1}^{(n+1)} + u_{i,j+1}^{(n)}\right] \tag{5.69}$$

**Example 5.5** Solve Laplace's equation for Figure Ex. 5.5.

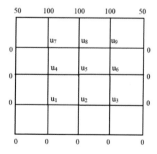

Fig. Ex. 5.5

**Solution**

First compute the quantities $u_5$, $u_7$, $u_9$, $u_1$, and $u_3$ by using the diagonal five-point formula:

$$u_5^{(1)} = \frac{1}{4}(0 + 0 + 50 + 50) == \frac{100}{2} = 25.0$$

$$u_7^{(1)} = \frac{1}{4}(0 + u_5 + 100 + 50) = \frac{150 + u_5}{4} = \frac{175}{4} = 43.75$$

$$u_9^{(1)} = \frac{1}{4}(0 + u_5 + 100 + 50) = \frac{175}{4} = 43.75$$

$$u_1^{(1)} = \frac{1}{4}(0 + 0 + 0 + u_5) = \frac{25}{4} = 6.25$$

$$u_3^{(1)} = \frac{1}{4}(0 + 0 + u_5 + 0) = \frac{25}{4} = 6.25$$

Now compute the remaining quantities, i.e. $u_8$, $u_4$, $u_6$, and $u_2$ by the standard five-point formula $u_8^{(1)} = \frac{1}{4}(u_5 + 100 + u_7 + u_9) = \frac{1}{4}(25 + 100 + 43.75 + 43.75) = 53.125$

$$u_4^{(1)} = \frac{1}{4}(u_1 + u_7 + 0 + u_5) = \frac{1}{4}(6.25 + 43.75 + 0 + 25) = 18.75$$

$$u_6^{(1)} = \frac{1}{4}(u_3 + u_9 + 0 + u_5) = \frac{1}{4}(6.25 + 43.75 + 0 + 25) = 18.75$$

$$u_2^{(1)} = \frac{1}{4}(0 + u_5 + u_1 + u_3) = \frac{1}{4}(25 + 6.25 + 6.25) = 9.375$$

By the above method, the first approximation of all nine mesh points is obtained. Now by using Gruss–Seidel formula for first four iterates, we obtain as follows:
First iterate: $u_1^{(1)} = \frac{1}{4}(0 + u_4 + 0 + u_2) = \frac{1}{4}(18.75 + 9.38) = 7.03$

$$u_2^{(1)} = \frac{1}{4}(0 + u_5 + u_1 + u_3) = \frac{1}{4}(25 + 7.03 + 6.25) = 9.57$$

$$u_3^{(1)} = \frac{1}{4}(0 + u_6 + 0 + u_2) = \frac{1}{4}(18.75 + 9.57) = 7.08$$

$$u_4^{(1)} = \frac{1}{4}(u_1 + u_7 + 0 + u_5) = \frac{1}{4}(7.03 + 43.75 + 25) = 18.95$$

$$u_5^{(1)} = \frac{1}{4}(u_2 + u_8 + u_4 + u_6) = \frac{1}{4}(9.57 + 53.13 + 18.95 + 18.75) = 25.1$$

$$u_6^{(1)} = \frac{1}{4}(u_3 + u_9 + 0 + u_5) = \frac{1}{4}(7.08 + 43.75 + 25.1) = 18.98$$

$$u_7^{(1)} = \frac{1}{4}(u_4 + 100 + 0 + u_8) = \frac{1}{4}(18.95 + 100 + 53.13) = 43.02$$

$$u_8^{(1)} = \frac{1}{4}(u_5 + 100 + u_7 + u_9) = \frac{1}{4}(25.1 + 100 + 43.02 + 43.75) = 52.97$$

$$u_9^{(1)} = \frac{1}{4}(u_6 + 100 + u_8 + 0) = \frac{1}{4}(18.98 + 100 + 52.97) = 42.99$$

Second iterate: $u_1^{(1)} = \frac{1}{4}(0 + u_4 + 0 + u_2) = \frac{1}{4}(18.95 + 9.57) = 7.13$

$$u_2^{(1)} = \frac{1}{4}(0 + u_5 + u_1 + u_3) = \frac{1}{4}(25.1 + 7.13 + 7.08) = 9.83$$

$$u_3^{(1)} = \frac{1}{4}(0 + u_6 + 0 + u_2) = \frac{1}{4}(18.98 + 9.83) = 7.20$$

$$u_4^{(1)} = \frac{1}{4}(u_1 + u_7 + 0 + u_5) = \frac{1}{4}(7.13 + 43.02 + 25.1) = 18.81$$

$$u_5^{(1)} = \frac{1}{4}(u_2 + u_8 + u_4 + u_6) = \frac{1}{4}(9.83 + 52.97 + 18.81 + 18.98) = 25.15$$

$$u_6^{(1)} = \frac{1}{4}(u_3 + u_9 + 0 + u_5) = \frac{1}{4}(7.20 + 42.99 + 25.15) = 18.84$$

$$u_7^{(1)} = \frac{1}{4}(u_4 + 100 + 0 + u_8) = \frac{1}{4}(18.81 + 1100 + 52.97) = 42.95$$

$$u_8^{(1)} = \frac{1}{4}(u_5 + 100 + u_7 + u_9) = \frac{1}{4}(25.15 + 100 + 42.95 + 42.99) = 52.77$$

$$u_9^{(1)} = \frac{1}{4}(u_6 + 100 + u_8 + 0) = \frac{1}{4}(18.84 + 100 + 52.77) = 42.90$$

Third iterate: $u_1^{(1)} = \frac{1}{4}(0 + u_4 + 0 + u_2) = \frac{1}{4}(18.81 + 9.83) = 7.16$

$$u_2^{(1)} = \frac{1}{4}(0 + u_5 + u_1 + u_3) = \frac{1}{4}(25.15 + 7.16 + 7.2) = 9.88$$

$$u_3^{(1)} = \frac{1}{4}(0 + u_6 + 0 + u_2) = \frac{1}{4}(18.84 + 9.88) = 7.18$$

$$u_4^{(1)} = \frac{1}{4}(u_1 + u_7 + 0 + u_5) = \frac{1}{4}(7.16 + 42.95 + 25.15) = 18.82$$

$$u_5^{(1)} = \frac{1}{4}(u_2 + u_8 + u_4 + u_6) = \frac{1}{4}(9.88 + 52.77 + 18.82 + 18.84) = 25.08$$

$$u_6^{(1)} = \frac{1}{4}(u_3 + u_9 + 0 + u_5) = \frac{1}{4}(7.18 + 42.9 + 25.08) = 18.79$$

$$u_7^{(1)} = \frac{1}{4}(u_4 + 100 + 0 + u_8) = \frac{1}{4}(18.82 + 100 + 52.77) = 42.90$$

$$u_8^{(1)} = \frac{1}{4}(u_5 + 100 + u_7 + u_9) = \frac{1}{4}(25.08 + 100 + 42.9 + 42.9) = 52.72$$

$$u_9^{(1)} = \frac{1}{4}(u_6 + 100 + u_8 + 0) = \frac{1}{4}(18.79 + 100 + 52.72) = 42.88$$

Fourth iterate: $u_1^{(1)} = \frac{1}{4}(0 + u_4 + 0 + u_2) = \frac{1}{4}(18.82 + 9.88) = 7.18$

$$u_2^{(1)} = \frac{1}{4}(0 + u_5 + u_1 + u_3) = \frac{1}{4}(25.08 + 7.18 + 7.18) = 9.86$$

$$u_3^{(1)} = \frac{1}{4}(0 + u_6 + 0 + u_2) = \frac{1}{4}(18.79 + 9.86) = 7.16$$

$$u_4^{(1)} = \frac{1}{4}(u_1 + u_7 + 0 + u_5) = \frac{1}{4}(7.18 + 42.9 + 25.08) = 18.79$$

$$u_5^{(1)} = \frac{1}{4}(u_2 + u_8 + u_4 + u_6) = \frac{1}{4}(9.86 + 52.72 + 18.79 + 18.79) = 25.04$$

$$u_6^{(1)} = \frac{1}{4}(u_3 + u_9 + 0 + u_5) = \frac{1}{4}(7.16 + 42.88 + 25.04) = 18.77$$

$$u_7^{(1)} = \frac{1}{4}(u_4 + 100 + 0 + u_8) = \frac{1}{4}(18.79 + 100 + 52.72) = 42.88$$

$$u_8^{(1)} = \frac{1}{4}(u_5 + 100 + u_7 + u_9) = \frac{1}{4}(25.04 + 100 + 42.88 + 42.88) = 52.7$$

$$u_9^{(1)} = \frac{1}{4}(u_6 + 100 + u_8 + 0) = \frac{1}{4}(18.77 + 100 + 52.7) = 42.87$$

## 5.7  Exercises

**Problem 5.1** Given $\frac{dy}{dx} = 1 + xy$ and $y(0) = 1$ obtain the Taylor series for $y(x)$, compute $y(0.15)$, and correct to four decimal places.
[Ans: 1.1623].

**Problem 5.2** Using Picard's method, solve the equation: $\frac{dy}{dx} = x(1 + x^3 y)$, $y(0) = 3$.

Find out the values of $y(0.25)$ and $y(0.35)$.

**Problem 5.3** Solve Laplace's equation for the following figure (use up to third iterate).

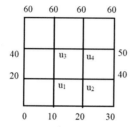

**Problem 5.4** The function u satisfies Laplace's equation at all points within the squares given in the following figure and has the boundary values indicated. Compute a solution and correct up to two decimals by any of the methods.

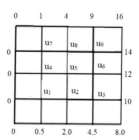

## 5.8  Questions

Q1. What do you understand by modeling? What is its aim? What are the types of modeling?

Q2. What are the advantages of mathematical models?

Q3. What is the classification of mathematical models? Discuss.

Q4. What are the stages for developing a mathematical model? Discuss each stage briefly.

Q5. What is the role of a mathematical model in process analysis?

Q6. What do you understand by physical model? What are the types of physical models?

Q7. What do you mean by the principle of similarity? What are the types of similarities?

Q8. What do you understand by geometric similarity?

Q9. Discuss about "diagonal five points" formula and "standard five points" formula for Laplace's equation.

# References

1. Lahari AK (1999) Workshop on "Process modeling application for steel industry", IISc, Bangalore
2. Mazumdar D (2004) IIM Metal News, p 16
3. Szekely J, Evans JW, Brimacombe JK (eds) (1988) The Mathematical and physical modeling of primary metals processing operations. Wiley, New York
4. Govindarajan S, Chatterjee A (1993) IIM Metal News 15(3), p 1
5. Mazumdar D (2004) IIM Metal News 7(2), p 5
6. Sastry SS (1995) Methods of numerical analysis. 2nd Ed, Prentice Hall of India Pvt Ltd

# Chapter 6
# Applications of Models

Applications of process modeling, such as, gas–solid reaction, gasification of carbon are discussed. Production of sponge iron by rotary kiln, oxygen jet momentum in BOF, hydrostatic pressure at the bottom of ladle, growth and detachment of bubbles nucleating in liquid bath are described. Continuous casting and production of duplex stainless steel by CONARC (material and heat balances) are also discussed.

## 6.1 Gas–Solid Reaction

Most of the fundamental kinetic information came from studies of reduction of single spherical pellet. The rates were mostly followed by noting the weight loss of the iron oxide at intervals of time in thermo-gravimetric setup. Spherical iron oxide pellet is surrounded by the reducing gas, hydrogen, or carbon monoxide (as shown in Fig. 6.1a).

Edstrom [1] first put forward a simple workable postulate of reaction mechanism applicable to dense samples of iron oxide in 1953. According to that, if a partially reduced hematite sample is examined, it would be found to consist of layers. The outermost layer would consist of metallic iron followed by wustite, magnetite, and a core of hematite. This was known as *topochemical pattern of reduction*. The outermost iron layer would be porous.

Mckewan [2] carried out experiment, first time in 1965, for the rate measurement extensively under controlled conditions. Mckewan assumed the following points to develop a mathematical model:

1. Topochemical pattern of reduction,
2. Negligible thickness of iron and wustite layers,
3. Reaction is chemical control, and
4. Reaction is first order reversible.

S. K. Dutta, *Fundamental of Transport Phenomena and Metallurgical Process Modeling*,
https://doi.org/10.1007/978-981-19-2156-8_6

**Fig. 6.1** Reduction of oxide
by reducing gas

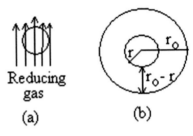

Considering the shape of metal oxide is spherical and that is surrounded by reducing hydrogen or carbon monoxide gas. Then reaction takes place on the surface of the oxide pellet with the hydrogen or carbon monoxide gas (Fig. 6.1b).

$$MO(s) + \{H_2\}/\{CO\} = M(s) + \{H_2O\}/\{CO_2\} \tag{6.1}$$

If the above reaction (Eq. (6.1)) is controlling the kinetic, then it is called *chemical reaction control*. Iron oxide reduction has been shown to be a *surface-controlled reaction*. To measure the reaction rate, the surface area of the sample must be known; but the way, the surface area changes with the reduction proceeds that must be known. If the sample is dense and of regular shape, the reaction surface area changes as a function of weight loss can be determined geometrically. Since weight loss easily measured, the reaction rate can also be easily measured for any regular shape particle such as a sphere, slab, or cylinder.

Let the rate of reaction is controlled by the reaction interface, i.e. rate of a reaction depends on interfacial area; a rate equation can be derived that will fit any shape of particle. Consider the initial radius of the sphere pellet is $r_o$, oxygen density is $d_o$, and total weight of the pellet is $W_o$.

Assume that the rate of formation of a uniform reaction product layer is proportional to the surface area (A) of the remaining oxide at time, t. If W is the weight of the material reacted, then

$$\left(\frac{dW}{dt}\right) \alpha\, A \quad \text{or} \quad \left(\frac{dW}{dt}\right) = k_M\, A \tag{6.2}$$

where $k_M$ is the proportionality constant or rate constant ($ML^{-2}t^{-1}$), which is a function of temperature, pressure, and gas composition.

If r is radius of the unreacted oxide core, then the fractional thickness of the reaction product layer (f) is defined as

$$\left(\frac{r_o - r}{r_o}\right) \tag{6.3}$$

or $f r_o = r_o - r$ or $r = r_o(1 - f)$

So surface area of remaining oxide (i.e. unreacted oxide core),

$$A = 4\pi r^2 = 4\pi \{r_o(1-f)\}^2 \tag{6.4}$$

Since material reacted, i.e. weight of oxygen removed from sample

$$= \text{Initial weight} - \text{Final weight}$$

So,

$$W = \left(\frac{4}{3}\right)\pi r_o^3 d_o - \left(\frac{4}{3}\right)\pi r^3 d_o$$

$$= \left(\frac{4}{3}\right)\pi r_o^3 d_o - \left(\frac{4}{3}\right)\pi \{r_o(1-f)\}^3 d_o \tag{6.5}$$

Since $r = r_o(1-f)$,

therefore, differentiating Eq. (6.5) with respect to time (t):

$$\left(\frac{dW}{dt}\right) = 4\pi r_o^3 d_o (1-f)^2 \left(\frac{df}{dt}\right) \tag{6.6}$$

since $r_o$ and $d_o$ are constant, they are not varying with time.

Substituting values of $\left(\frac{dW}{dt}\right)$ and A [from Eqs. (6.6) and (6.4)] in Eq. (6.2):

$$4\pi r_o^3 d_o (1-f)^2 \left(\frac{df}{dt}\right) = k_M 4\pi \{r_o(1-f)\}^2$$

Therefore,

$$r_o d_o \left(\frac{df}{dt}\right) = k_M \tag{6.7}$$

By integrating of Eq. (6.7):

$$r_o d_o \int df = k_M \int dt \quad \text{or} \quad \mathbf{r_o d_o f = k_M t} \tag{6.8}$$

Equation (6.8) means that the reaction product layer grows linearly with time (as shown in Fig. 6.2a), i.e. this reduction reaction is *interfacial reaction control*. To use Eq. (6.8), it is necessary to correlate weight change data to the thickness of the reaction product layer.

The *fractional reduction* (R) is defined as the oxygen removed from the oxide, divided by the total removable oxygen originally present in oxide sample.

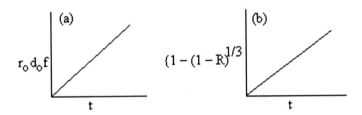

**Fig. 6.2** Kinetic plots

$$R = \left(\frac{W}{W_o}\right) = \left(\frac{\frac{4}{3}\pi r_o^3 d_o - \frac{4}{3}\pi r^3 d_o}{\frac{4}{3}\pi r_o^3 d_o}\right) = \left\{1 - (1 - f)^3\right\} \quad (6.9)$$

Therefore,

$$f = \left\{1 - (1 - R)^{1/3}\right\} \quad (6.10)$$

Again, from Eq. (6.8):

$$r_o d_o f = k_M t \quad \text{or} \quad r_o d_o \left\{1 - (1 - R)^{1/3}\right\} = k_M t$$

or

$$\left\{1 - (1 - R)^{1/3}\right\} = \left(\frac{k_M}{r_o d_o}\right) t \quad (6.11)$$

Equation (6.11) is known as *Mckewan Equation* or *Mckewan Model*.

From the experimental data, $\{1 - (1 - R)^{1/3}\}$ vs t plot (Fig. 6.2b) should be straight line, that means reduction reaction (Eq. (6.1)) obeys Mckewan's equation. Mckewan concluded that reduction of oxide was controlled by the interfacial chemical reaction.

**Example 6.1** Initial weight of pure iron oxide ($Fe_2O_3$) fines is 296.5 mg. Oxide is reduced by hydrogen gas at 1073 K. Find out: (i) fraction of reduction with respect to time; and (ii) whether these data fit to the Mckewan equation?

The following data are obtained:

| Time (s) | 36 | 60 | 120 | 180 | 240 | 300 | 420 | 600 |
|----------|------|-------|-------|-------|-------|-------|-------|-------|
| Wt. loss (mg) | 6.69 | 14.05 | 31.95 | 48.14 | 55.14 | 59.34 | 64.94 | 70.24 |

**Solution**

Total removable oxygen present in iron oxide $= 296.5 \times 0.3 = 88.95$ mg.

| Time (s) | Weight loss (mg) | Fraction of reduction (R) | $(1-R)^{1/3}$ | $[1-(1-R)^{1/3}]$ |
|---|---|---|---|---|
| 36 | 6.69 | 0.075 | 0.974 | 0.026 |
| 60 | 14.05 | 0.158 | 0.944 | 0.056 |
| 120 | 31.95 | 0.359 | 0.862 | 0.138 |
| 180 | 48.14 | 0.541 | 0.771 | 0.229 |
| 240 | 55.14 | 0.620 | 0.724 | 0.276 |
| 300 | 59.34 | 0.667 | 0.693 | 0.307 |
| 420 | 64.94 | 0.730 | 0.646 | 0.354 |
| 600 | 70.24 | 0.790 | 0.594 | 0.406 |

From Eq. (6.11), by plotting $[1-(1-R)^{1/3}]$ vs time, line should be straight for Mckewan equation, below figure shows that line is not straight; hence, the given data for reduction of pure iron oxide ($Fe_2O_3$) fine does not obey the Mckewan equation.

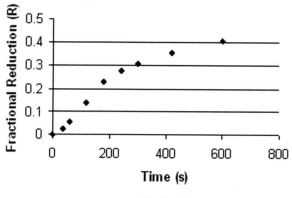

Fig. Ex. 6.1

## 6.2 Gasification of Carbon

Reduction of iron ore by solid carbon takes place (via the gaseous phase) in two stages as follows:

(i) Gasification reaction or Boudouard's reaction:

$$C(s) + \{CO_2\} = 2\{CO\} \qquad (6.12)$$

(ii)    Reduction of iron oxide:

$$Fe_xO_y(s) + \{CO\} = 2Fe_xO_{y-1}(s) + \{CO_2\} \qquad (6.13)$$

where $Fe_xO_y$ denotes $Fe_2O_3$, $Fe_3O_4$, $Fe_xO$; and $Fe_xO_{y-1}$ denotes $Fe_3O_4$, $Fe_xO$, Fe.
    Mechanism of gasification reaction:
    A two-step mechanism proposed by Reif [3] in 1952 (it is also known as Reif's mechanism):
    First step: Reversible oxygen exchange between gas phase and carbon surface:

$$C^1(s) + \{CO_2\} \overset{k_f^1}{\underset{k_b^1}{\leftrightarrow \rightleftharpoons}} C^{1^o} + \{CO\} \qquad (6.14)$$

Second step: carbon gasification stage:

$$C^{1^o} \overset{k_f^2}{\rightarrow} \{CO\} \qquad (6.15)$$

where $C^1$ represents a free carbon site on the surface and $C^{1^o}$ denotes a site where oxygen atom is chemisorbed. $k_f$, $k_b$ are forward and backward reaction rate constants, respectively.
    Step-1 is reversible, so it is fast. Step-2 proceeds relatively slow, so it is rate-controlling step.
    Based on Reif's mechanism, Ergun [4] in 1956 derived the rate expression for gasification of carbon by $CO_2$ gas, which is known as Ergun's rate equation.

$$\text{Overall rate (r)} = \text{rate of step-2} \simeq k_f^2 C_O \qquad (6.16)$$

where $C_O$ is the concentration of $C^{1^o}$ and r is rate of reaction per unit surface area.
    The value of $C_O$ can be determined by applying steady-state approximation as follows:
    At steady state, i.e. concentration of intermediate, e.g. CO and $C^{1^o}$ are not changed with time, i.e. $C_O$ is constant.
    Therefore, rate of reaction (r) = rate of step-1 = rate of step-2.
    Hence,

$$r = k_f^2 C_O = k_f^1 p_{CO_2} \cdot C_f - k_b^1 p_{CO} C_O \qquad (6.17)$$

where $C_f$ is the concentration of free carbon site on the surface.
    Now,

$$C_O + C_f = C_T = \text{constant}$$

or

$$C_f = C_T - C_O \tag{6.18}$$

where $C_T$ is the total number of carbon sites on the surface.
Combining Eqs. (6.17) and (6.18):

$$k_f^2 C_O = k_f^1 p_{CO_2} \cdot (C_T - C_O) - k_b^1 p_{CO} C_O$$

or,

$$\left(k_f^2 + k_f^1 p_{CO_2} + k_b^1 p_{CO}\right) C_O = k_f^1 p_{CO_2} \cdot C_T$$

Therefore,

$$C_O = \left( \frac{k_f^1 C_T p_{CO_2}}{\left(k_f^2 + k_f^1 p_{CO_2} + k_b^1 p_{CO}\right)} \right) \tag{6.19}$$

Again, combining Eqs. (6.16) and (6.19):

$$r = k_f^2 C_O = k_f^2 \left( \frac{k_f^1 C_T p_{CO_2}}{\left(k_f^2 + k_f^1 p_{CO_2} + k_b^1 p_{CO}\right)} \right) \tag{6.20}$$

or,

$$r = \frac{k_f^1 \cdot C_T \cdot p_{CO_2}}{\left[1 + \left(\frac{k_f^1}{k_f^2}\right) p_{CO_2} + \left(\frac{k_b^1}{k_f^2}\right) p_{CO}\right]} = \frac{k_1 \cdot p_{CO_2}}{1 + k_2 \cdot p_{CO_2} + k_3 \cdot p_{CO}} \tag{6.21}$$

where

$$k_1 = k_f^1 \cdot C_T = \text{constant}$$
$$k_2 = \left(\frac{k_f^1}{k_f^2}\right) = \text{constant}$$
$$k_3 = \left(\frac{k_b^1}{k_f^2}\right) = \text{constant}$$

Now $k_2$ and $k_3$ do not depend on nature of carbon, but $k_1$ depends on nature of carbon.

The reaction (6.12) has been found to be controlled by the interfacial chemical reaction step below 1100 °C if the size is not too large. Such a conclusion has been drawing from evidences such as (i) high activation energy (300–350 kJ/mol) and (ii) strong influence of even trace amounts of solid and gaseous impurities on reaction rate.

The gasification reaction (6.12) is also of significant interest in carbothermic reduction. The term reactivity is commonly used to denote the speed of the gasification reaction. The higher the reactivity, the faster is the gasification. The reactivity ($r_C$) can be defined as

$$r_C = \left( \frac{dF_C}{dt} \right) \tag{6.22}$$

where $F_C$ is the degree of gasification of carbon and t is the time of reaction.

The kinetic steps involved in the gasification of a carbon are as follows:

- Transfer of $CO_2$ across the gas boundary layer to the surface of the particle,
- Inward diffusion of $CO_2$ through the pores,
- Chemical reaction on the pore surface,
- Outward diffusion of CO through the pores, and
- Transfer of CO into the bulk gas by mass transfer across the boundary layer.

Laboratory experiments have demonstrated that the chemical reaction is much slower compared with the other steps; hence, it constitutes the principal rate-controlling step. Such a conclusion has been arrived at based on the large activation energy involved, strong retarding influence of CO on the rate.

## 6.3  Production of Sponge Iron by Rotary Kiln

The product output of any process depends on the amount of materials consumed as raw materials and the chemical reactions taking place. Steps involved in material balance calculation are as follows [5]:

1.  Represent the process by a rectangle and indicate each feed and product materials by arrows,
2.  Putting the data on the flow diagram,
3.  Considering suitable assumption if necessary,
4.  Considering a convenient basis for calculation,
5.  Writing the overall chemical reaction that take place in the process,
6.  Writing the element balance equations,
7.  Solving the equations, and
8.  Making a material balance table.

It is desirable to first obtain a block diagram of rotary kiln process (Fig. 6.3) for

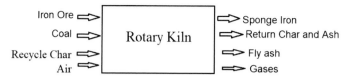

**Fig. 6.3** Block diagram of rotary kiln

material balance.

Material balance is essentially an application of law of conservation of mass, i.e. *the mass of an isolated system remains constant irrespective of the charges occurring within the system.* Rotary kiln process is operated continuously, so continuous streams of materials enter the unit's feed end and continuous streams of materials leave the same on the other discharged end. Hence, in such a process, the rate of input is equal to the rate of output of materials.

As steady state, the chemical compositions of the input materials and output materials remain unchanged and there is no accumulation of materials within the system.

Hence, it can be written as

$$[\text{Rate of input materials into the system}] = [\text{Rate of output materials from the system}] \tag{6.23}$$

Basis of calculation: Rate of production in tonne/hour (t/h).

**Fe Balance:**

$$\text{Fe input} = \text{Fe output}$$

$$(\text{Fe from iron ore}) = [(\text{Total Fe present in sponge iron}) + (\text{Fe present in fly ash})] \tag{6.24}$$

$$\rightarrow \quad (W_{Ore} \times f_1 \times f_2) = \left[(W_{Sp} \times f_3) + (W_{Fly} \times f_4 \times f_2)\right] \tag{6.25}$$

where

$W_{Ore}$ is the amount of iron ore require, t/h,
$f_1$ is the fraction of $Fe_2O_3$ present in iron ore,
$f_2$ is the fraction of Fe present in $Fe_2O_3$,
$W_{Sp}$ is the amount of sponge iron produced, t/h,
$f_3$ is the fraction of total Fe present in sponge iron,
$W_{Fly}$ is the amount of fly ash discharged, t/h,
$f_4$ is the fraction of of $Fe_2O_3$ present in fly ash.

**C Balance:**

$$C \text{ input} = C \text{ output}$$

$$[(C \text{ from Coal}) + (C \text{ from recycle char})]$$
$$= [(C \text{ present in sponge iron}) + (C \text{ in discharge char}) + (C \text{ in fly ash})$$
$$+ (C \text{ in coal ash}) + (C \text{ in gases})]$$

$$[(W_{Coal} \times f_5) + (W_{RC} \times f_6)] = [(W_{Sp} \times f_7) + (W_{Char} \times f_6)$$
$$+ (W_{Fly} \times f_8) + (W_{Ash} \times f_9) + W_C] \quad (6.26)$$

where

$W_{Coal}$ is the amount of coal used, t/h,
$f_5$ is the fraction of C present in coal,
$W_{RC}$ is the amount of recycle char used, t/h,
$f_6$ is the fraction of C present in char,
$f_7$ is the fraction of C present in sponge iron,
$W_{Char}$ is the amount of char discharge, t/h,
$W_{Fly}$ is the amount of fly ash discharge, t/h,
$f_8$ is the fraction of C present in fly ash,
$W_{Ash}$ is the amount of coal ash discharge, t/h,
$f_9$ is the fraction of C present in coal ash,
$W_C$ is the amount of C goes into gases, t/h.

**H Balance**:

$$H \text{ input} = H \text{ output}$$

$$[(H \text{ from coal}) + (H \text{ from recycle char})]$$
$$= [(H \text{ in discharge char}) + (H \text{ in form of } H_2O \text{ in gases})]$$

$$[(W_{Coal} \times f_{10}) + (W_{RC} \times f_{11})] = (W_{Char} \times f_{11}) + W_H] \quad (6.27)$$

where

$f_{10}$ is the fraction of H present in coal,
$f_{11}$ is the fraction of H present in char,
$W_H$ is the amount of H goes into gases in form of $H_2O$, t/h.

**O Balance**:

$$O \text{ input} = O \text{ output}$$

$$[(O \text{ from iron ore}) + (O \text{ from coal}) + (O \text{ from recycle char}) + (O \text{ from air})]$$

$$= \left[(\text{O present in sponge iron}) + (\text{O in discharge char}) + (\text{O in fly ash}) + (\text{O in gases})\right]$$

$$[(W_{Ore} \times f_1 \times f_{12}) + (W_{Coal} \times f_{13}) + (W_{RC} \times f_{14}) + (W_{Air} \times f_{15})$$
$$= [(W_{Sp} \times f_{16} \times f_{17}) + (W_{Char} \times f_{14}) + (W_{Fly} \times f_4 \times f_{12}) + W_O] \quad (6.28)$$

where

$f_{12}$ is the fraction of O present in $Fe_2O_3$,
$f_{13}$ is the fraction of O present in coal,
$f_{14}$ is the fraction of O present in recycle char,
$W_{Air}$ is the amount of air require, t/h,
$f_{15}$ is the fraction of O present in air,
$f_{16}$ is the fraction of FeO present in sponge iron,
$f_{17}$ is the fraction of O present in FeO,
$W_O$ is the amount of O goes into gases, t/h.

Therefore,

$$\text{amount of flue gases } (W_{FG}) = W_{CO2} + W_{CO} + W_H + (W_{Air} \times f_{18}) \quad (6.29)$$

where

$W_{CO_2}$ is the amount of $CO_2$ present in flue gases, t/h,
$W_{CO}$ is the amount of CO present in flue gases, t/h,
$f_{18}$ is the fraction of N present in air.

Again, amount of oxygen goes into gases:

$$W_O = \left[\left\{\left(\frac{2M_O}{M_{CO2}}\right) \times W_{CO2}\right\} + \left\{\left(\frac{M_O}{M_{CO}}\right) \times W_{CO}\right\} + \left\{\left(\frac{M_O}{M_{H2O}}\right) \times W_H\right\}\right]$$
$$(6.30)$$

and the amount of carbon goes into gases:

$$W_C = \left[\left\{\left(\frac{M_C}{M_{CO2}}\right) \times W_{CO2}\right\} + \left\{\left(\frac{M_C}{M_{CO}}\right) \times W_{CO}\right\}\right] \quad (6.31)$$

**Example 6.2** In a rotary kiln for DRI production, how much ore (t/h) is charged for 4 t/h production? Also calculate and draw the consumption of ore varies with variation of purity of ore.

Given: $Fe_2O_3$ in ore is 93%, Fe in fly ash is 4%, amount of fly ash is 0.25 t/h, and total Fe contain in DRI is 90%.

**Solution**

**Fe Balance**:

$$\text{Fe input} = \text{Fe output}$$

$$(\text{Fe from iron ore}) = \left[(\text{Total Fe present in sponge iron}) + (\text{Fe present in fly ash})\right]$$

$$\rightarrow (W_{Ore} \times f_1 \times f_2) = \left[(W_{Sp} \times f_3) + (W_{Fly} \times f_4)\right]$$

where

$W_{Ore}$ is the amount of iron ore require, t/h,
$f_1$ is the fraction of $Fe_2O_3$ present in iron ore, (0.93).
$f_2$ is the fraction of Fe present in $Fe_2O_3$, (0.70).
$W_{Sp}$ is the amount of sponge iron produced, t/h, (4 t/h).
$f_3$ is the fraction of total Fe present in sponge iron, (0.9).
$W_{Fly}$ is the amount of fly ash discharged, t/h, (0.25 t/h).
$f_4$ is the fraction of of Fe present in fly ash, (0.04).

$$\rightarrow (W_{Ore} \times 0.93 \times 0.7) = [(4 \times 0.9) + (0.25 \times 0.04)] = 3.6 + 0.01 = 3.61$$

Therefore, $W_{Ore} = \left[\left(\frac{3.61}{0.7}\right)\left(\frac{1}{0.93}\right)\right] = \mathbf{5.545\,t/h}$.
So, $W_{Ore} = \left(\frac{5.157}{f_1}\right)$.

| Sr. no | Variation of purity of ore ($f_1$) | Consumption of ore (t/h) |
|--------|-----------------------------------|--------------------------|
| 1      | 0.85                              | 6.067                    |
| 2      | 0.875                             | 5.894                    |
| 3      | 0.9                               | 5.73                     |
| 4      | **0.93**                          | **5.545**                |
| 5      | 0.955                             | 5.40                     |
| 6      | 0.98                              | 5.262                    |

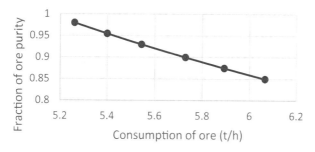

Fig. Ex. 6.2

Above table and figure show that as the purity of ore increases, the consumption of ore also decreases.

## 6.4 Oxygen Jet Momentum in BOF

Oxygen is introduced into the Basic Oxygen Furnace (BOF) converter from the top through a water-cooled lance. Oxygen is blown at a pressures of 10–12 kg/cm$^2$ through a nozzle so that the jet issuing at the nozzle exit is supersonic and generally has a speed between 2.0 and 2.5 times the speed of sound (i.e. 331.45 m/s in dry air). Previously, a single nozzle lance is used for small size converter. Modern big converter used multiple-nozzles lance. Six to eight nozzles are used for 350–400 t vessels [6]. By using multiple-nozzles lance, increase the surface area of jet impact area.

Total energy of the jet from the multiple-nozzles lance is distributed over a larger surface area of the bath. This improves slag–metal interaction. The important parameters are jet momentum, jet height, and fluid properties. Jet momentum, means quantity of motion, can be calculated as follows [7]:

$$\text{Momentum} = \text{mass} \times \text{velocity}$$

Or

$$M = m \times u \tag{6.32}$$

Again,

$$\text{mass flow rate of gas} \left(kg \cdot s^{-1}\right), m = \rho_G \cdot A \cdot u \tag{6.33}$$

where

$\rho_G =$ density of gas (kg · m$^{-3}$),

A = cross-sectional area of nozzle opening (m$^2$) = $\left\{ \frac{\pi d^2}{4} \right\}$,

d = throat diameter of each nozzle (m),

u = gas velocity (m · s$^{-1}$).

For a lance with n nozzles and total gas flow rate (Q, m$^3$ · s$^{-1}$), the flow rate of gas through each nozzle:

$$Q'_n = \frac{Q}{n} \tag{6.34}$$

Therefore, mass flow rate of gas (kg · s$^{-1}$) through each nozzle:

$$m'_n = \left\{ \frac{(Q \cdot \rho_G)}{n} \right\} \tag{6.35}$$

Again, mass flow rate of gas (kg.s$^{-1}$), through each nozzle:

$$m'_n = \rho_G \cdot A \cdot u'_n \tag{6.36}$$

Now combining Eqs. (6.35) and (6.36):

Velocity of gas at each nozzle: $u'_n = \left\{ \frac{Q}{(n \cdot A)} \right\} = \left\{ \frac{4Q}{(n \cdot \pi d^2)} \right\}$ \tag{6.37}

When throat diameter of each nozzle (d) and total flow rate (Q) are constant, then Eq. (6.35) becomes:

$$m'_n = \frac{K_1}{n} \tag{6.38}$$

and Eq. (6.37) becomes:

$$u'_n = \frac{K_2}{n} \tag{6.39}$$

where K is the constant, $K_1 = Q \cdot \rho_G$ and $K_2 = \left\{ \frac{4Q}{(\pi d^2)} \right\}$

Therefore, momentum of jet in each nozzle:

$$M'_n = m'_n \cdot u'_n = \frac{K_1}{n} \cdot \frac{K_2}{n} = \frac{K}{n^2} \tag{6.40}$$

where $K = K_1 \cdot K_2$.

Hence, the momentum of jet in each nozzle is inversely proportional to the square of the number of nozzles in the lance.

Therefore, total momentum for all the jets:

$$M = \sum M'_n = n \cdot M'_n = \frac{K}{n} \tag{6.41}$$

Total momentum for all the jets is inversely proportional to the number of nozzles.
Less momentum or force of jet means less splashing of metal, i.e. less loss of iron
or more iron yield.

## 6.5 Hydrostatic Pressure at the Bottom of Ladle

Pressure developed at the bottom surface of the ladle which is full of liquid metal
(Fig. 6.4).

$$\int_{p_0}^{p_h} dp = \int_{z=0}^{z=h} \rho g \cdot dz \tag{6.42}$$

Therefore,

$$p_h - p_0 = \rho gh \tag{6.43}$$

where $\rho$ is the density of liquid metal and h is the depth of liquid bath.
    Hence, resultant force:

$$F = p_h \cdot A - p_0 \cdot A = (p_h - p_0)A = (p_h - p_0)\left(\frac{\pi d^2}{4}\right) \tag{6.44}$$

where A is area, d is internal diameter of the ladle.

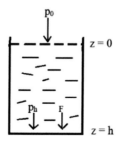

**Fig. 6.4** Ladle with full of liquid metal

**Example 6.3** A ladle (3 m internal diameter and 3.5 m height) is filled with liquid steel leaving a freeboard of 0.15 m. Calculate the pressure developed on its bottom surface and resultant force.

Given: Density of liquid steel 7000 kg/m$^3$ and g $= 9.81$ m/s$^2$.

**Solution**

Since $p_h = p_0 + \rho g h$ (from Eq. 6.43).

Here, $p_0 =$ atm pressure $= 1.01 \times 10^5$ Pa, h $= 3.5 - 0.15 = 3.35$ m.

Therefore, $p_h = 1.01 \times 10^5 + (7000 \times 9.81 \times 3.35) = 3.31 \times 10^5$ Pa $=$ **3.28 atm**.

Again F $= (p_h - p_0)\left(\frac{\pi d^2}{4}\right) = \{(3.31 - 1.01) \times 10^5\}\left(\frac{3.14 \times (3)^2}{4}\right) =$ **16.25 $\times$ 10$^5$ N**.

## 6.6  Growth and Detachment of Bubbles Nucleating in Liquid Bath

The gas bubbles can nucleate at gas-filled crevices within the surface of a refractory or container vessel; i.e. the mechanism of CO bubbles nucleation in a BOF vessel during later stages of refining when bath carbon is low ($\leq 0.1\%$).

Carrying out a static force balance between surface tension forces around the equator of a growing hemisphere, and the surface tension over-pressure ($\Delta p_\sigma$) within the bubble:

$$2\pi r\sigma = \Delta p_\sigma \pi r^2 \qquad (6.45)$$

or

$$\Delta p_\sigma = \frac{2\sigma}{r} \qquad (6.46)$$

Therefore, gas over-pressure,

$$\Delta p_{gas} = \Delta p_{static} + \Delta p_\sigma = \rho g h + \frac{2\sigma}{r} \qquad (6.47)$$

Since the aqueous system is wetting, the rim of the forming bubble remains attached to the periphery of the crevice, and detachment will occur once buoyancy forces exceed adhesion forces.

For releasing the gas bubble:

$$(\rho_L - \rho_g)g\,V_b \geq 2\pi r_\varphi \sigma \tag{6.48}$$

where $\rho_L$ and $\rho_g$ are the densities of liquid and gas, respectively, $V_b$ is the volume of bubble, $r_\varphi$ is the radius of the crevice.

Therefore,

$$V_b \geq \frac{2\pi r_\varphi \sigma}{\rho_L g} \quad \text{(since the value of } \rho_g \text{ is neglected)} \tag{6.49}$$

For the liquid metal systems, which are non-wetting, gas will spread across the refractory surface since the bubble rim will not be anchored to the crevice's periphery. Therefore, maximum radius ($r_m$) for a growing hemispherical bubble given by

$$(\rho_L - \rho_g)g\left\{ \frac{4\pi r_m^3}{3} x \frac{1}{2} \right\} = 2\pi r_m \sigma \tag{6.50}$$

or

$$\rho_L \cdot g\left\{ \frac{2\pi r_m^3}{3} \right\} = 2\pi r_m \sigma \quad \text{or} \quad \frac{r_m^3}{r_m} = \frac{3\sigma}{\rho_L g}$$

Therefore,

$$r_m = \left( \frac{3\sigma}{\rho_L g} \right)^{1/2} \tag{6.51}$$

and

$$V_m = \frac{4\pi r_m^3}{3} = \frac{4\pi}{3}\left( \frac{3\sigma}{\rho_L g} \right)^{3/2} \tag{6.52}$$

**Example 6.4** Calculate (i) the dissolved over-pressure (with respect to ambient pressure) needed for continued nucleation and growth of bubbles on the bottom of a 1-m-deep body of liquid: in (a) liquid Fe, (b) liquid Al, (c) water; (ii) the size of bubble that detaches.

Given: Diameter of crevice is 1 mm.

| Material | $\sigma$ (N/m) | Contact angle ($\theta$) | $\rho$ (kg/m$^3$) |
|----------|----------------|--------------------------|-------------------|
| $H_2O$   | 0.07           | 0                        | 1000              |
| Al       | 0.90           | 90                       | 2300              |
| Fe       | 1.76           | 90                       | 7000              |

**Solution**

(i)   Since $\Delta p_{gas} = \rho g h + \frac{2\sigma}{r}$.

For (a) liquid Fe: $r = (1/2)$ mm $= 0.5 \times 10^{-3}$ m.
So $\Delta p_{gas} = (7000 \times 9.81 \times 1) + \frac{2 \times 1.76}{0.5 \times 10^{-3}} = \mathbf{0.76 \times 10^5}$ **Pa**.
For (b) liquid Al: $\Delta p_{gas} = (2300 \times 9.81 \times 1) + \frac{2 \times 0.9}{0.5 \times 10^{-3}} = \mathbf{0.26 \times 10^5}$ **Pa**.
For (c) water: $\Delta p_{gas} = (1000 \times 9.81 \times 1) + \frac{2 \times 0.07}{0.5 \times 10^{-3}} = \mathbf{0.1 \times 10^5}$ **Pa**.

(ii)   Since $r_m = \left(\frac{3\sigma}{\rho_L g}\right)^{1/2}$.

and $V_m = \frac{4\pi r_m^3}{3}$.
For (a) liquid Fe:

$$r_m = \left(\frac{3 \times 1.76}{7000 \times 9.81}\right)^{1/2} = 8.77 \times 10^{-3} m = 8.77\,\text{mm},$$

$$V_m = \frac{4\pi r_m^3}{3} = \frac{4\pi}{3}(r_m)^3 = \frac{4 \times 3.14}{3}\left(8.77 \times 10^{-3}\right)^3$$
$$= 2.83 \times 10^{-6}\,m = \mathbf{2.83\,cm^3}$$

For (b) liquid Al:

$$r_m = \left(\frac{3 \times 0.9}{2300 \times 9.81}\right)^{1/2} = 10.94 \times 10^{-3} m = 10.94\,\text{mm},$$

$$V_m = \frac{4\pi r_m^3}{3} = \frac{4\pi}{3}(r_m)^3 = \frac{4 \times 3.14}{3}\left(10.94 \times 10^{-3}\right)^3$$
$$= 5.49 \times 10^{-6}\,m = \mathbf{5.49\,cm^3}.$$

For (c) water:

$$r_m = \left(\frac{3 \times 0.07}{1000 \times 9.81}\right)^{1/2} = 4.63 \times 10^{-3}\,m = 4.63\,\text{mm},$$

$$V_m = \frac{4\pi r_m^3}{3} = \frac{4\pi}{3}(r_m)^3 = \frac{4 \times 3.14}{3}\left(4.63 \times 10^{-3}\right)^3$$
$$= 4.16 \times 10^{-7} m = \mathbf{0.416\,cm^3}.$$

## 6.7 Continuous Casting (Concast)

Continuous casting (or Concast) is replacing the ingot casting because of (i) improved yield, (ii) lower capital cost due to elimination of many downstream facilities, (iii) fuel saving due to not reheating of ingot before rolling to produce slab/billet, and (iv) improved quality of product [8].

Liquid steel with a superheat of 10–40 °C is poured continuously in the oscillating mold from the tundish at one end and the solid strand in pulled out at the other end. When the molten metal encounters the water cooled copper mold, a thin solid skin is formed. By the time cast metal comes out from the mold, the skin becomes sufficiently thick to withstand the ferro-static pressure of liquid metal in core. Beyond the mold region, the cast is cooled by direct spraying of water and, finally, it is cooled by radiation mode of heat transfer only. Although the process appears to be a very simple one, it is a very complex system.

Since the actual phenomena is very complex, for the purpose of modeling, the following simple assumptions are taken:

1. Metal is poured at a constant temperature and flow rate, so the continuous casting process is at steady state.
2. Cooling of the cast is symmetrical, i.e. cooling rate of the opposite faces is same.
3. Heat transfer in the liquid core is by conduction only.
4. Heat conduction is in axial direction only.
5. Density of liquid and solid are same.

The first two assumptions are reasonably valid in the actual process. The heat transfer in liquid core is both by conduction and convection. In the mold region, convection plays the major role but further down gradually conduction predominates. However, for mathematical simplicity, it is assumed that the heat transfer in the liquid core is by conduction.

The heat flow in the axial direction is both due to downward movement of the cast and conduction. But since the contribution of former is much greater than the later, assumption (4) is made. The volume contraction due to solidification is very small, so it is neglected.

To further simplify the problem, let assume that for a cast of rectangular cross section, heat transfer along the thickness of slab is important. This assumption is reasonably valid except near the corner of the slab. Figure 6.5 shows the heat balance across an elementary volume of $\Delta z \cdot \Delta x \cdot w$, where w is the width of the slab, x is the direction of thickness, and z is the direction of casting.

[(Heat entering through the face of z−that leaving the face at $z + \Delta z$)

$+$ (heat entering through the face of x

$-$ that leaving the face at $x + \Delta x$)] $= 0$ \hfill (6.53)

**Fig. 6.5** Heat balance over a
control volume

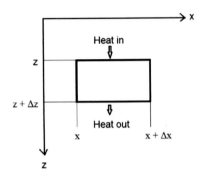

$$\rightarrow \quad w\Delta x \,(u\,\Delta H_z - u\,\Delta H_{z+\Delta z}) + w\Delta z\left[k_{eff}\left(\frac{\delta T}{\delta x}\right)_x - k_{eff}\left(\frac{\delta T}{\delta x}\right)_{x+\Delta x}\right] = 0 \tag{6.54}$$

where

    u is the casting speed,
    $\Delta H$ is the change of enthalpy per unit volume,
    T is the temperature, and
    $k_{eff}$ is the effective thermal conductivity.

Dividing Eq. (6.54) by $w\Delta x\Delta z$ and since $M_{x+\Delta x} = M_x + \left(\frac{\delta M_x}{\delta x}\right)\Delta x$:

$$\frac{1}{\Delta z}\left[u\,\Delta H_z - u\,\Delta H_z - \left(\frac{\delta(u\Delta H_z)}{\delta z}\right)\Delta z\right]$$

$$+ \frac{1}{\Delta x}\left[k_{eff}\left(\frac{\delta T}{\delta x}\right)_x - k_{eff}\left(\frac{\delta T}{\delta x}\right)_x - \left(\frac{\delta\{k_{eff}(\frac{\delta T}{\delta x})\}}{\delta x}\right)\Delta x\right] = 0$$

$$\rightarrow \quad \left(\frac{\delta(u\Delta H_z)}{\delta z}\right) = -\left(\frac{\delta\{k_{eff}(\frac{\delta T}{\delta x})\}}{\delta x}\right) \tag{6.55}$$

Since casting speed, u is constant, Eq. (6.55) is can be written as

$$\left(\frac{\delta\Delta H}{\delta t}\right) = -\left(\frac{\delta\{k_{eff}(\frac{\delta T}{\delta x})\}}{\delta x}\right) \tag{6.56}$$

Since $u = \frac{z}{t}$, so $t = \frac{z}{u}$, therefore, $dt = \frac{dz}{u}$ or $dz = u \cdot dt$.
The enthalpy and effective thermal conductivity of metal is defined as

$$\Delta H = \rho\left(C_{p,l}T + \Delta H_f\right) \quad \text{and} \quad k_{eff} = \lambda k_l, \quad \text{when} \quad T > T_l \tag{6.57}$$

where

$\rho$ is the density of metal,
$C_{p,l}$ is the specific heat of liquid,
$\Delta H_f$ is the enthalpy of fusion,
$\lambda$ is the correction factor of thermal conductivity, and
$k_l$ is the thermal conductivities of liquid.

The effect of heat transfer by convection is considered by the correction factor of thermal conductivity ($\lambda$).

$$\Delta H = \rho(C_{p,l}T + f_s\,\Delta H_f) \quad \text{and} \quad k_{eff} = (1 - f_s)\,\lambda k_l + f_s\,k_s, \quad \text{when} \quad T_s < T < T_l \tag{6.58}$$

where

$f_s$ is the fraction of solidification,
$k_s$ is the thermal conductivities of solid.

Since

$$\Delta H = \rho(C_{p,s}\,T) \text{ and } k_{eff} = k_s, \text{ when } T < T_s \tag{6.59}$$

where

$C_{p,s}$ is the specific heat of solid,
$T_s, T_l$ are the solidus and liquidus temperature.

Fraction of solid, $f_s$ can be calculated as:

$$f_s = \frac{T_l - T}{T_l - T_s} \tag{6.60}$$

The initial and boundary condition of Eq. (6.60) are:

at $t = 0$, $T = T_p$ for $0 \le x \le d$ (where $T_p$ is the pouring temperature)  (6.61)

$$\text{at } x = \left(\frac{d}{2}\right), \quad \left(\frac{\delta T}{\delta x}\right) = 0 \qquad \text{(at center)} \tag{6.62}$$

Therefore,

$$x = 0, \quad -k_s\left(\frac{\delta T}{\delta x}\right) = q \tag{6.63}$$

where q is the heat flux from the strand surface.

The Eq. (6.61) states that the metal temperature on the top of mold is pouring temperature, $T_p$ and Eq. (6.62) is the symmetry condition of temperature (assumption 2). The other boundary condition, Eq. (6.63) is heat flux condition on the surface of the cast. The value of surface heat flux q depends on cooling zone. In the mold, q is mostly calculated by Savage and Pickard correlation:

$$q = 640 - 80\sqrt{t} \tag{6.64}$$

where q is the heat flux in kcal/m$^2$s and t is time in s.

The value of q in the mold depends on water flow rate, casting speed, type of lubricant or flux used, mold design, and the steel chemistry. It can be different from that given in Eq. (6.64) by more than 40%. The surface heat flux in the spray cooling zone is given by

$$q = h(T - T_w) \tag{6.65}$$

where h is heat transfer coefficient, and $T_w$ is temperature of water.

The heat transfer coefficient depends on the type of nozzle, water pressure, nozzle stand-off, and even on support roll configuration. In the radiation cooling zone, surface heat flux is

$$q = \sigma\varepsilon\left(T^4 - T_a^4\right) \tag{6.66}$$

where $\sigma$ is Stefan–Boltzmann constant ($5.669 \times 10^{-8}$ W/m$^2$ K$^4$), $\varepsilon$ is emissivity of the strand surface, and $T_a$ is the ambient temperature. The emissivity for oxidized steel surface is 0.85.

The solution of Eqs. (6.56)–(6.66) gives the temperature profile in the cast. In case of billet the one-dimensional Eq. (6.56) is valid for the mold region but to simulate the entire strand the heat transfer in y direction also should be taken.

**Example 6.5**  Calculate radiation loss per tonne of liquid steel at 1750 °C in an electric arc furnace. Given: (i) diameter of furnace = 7.6 m, capacity = 180 t, (ii) emissivity of liquid steel = 0.28, Stefan–Boltzmann constant = $5.669 \times 10^{-8}$ W/m$^2$ K$^4$.

**Solution**

From Eq. (6.55): $q = \sigma\varepsilon\ (T^4 - T_a^4) = \sigma\varepsilon\ T^4$ (since $T \gg T_a$, so value of $T_a^4$ is neglected).

Here, $\sigma = 5.669 \times 10^{-8}$ W/m$^2 \cdot$ K$^4 \cdot$ s, $\varepsilon = 0.28$, T = 1750 + 273 = 2023 K, d = 7.6 m,

$$q = (5.669 \times 10^{-8}) \times 0.28(2023)^4 = 1.587 \times 10^{-8}\left(1.675 \times 10^{13}\right)$$
$$= 2.6588 \times 10^5 = 265.88 \text{ kW/m}^2$$

Area of EAF $= \pi \, r^2 = 3.14 \times (3.8)^2 = 45.34 \text{ m}^2$.
Radiation loss for $1 \text{m}^2$ is 265.88 kW.
So, for radiation loss for 45.34 $\text{m}^2$ is $(45.34 \times 265.88) = 12{,}055$ kW.
Since furnace capacity is 180 t,
     Therefore, radiation loss per tonne of liquid steel $= (12{,}055/180) = 66.97 =$
**67 kW**.

## 6.8   Addition of Slag Powder on Concast

Develop an equation to show how much slag powder should be added to a continuous
casting (i.e. Concast) to maintain a coherent lubricating film of thickness δ, between
the mold and the slab. Figure 6.6a shows a section of the water cooled copper mold,
together with the slab and on overlaying fluid slag lubricant [9]. Depending on the
thickness of the film, it is expected that it is drained out by itself due to gravity,
i.e. this is the drag-out effect of the downward moving slab, dragging slag down
towards the bottom of the cooled copper mold. This happens because of the non-slip
condition between a solid surface and the adjacent molecules of fluid. Similarly,
since the solidified steel slab is still at a high temperature, and the slag composition
is chosen for its fluidity and low freezing range. Steel makers are ensuring that most
of the film remains liquid during its descent through the water cooled copper mold.
     In Fig. 6.6b, a suitable coordinate system has been chosen for solving the problem.
An x–y coordinate system was chosen with x as the vertical axis, taken as positive
in the downward direction. The general approach required for solving fluid flow
phenomena is to write down the governing differential equations for continuity of
mass, and for momentum, for a volume element of fluid. For simple problems, it can
demonstrate the use of the shell momentum balance. It will apply this to a typical

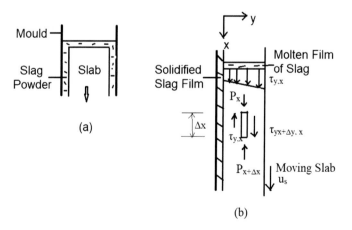

**Fig. 6.6   a** A section of slab with slag powder in copper mold, **b** a section of moving slab with
solidified slag film

volume element of slag to deduce the differential equations of motion governing its flow. Suitable boundary conditions then allow velocity profiles for the specific problem to be solved, and hence the flow rates.

For a steady flow, for momentum balance:

$$[(\text{Rate of momentum in}) - (\text{Rate of momentum out})$$
$$+ (\text{Sum of forces acting on system})] = 0 \tag{6.67}$$

It is sufficient to note that since the flow is one-dimensional, i.e. $u_y$ and $u_z$ are zero. Similarly, since slag is incompressible, so $\frac{\delta \rho}{\delta t} = 0$ and $\frac{\delta(\rho u_x)}{\delta x} = 0$.

Now to the momentum equation and choosing a volume element of thickness $\Delta y$, length $\Delta x$, and unit perimeter length, then shell momentum balances as follows:

(i)     Rate of x momentum in across top surface at x: $(L\Delta y)\, \rho u_x \cdot u_x$,
(ii)    Rate of x momentum out from bottom surface at $x + \Delta x$: $(L\Delta y)\, \rho u_{x+\Delta x} \cdot u_{x+\Delta x}$,
(iii)   Rate of x momentum in across side surface of element between
        x and $x + \Delta x$: $(L\Delta x)\tau_{y,x}$,
(iv)    Rate of x momentum out across side surface of element between
        x and $x + \Delta x$: $(L\Delta x)\tau_{yx+\Delta y,\, x}$,
(v)     Gravity force acting on fluid element (i.e. component of weight):
        $(L\Delta x \Delta y)\rho g_x$,
(vi)    Pressure force acting on top surface of element: $(L\Delta y)P_x$,
(vii)   Pressure force acting on bottom surface of element:

$$(L\Delta y)P_{x+\Delta x}. \tag{6.68}$$

Now putting these above values in Eq. (6.67):

$$\rightarrow L\Delta y \rho (u_x \cdot u_x - u_{x+\Delta x} \cdot u_{x+\Delta x}) + L\Delta x \left(\tau_{y,x} - \tau_{y,x+\Delta y,x}\right)$$
$$+ \left\{(L\Delta x \Delta y)\rho g_x\right\} + L\Delta y (P_x - P_{x+\Delta x}) = 0 \tag{6.69}$$

Since $\left\{\left\{\frac{u_{x+\Delta x} - u_x}{\Delta x}\right\} = \frac{\delta u_x}{\delta x}\right.$ or $u_{x+\Delta x} = u_x + \frac{\delta u_x}{\delta x} \cdot \Delta x$; similarly $\tau_{y,x+\Delta y} = \tau_{y,x} + \frac{\delta \tau_{y,x}}{\delta y} \cdot \Delta y$

$$\rightarrow \; L\Delta y \rho \left[u_x \cdot u_x - u_x \cdot u_x - \frac{\delta(u_x u_x)}{\delta x} \cdot \Delta x\right] + L\Delta x \left[\tau_{y,x} - \tau_{y,x} - \frac{\delta \tau_{y,x}}{\delta y} \cdot \Delta y\right]$$
$$+ \left\{(L\Delta x \Delta y)\rho g_x\right\} - L\Delta y \left(\frac{\delta P_x}{\delta x}\right) \cdot \Delta x = 0 \tag{6.70}$$

Equation (6.70) divided by $(L\Delta x \Delta y)$:

$$\rightarrow \; -\left[\rho \cdot \frac{\delta(u_x u_x)}{\delta x} + \frac{\delta \tau_{y,x}}{\delta y}\right] + \rho g_x - \left(\frac{\delta P_x}{\delta x}\right) = 0 \tag{6.71}$$

Since $\frac{\delta u_x}{\delta x} = 0$, for one-dimensional steady state, incompressible flow, and realizing that $(\frac{\delta P}{\delta x}) = 0$ (since $P_x = P_{at} =$ both at the top of the mold and at the bottom where the molten slag is existing).

Therefore, Eq. (6.71) becomes:

$$\frac{\delta \tau_{y,x}}{\delta y} = \rho g_x \tag{6.72}$$

Integrating Eq. (6.72):

$$\tau_{y,x} = \rho g_x \cdot y + I_1 \tag{6.73}$$

where $I_1$ is integration constant.

Substituting in Newton's law of viscosity for laminar, one-dimensional flow:

$$-\mu \frac{\delta u_x}{\delta y} = \rho g_x \cdot y + I_1 \tag{6.74}$$

Again integrating Eq. (6.74):

$$-\mu\, u_x = \rho g_x \cdot \left(\frac{y^2}{2}\right) + I_1 y + I_2 \tag{6.75}$$

where $I_2$ is another integration constant.

Applying non-slip boundary condition: $u_x = 0$ at $y = 0$.

Therefore,

$$I_2 = 0 \tag{6.76}$$

Again, $u_x = u_s$ at $y = \delta_s$

Therefore,

$$-\mu\, u_s = \rho g_x \cdot \left(\frac{\delta_s^2}{2}\right) + I_1 \delta_s \quad \text{or } I_1 = \left\{-\left(\frac{\mu u_s}{\delta_s}\right) - \left(\frac{\rho g_x \delta_s}{2}\right)\right\} \tag{6.77}$$

Substituting Eq. (6.77) in Eq. (6.75), finally getting an expression for the way in which the velocity $u_x$, varies with distance, across the slag film:

$$-\mu\, u_x = \rho g_x \cdot \left(\frac{y^2}{2}\right) + \left\{-\left(\frac{\mu u_s}{\delta_s}\right) - \left(\frac{\rho g_x \delta_s}{2}\right)\right\} y$$

Therefore,

$$u_x = u_s\left(\frac{y}{\delta_s}\right) + \left(\frac{\rho g_x}{2\mu}\right)(\delta_s\, y - y^2) \qquad (6.78)$$

This expression shows that, for very high viscosities, the second term vanishes and a linear velocity profile across the slag film will be generated by the drag effect of the moving slab.

i.e.

$$u_x = u_s \text{ at } y = \delta_s \qquad (6.79)$$

At the other extreme, for low viscosity larger slag thickness and low casting speeds, the right-hand side term dominates:

$$u_x \approx \left(\frac{\rho g_x y^2}{2\mu}\right)\left(\frac{\delta_s}{y} - 1\right) \qquad (6.80)$$

Differentiating and setting $\left(\frac{\delta u_x}{\delta y}\right) = 0$, getting a maximum in the self-draining velocity at $y = \frac{\delta_s}{2}$, equals to $\left(\frac{\rho g_x y^2}{2\mu}\right)$. Consequently, as seen in Fig. 6.6b, a set of velocity profiles is possible depending upon the relative magnitudes of the parameters appearing in Eq. (6.80).

Now move on to show how much slag needs to be fed per unit length of slab periphery.

The slag flow rate ($Q'$) is given by

$$Q' = \int_0^{\delta_s} u_x dy \qquad (6.81)$$

$$= \int_0^{\delta_s}\left\{u_s\left(\frac{y}{\delta_s}\right) + \left(\frac{\rho g_x}{2\mu}\right)(\delta_s\, y - y^2)\right\}dy \quad \text{[from Eq. (6.78)]}$$

$$= \left\{u_s\left(\frac{\delta_s}{2}\right) + \frac{\rho g_x \delta_s^3}{12\mu}\right\} \qquad (6.82)$$

The expression for the mass flow of slag per linear meter finally becomes

$$m' = \rho\left(\frac{u_s \delta_s}{2} + \frac{\rho g_x \delta_s^3}{12\mu}\right) \qquad (6.83)$$

As it would be expected, the mass flow of slag is proportional to slag density and slab withdrawal velocity, and inversely proportional to slag viscosity. It has a relatively complex dependence on slag film thickness, $\delta_s$.

**Example 6.6** Calculate the mass flow of slag per meter. If density of slag is 2900 kg/m³, velocity of slab is 45 cm/s, viscosity of slag is 0.3 kg/ms, and thickness of slag is 0.1 mm.

**Solution**

Here, $\rho = 2900$ kg/m³, $u_s = 45$ cm/s $= 0.045$ m/s, $\mu = 0.3$ kg/ms, $\delta_s = 0.1$ mm $= 10^{-4}$ m.

From Eq. (6.83):

$$m' = \rho\left(\frac{u_s\delta_s}{2} + \frac{\rho g_x\delta_s^3}{12\mu}\right) = 2900\left(\frac{0.045 \times 10^{-4}}{2} + \frac{2900 \times 9.81 \times 10^{-12}}{12 \times 0.3}\right)$$
$$= 6.5479 \times 10^{-3}\,\text{kg/ms} = \mathbf{6.55\,g/ms}.$$

## 6.9  Production of Duplex Stainless Steel by CONARC

### 6.9.1  Material Balance

Duplex stainless steel is one of the members of stainless steel family, which consists a mixture of austenitic and ferritic structures, developed with a view to combine the toughness and weldability of the austenite with the strength, and resistance to localized corrosion. CONARC technology is combination of half-EAF and half-BOF processes, where scrap and hot metal can be charged as a feed materials [10].[1]

The product output of any process depends on the amount of materials consumed as raw materials and the chemical reactions taking place. Material balance is essentially an application of law of conservation of mass, i.e. *the mass of an isolated system remains constant irrespective of the charges occurring within the system.* The following equation is applicable for material balance:

$$\text{Accumulation of mass within the system} = (\text{Input}-\text{output}) \qquad (6.84)$$

If there is no accumulation of mass within the system, then the above equation can be written as

$$\text{Input} = \text{output} \qquad (6.85)$$

**Fe Balance**:

$$\text{Fe input} = \text{Fe output}$$

---

[1] With permission from "*JPC Bulletin on Iron & Steel*", Joint Plant Committee, Ministry of Steel, Kolkata.

(Fe from HM + Fe from MS scrap + Fe from DRI + Fe from Fe−Cr

   + Fe from Fe − Mn + Fe from Fe−Mo)

= (Fe in Steel + Fe losses in slag + Fe losses in flumes)

$$\rightarrow [(W_{HM} \times f_1) + (W_{Sc} \times f_2) + \{(W_{DRI} \times f_3) + (W_{DRI} \times f_4 \times f_5)\}$$
$$+ (W_{Fe-Cr} \times f_6) + (W_{Fe-Mn} \times f_7) + (W_{Fe-Mo} \times f_8)]$$
$$= \{(W_{Steel} \times f_9) + (W_{Slag} \times f_{10}) + (W_{Flu} \times f_{11}) \qquad (6.86)$$

where

$W_{HM}$ is weight of hot metal charged,
$f_1$ is fraction of Fe present in hot metal,
$W_{Sc}$ is weight of MS scrap charged,
$f_2$ is fraction of Fe present in ms scrap,
$W_{DRI}$ is weight of DRI charged,
$f_3$ is the fraction of metallized Fe present in DRI,
$f_4$ is fraction of FeO present in DRI,
$f_5$ is fraction of Fe present in FeO,
$W_{Fe-Cr}$ is weight of Fe–Cr charged,
$f_6$ is fraction of Fe present in Fe–Cr,
$W_{Fe-Mn}$ is weight of Fe–Mn charged,
$f_7$ is fraction of Fe present in Fe–Mn,
$W_{Fe-Mo}$ is weight of Fe–Mo charged,
$f_8$ is fraction of Fe present in Fe–Mo,
$W_{Steel}$ is weight of steel produced,
$f_9$ is fraction of Fe present in steel,
$W_{Slag}$ is weight of slag produced,
$f_{10}$ is fraction of Fe present in slag,
$W_{Flu}$ is weight of flumes produced, and
$f_{11}$ is fraction of Fe present in flumes.

$$\rightarrow \quad Fe + 1/2\ O_2 = FeO$$
$$\quad 56 \quad 16 \qquad 72$$

72 kg FeO content 56 kg of Fe

$$X\ kg \ldots\ldots\ldots\ldots \left\{ \left(\frac{56}{72}\right) \times X \right\} kg\ Fe \qquad (6.87)$$

Again, 56 kg Fe form 72 kg FeO
So,

$$Y \text{ kg} \ldots \ldots \left\{ \left( \frac{72}{56} \right) \times Y \right\} \text{ kg of FeO goes to slag.} \qquad (6.88)$$

Again, 56 kg Fe react with 16 kg oxygen to form 72 kg FeO
So,

$$Y \text{ kg} \ldots \ldots \ldots \left\{ \left( \frac{16}{56} \right) \times Y \right\} \text{ kg oxygen requires for FeO formation.}$$

$$(6.89)$$

**Si Balance**:

$$Si \text{ input} = Si \text{ output}$$

$$(\text{Si from HM} + \text{Si from MS scrap} + \text{Si from DRI} + \text{Si from Fe} - \text{Cr}$$
$$+ \text{ Si from Fe}-\text{Mn} + \text{Si from lime} + \text{Si from spar})$$

$$= (\text{Si in Steel} + \text{Si losses in slag})$$

$$\rightarrow \{(W_{HM} \times f_{12}) + (W_{Sc} \times f_{13}) + (W_{DRI} \times f_{14} \times f_{15}) + (W_{Fe-Cr} \times f_{16})$$
$$+ (W_{Fe-Mn} \times f_{17}) + (W_{Lime} \times f_{18} \times f_{15})$$
$$+ (W_{Spar} \times f_{19} \times f_{15})\} = \{(W_{Steel} \times f_{20}) + W_{Si \text{ in slag}}\} \qquad (6.90)$$

where

$f_{12}$ is fraction of Si present in hot metal,
$f_{13}$ is fraction of Si present in ms scrap,
$f_{14}$ is fraction of $SiO_2$ present in DRI,
$f_{15}$ is fraction of Si present in $SiO_2$,
$f_{16}$ is fraction of Si present in Fe–Cr,
$f_{17}$ is fraction of Si present in Fe–Mn,
$W_{Lime}$ is weight of lime charged,
$f_{18}$ is fraction of $SiO_2$ present in lime,
$W_{Spar}$ is weight of spar charged,
$f_{19}$ is fraction of $SiO_2$ in spar,
$f_{20}$ is fraction of Si present in steel, and
$W_{Si \text{ in slag}}$ is weight of Si losses in slag.

$$\rightarrow \quad Si + O_2 = SiO_2$$
$$28 \quad 32 \quad 60$$

60 kg $SiO_2$ content 28 kg Si

So,

$$M \, kg \ldots\ldots\ldots\ldots\ldots \left\{ \left( \frac{28}{60} \right) \times M \right\} kg \, Si \qquad (6.91)$$

Again 28 kg Si form 60 kg $SiO_2$.
So,

$$N \, kg \, Si \, form \left\{ \left( \frac{60}{28} \right) \times N \right\} kg \, SiO_2 \, goes \, to \, slag \qquad (6.92)$$

Again, 28 kg Si react with 32 kg oxygen to form 60 kg $SiO_2$

$$\rightarrow \quad So \quad N \, kg \, Si \ldots\ldots\ldots \left\{ \left( \frac{32}{28} \right) \times N \right\} kg \, oxygen \, requires \, for \, SiO_2 \, formation \qquad (6.93)$$

**Mn Balance**:

$$Mn \, input = Mn \, output$$

$(Mn \, from \, HM + Mn \, from \, MS \, scrap + Mn \, from \, Fe{-}Cr + Mn \, from \, Fe - Mn)$
$= (Mn \, in \, Steel + Mn \, losses \, in \, slag)$

$$\rightarrow \quad \{(W_{HM} \times f_{21}) + (W_{Sc} \times f_{22}) + (W_{Fe-Cr} \times f_{23}) + (W_{Fe-Mn} \times f_{24})\}$$
$$= \{(W_{Steel} \times f_{25}) + W_{Mn \, in \, slag}\} \qquad (6.94)$$

where

$f_{21}$ is fraction of Mn present in hot metal,
$f_{22}$ is fraction of Mn present in ms scrap,
$f_{23}$ is fraction of Mn present in Fe–Cr,
$f_{24}$ is fraction of Mn present in Fe–Mn,
$f_{25}$ is fraction of Mn present in steel, and
$W_{Mn \, in \, slag}$ is weight of Mn losses in slag.

$$\rightarrow \quad Mn + 1/2 \, O_2 = MnO$$
$$55 \quad 16 \quad 71$$

55 kg Mn form 71 kg MnO
So,

$$Z \text{ kg} \ldots \ldots \left\{\left(\frac{71}{55}\right) \times Z\right\} \text{ kg MnO goes to slag} \qquad (6.95)$$

Again 55 kg Mn react with 16 kg oxygen to form MnO.
So,

$$Z \text{ kg} \ldots \ldots \left\{\left(\frac{16}{55}\right) \times Z\right\} \text{ kg oxygen requires to form MnO} \qquad (6.96)$$

**C Balance:**

$$C \text{ input} = C \text{ output}$$

(C from HM + C from MS scrap + C from DRI + C from
Fe−Cr + C from Fe − Mn + C from Fe − Mo)

$$= (C \text{ in Steel } + C \text{ losses in gases})$$

$$\rightarrow \left[(W_{HM} \times f_{26}) + (W_{Sc} \times f_{27}) + \{(W_{DRI} \times f_{28}) + (W_{Fe-Cr} \times f_{29}) \right.$$
$$\left. + (W_{Fe-Mn} \times f_{30}) + (W_{Fe-Mo} \times f_{31})\right]$$
$$= \left\{(W_{Steel} \times f_{32}) + W_{C \text{ in gas}}\right\} \qquad (6.97)$$

where

$f_{26}$ is fraction of C present in hot metal,
$f_{27}$ is fraction of C present in ms scrap,
$f_{28}$ is fraction of C present in DRI,
$f_{29}$ is fraction of C present in Fe–Cr,
$f_{30}$ is fraction of C present in Fe–Mn,
$f_{31}$ is fraction of C present in Fe–Mo,
$f_{32}$ is fraction of C present in steel, and
$W_{C \text{ in gas}}$ is weight of C losses in gases.

$$\rightarrow FeO + C = Fe + CO$$
$$72 \quad 12 \quad 56 \quad 28$$

72 kg FeO reduced by 12 kg C

$$(W_{DRI} \times f_4) \text{ kg FeO (in DRI) reduced by} \left\{\left(\frac{12}{72}\right) \times (W_{DRI} \times f_4)\right\} = X_{C,FeO} \text{ kg of C} \quad (6.98)$$

72 kg FeO reduced by C to form 28 kg CO

$$(W_{DRI} \times f_4)\, kg \dots\dots\dots\dots\dots\dots\dots\dots\left\{\left(\frac{28}{72}\right) \times (W_{DRI} \times f_4)\right\} \textbf{kg CO}$$

$$(6.99)$$

Hence,

$$\text{C available for CO and } CO_2 \text{ formation} = \left(W_{C\,in\,gas}\right) - X_{C,FeO} = X_{C,gas}\, \textbf{kg} \quad (6.100)$$

Assume, 90% CO and 10% $CO_2$ will be formed.
So, C consume for CO formation $= 0.9 \times X_{C,gas} = X_{C,CO}$ kg

$$C + 1/2\, O_2 = CO$$
$$12 \quad 16 \qquad 28$$

12 kg C form 28 kg CO gas
So,

$$X_{C,CO}\, kg \dots\dots\left\{\left(\frac{28}{12}\right) \times X_{C,CO}\right\} = X_{CO} \textbf{ kg of CO} \qquad (6.101)$$

Again 12 kg C react with 16 kg oxygen to form CO

$$X_{C,CO}\, kg \dots\dots\left\{\left(\frac{16}{12}\right) \times X_{C,CO}\right\} = X_{O,CO} \textbf{ kg oxygen} \text{ to form CO} \quad (6.102)$$

Total CO formation $= [\{(\frac{28}{72}) \times (W_{DRI} \times f_4)\} + X_{CO}]\, \textbf{kg}$.
Again C consume for $CO_2$ formation $= 0.1 \times X_{C,gas} = X_{C,CO_2}\, E$

$$C + O_2 = CO_2$$
$$12 \quad 32 \qquad 44$$

12 kg C form 44 kg $CO_2$ gas
So,

$$X_{C,CO_2}\, kg \dots\dots\left\{\left(\frac{44}{12}\right) \times X_{C,CO_2}\right\} = X_{CO_2} \textbf{ kg of } CO_2 \qquad (6.103)$$

Again 12 kg C react with 32 kg oxygen

$$X_{C,CO_2}\, kg \dots\dots\left\{\left(\frac{32}{12}\right) \times X_{C,CO_2}\right\} = X_{O,CO_2} \textbf{ kg oxygen} \text{ for } CO_2 \text{ formation}$$

$$(6.104)$$

**Oxygen consumption for gas formation** $=$ Oxy for CO formation

$$+ \text{ Oxy for } CO_2 \text{ formation}$$
$$= X_{O,CO} + X_{O,CO_2} = \mathbf{X_{O,gas}} \text{ kg}  (6.105)$$

**P Balance**:

$$P \text{ input} = P \text{ output}$$

$(P \text{ from } HM + P \text{ from } MS \text{ scrap} + P \text{ from } DRI + P \text{ from } Fe-Cr + P \text{ from } Fe - Mn)$
$= (P \text{ in Steel} + P \text{ goes in slag})$

$$\left[ (W_{HM} \times f_{33}) + (W_{Sc} \times f_{34}) + \{ (W_{DRI} \times f_{35}) + (W_{Fe-Cr} \times f_{36}) + (W_{Fe-Mn} \times f_{37}) \right]$$
$$= \{ (W_{steel} \times f_{38}) + W_{P \text{ in slag}} \}  (6.106)$$

where

$f_{33}$ is fraction of P present in hot metal,
$f_{34}$ is fraction of P present in ms scrap,
$f_{35}$ is fraction of P present in DRI,
$f_{36}$ is fraction of P present in Fe–Cr,
$f_{37}$ is fraction of P present in Fe–Mn,
$f_{38}$ is fraction of C present in steel, and
$W_{P \text{ in slag}}$ is weight of P losses in slag.

$$2P + 5/2\,O_2 = P_2O_5$$
$$62 \quad 80 \quad 142$$

62 kg P forms 142 kg $P_2O_5$
So,

$$W_{P \text{ in slag}} \text{kg} \ldots\ldots\ldots \left\{ \left( \frac{142}{62} \right) \times W_{P \text{ in slag}} \right\} \textbf{kg } \mathbf{P_2O_5} \text{ goes to slag}  (6.107)$$

Again, 62 kg P reacts with 80 kg oxygen.
So,

$$W_{P \text{ in slag}} \text{kg} \ldots\ldots\ldots \left\{ \left( \frac{80}{62} \right) \times W_{P \text{ in slag}} \right\} \textbf{kg oxygen} \text{ require for } P_2O_5  (6.108)$$

**S Balance**:

$$S \text{ input} = S \text{ output}$$

$(S \text{ from } HM + S \text{ from } MS \text{ scrap} + S \text{ from } DRI + S \text{ from } HC \text{ } Fe-Cr$

+ S from HC Fe − Mn + S from lime + S from spar)
= (S in Steel + S goes in slag)

$$\rightarrow \{(W_{HM} \times f_{39}) + (W_{Sc} \times f_{40}) + (W_{DRI} \times f_{41}) + (W_{Fe-Cr} \times f_{42})$$
$$+ (W_{Fe-Mn} \times f_{43}) + (W_{Lime} \times f_{44}) + (W_{Spar} \times f_{45})\}$$
$$= \{(W_{Steel} \times f_{46}) + W_{S \text{ in slag}}\} \tag{6.109}$$

where

$f_{39}$ is fraction of S present in hot metal,
$f_{40}$ is fraction of S present in ms scrap,
$f_{41}$ is fraction of S present in DRI,
$f_{42}$ is fraction of S present in Fe–Cr,
$f_{43}$ is fraction of S present in Fe–Mn,
$f_{44}$ is fraction of S present in lime,
$f_{45}$ is fraction of S present in spar,
$f_{46}$ is fraction of S present in steel, and
$W_{S \text{ in slag}}$ is weight of S losses in slag.

$$CaO + S = CaS + O$$
$$56 \quad 32 \quad 72$$

32 kg S forms 72 kg CaS

$$W_{S \text{ in slag}} kg \dots \dots \left\{\left(\frac{72}{32}\right) \times W_{S \text{ in slag}}\right\} \textbf{kg of CaS} \tag{6.110}$$

Again 72 kg CaS forms from 56 kg of CaO

$$\left\{\left(\frac{72}{32}\right) \times W_{S \text{ in slag}}\right\} kg \dots \dots \left[\left(\frac{56}{72}\right) \times \left\{\left(\frac{72}{32}\right) \times W_{S \text{ in slag}}\right\}\right]$$
**kg of CaO consume for CaS.** $\tag{6.111}$

**Cr Balance:**

$$Cr \text{ input} = Cr \text{ output}$$

$$(Cr \text{ from Fe}-Cr) = (Cr \text{ in Steel} + Cr \text{ goes in slag})$$
$$\rightarrow (W_{Fe-Cr} \times f_{47}) = \{(W_{Steel} \times f_{48}) + W_{Cr \text{ in slag}}\} \tag{6.112}$$

where

f$_{47}$ is fraction of Cr present in Fe–Cr,
f$_{48}$ is fraction of Cr present in steel, and
W$_{\text{Cr in slag}}$ is weight of Cr losses in slag.

$$2Cr + 3/2\, O_2 = Cr_2O_3$$
$$104 \quad\ 48 \qquad 152$$

104 kg Cr forms 152 kg Cr$_2$O$_3$

$$\left(W_{\text{Cr in slag}}\right) kg \ldots\ldots\ldots \left\{\left(\frac{152}{104}\right) \times \left(W_{\text{Cr in slag}}\right)\right\} \textbf{kg of } Cr_2O_3 \text{ in slag} \quad (6.113)$$

Again 104 kg Cr reacts with 48 kg oxygen to form

$$\left(W_{\text{Cr in slag}}\right) kg \ldots\ldots\ldots \left\{\left(\frac{48}{104}\right) \times \left(W_{\text{Cr in slag}}\right)\right\} \textbf{kg oxygen to } \text{form } Cr_2O_3$$
$$(6.114)$$

**Mo Balance**:

$$\text{Mo input} = \text{Mo output}$$

$$(\text{Mo from Fe}-\text{Mo}) = \{(\text{Mo in Steel}) + (\text{Mo losses in slag})\}$$
$$\rightarrow (W_{\text{Fe}-\text{Mo}} \times f_{49}) = \left\{(W_{\text{Steel}} \times f_{50}) + W_{\text{Mo in slag}}\right\} \quad (6.115)$$

where

f$_{49}$ is fraction of Mo present in Fe–Mo,
f$_{50}$ is fraction of Mo present in steel, and
W$_{\text{Mo in slag}}$ is weight of Mo losses in slag.

**CaO Balance**:

$$\text{CaO input} = \text{CaO output}$$

$$\{(\text{CaO from lime} + \text{CaO from spar} + \text{CaO from DRI})\}$$
$$= \{(\text{CaO in slag}) + (\text{CaO for CaS})\}$$
$$\rightarrow \left\{(W_{\text{Lime}} \times f_{51}) + \left(W_{\text{Spar}} \times f_{52}\right) + \left(W_{\text{DRI}} \times f_{53}\right)\right\}$$
$$= \left\{W_{\text{CaO in slag}} + W_{\text{CaO for CaS}}\right\} \quad (6.116)$$

where

$f_{51}$ is fraction of CaO present in lime,
$f_{52}$ is fraction of CaO present in spar,
$f_{53}$ is fraction of CaO present in DRI,
$W_{\text{Cao in slag}}$ is weight of CaO goes in slag, and
$W_{\text{CaO for CaS}}$ is weight of CaO consume for CaS.

**MgO Balance**:

$$\text{MgO input} = \text{MgO output}$$

$$\{(\text{MgO from lime} + \text{MgO from DRI})\} = (\text{MgO in slag})$$
$$\rightarrow \{(W_{\text{Lime}} \times f_{54}) + (W_{\text{DRI}} \times f_{55})\} = \left(W_{\text{MgO in slag}}\right) \tag{6.117}$$

where

$f_{54}$ is fraction of MgO present in lime,
$f_{55}$ is fraction of MgO present in DRI, and
$W_{\text{Mgo in slag}}$ is weight of MgO goes in slag.

**Al$_2$O$_3$ Balance**:

$$\text{Al}_2\text{O}_3 \text{ input} = \text{Al}_2\text{O}_3 \text{ output}$$

$$\rightarrow \quad \{(\text{Al}_2\text{O}_3 \text{ from spar} + \text{Al}_2\text{O}_3 \text{ from DRI})\} = (\text{Al}_2\text{O}_3 \text{ in slag})$$

$$\rightarrow \quad \{(W_{\text{Spar}} \times f_{56}) + (W_{\text{DRI}} \times f_{57})\} = \left(W_{\text{Al2O3 in slag}}\right) \tag{6.118}$$

where

$f_{56}$ is fraction of Al$_2$O$_3$ present in spar,
$f_{57}$ is fraction of Al$_2$O$_3$ present in DRI, and
W is weigh $W_{\text{Al}_2\text{O}_3 \text{ in slag}}$t of Al$_2$O$_3$ goes in slag.

**CaF$_2$ Balance**:

$$\text{CaF}_2 \text{ input} = \text{CaF}_2 \text{ output}$$

$$\rightarrow \quad (\text{CaF}_2 \text{ from spar}) = (\text{CaF}_2 \text{ in slag})$$

$$\rightarrow \quad \left(W_{\text{Spar}} \times f_{58}\right) = \left(W_{\text{CaF2 in slag}}\right) \tag{6.119}$$

where

f$_{58}$ is fraction of $CaF_2$ present in spar, and
W$_{CaF2\ in\ slag}$ is weight of CaF goes in slag.

$$\textbf{Amount of slag} = W_{FeO\ in\ slag} + W_{SiO_2\ in\ slag} + W_{MnO\ in\ slag} + W_{P_2O_5\ in\ slag}$$
$$+ \ W_{CaS\ in\ slag} + W_{Cr_2O_3\ in\ slag} + W_{Mo\ in\ slag} + W_{CaO\ in\ slag}$$
$$+ \ W_{MgO\ in\ slag} + W_{Al_2O_3\ in\ slag} + W_{CaF_2\ in\ slag} \qquad (6.120)$$

$$\textbf{Amount of oxygen require} = Oxy\ for\ FeO + Oxy\ for\ SiO_2 + Oxy\ for\ MnO$$
$$+ \ Oxy\ for\ P_2O_5 + Oxy\ for\ Cr_2O_3$$
$$+ \ Oxy\ for\ gases\ formation \qquad (6.121)$$

## 6.9.2 Heat Balance

The aim of heat balance calculation is to estimate in-flow and out-flow of heat. The basic equation is as follows [11][2]:

$$Accumulation\ of\ heat\ within\ the\ system = (Heat\ input - Heat\ output) \qquad (6.122)$$

If there is no accumulation of heat within the system, then the above equation can be written as

$$Heat\ input = Heat\ output \qquad (6.123)$$

(1) **Heat input**
(i) Sensible heat carries by hot metal [12]:

$$(H_{HM}) = (0.22T + 17)\ kWh/t$$
$$= \left(\frac{(0.22T + 17)}{1000}\right)\ kWh/kg$$

(where T is in °C)

(ii) Chemical reactions are takes place at 1750 °C (i.e. 2023 K)

Heat generated due to chemical reactions ($\Delta H_{f,T}^0$), calculated values of heat of reaction for oxide formation at 2023 K (as shown in Table 6.1).

**Total heat input** = (i) Sensible heat carries by hot metal ($H_{HM}$), kWh

---

[2] With permission from "*JPC Bulletin on Iron & Steel*", Joint Plant Committee, Ministry of Steel, Kolkata.

**Table 6.1** Calculated values of heat of reaction for oxide formation at 2023 K [13]

| Oxide formation | $\Delta H^0_{298}$ (kJ/mol) | $\Delta H^0_{2023}$ (kJ/mol) | $\Delta H^0_{2023}$ (kJ/g) | $\Delta H^0_{2023}$ (kJ/kg) |
|---|---|---|---|---|
| CO | $-110.54$ | $-53.47$ | $-1.9096$ | $-1909.70$ of CO |
| $CO_2$ | $-393.51$ | $-301.71$ | $-6.8570$ | $-6857.05$ of $CO_2$ |
| FeO | $-264.43$ | $-146.77$ | $-2.0385$ | $-2038.54$ of FeO |
| MnO | $-384.93$ | $-289.53$ | $-4.0779$ | $-4077.93$ of MnO |
| $P_2O_5$ | $-1492.01$ | $-1042.32$ | $-7.3402$ | $-7340.27$ of $P_2O_5$ |
| $SiO_2$ | $-910.44$ | $-794.17$ | $-13.236$ | $-13,236.08$ of $SiO_2$ |
| $Cr_2O_3$ | $-1129.68$ | $-909.81$ | $-5.9855$ | $-5985.60$ of $Cr_2O_3$ |
| FeO + C = Fe + CO | $+153.89$ | $+93.30$ | $+1.2958$ | $+1295.87$ of FeO |

$+$ (ii) Total heat of reactions ($\Delta H^0_{f,T}$), kJ.

$=$ (i)  Sensible heat carries by hot metal ($H_{HM}$), kWh

$+$ (ii)  Total heat of reactions $\left[ \Delta H^0_{f,T}, \text{kJ} \times \left( \dfrac{1}{3600} \right) \right.$, kWh.

$$(6.124)$$

(Since 1 kwh = 3600 kJ).

Therefore, **Total heat input, without electric power** ($H_{in}$)

$$= \left[ H_{HM} + \left\{ \Delta H^0_{f,T} \times \left( \frac{1}{3600} \right) \right\} \right], \text{kWh} \qquad (6.125)$$

(2)    **Heat output**

(i)    Melting of charge material ($H_{melting}$), kWh

$$= \left[ \{ W \times q \times (T_2 - T_1) \} + (W \times L_f) \right] \times \left( \frac{1}{3600} \right) \qquad (6.126)$$

where

W = weight of material, kg
q = specific heat of material, kJ/kg.K
$L_f$ = latent heat of materials, kJ/kg.K
$T_2$ = temperature of the molten bath (i.e. 2023 K), and
$T_1$ = room temperature (i.e. 298 K).

Thermal data for charge materials is shown in Table 6.2.

(ii)    The heat requires for increasing temperature of hot metal ($H_{HM\ to\ T3}$), kWh

**Table 6.2** Thermal data for charge materials [13]

| Charge materials | Specific heat, kJ/kg · K × $10^3$ | Latent heat, kJ/kg · K |
|---|---|---|
| Hot metal | 510.0 | – |
| Steel scrap | 681.97 | 271.95 |
| DRI | 837.0 | 271.95 |
| HC Fe–Cr | 670.0 | 324.52 |
| HC Fe–Mn | 700.0 | 534.654 |
| LC Si–Mn | 628.0 | 578.78 |
| Fe–Mo | 1005.0 | 486.78 |

[a]Heat require [14] to melt of 1 tonne Ni at 2023 K is 295 kWh, i.e. 0.295 kWh/kg
[b]Heat require [14] to form slag at 2023 K is 444.44 kWh/t, i.e. 0.444 kWh/kg

$$= \left\{ W_{HM} \times q_{HM} \times (T_2 - T_3) \right\} \times \left( \frac{1}{3600} \right) \qquad (6.127)$$

where $T_3$ is temperature of the hot metal (K), $q_{HM}$ is' specific heat of hot metal, kJ/kg.K.

(iii)  Heat require [14] to form slag ($H_{slag}$) at 2023 K is 444.44 kWh/t, i.e. 0.444 kWh/kg.

(iv)  Radiation and other heat losses

   (a)  The heat losses by radiation from the furnace:

$$q_r = \varepsilon. \, \sigma. \, A. \, (T_2^4 - T_1^4) \;\; = \varepsilon. \, \sigma. \, A. \, (T_2^4) \qquad (6.128)$$

   (since $T_2 \gg T_1$, then value of $T_1^4$ is neglected)

   where

   $q_r$ = Heat losses by radiation in watts;
   $\varepsilon$ = Emissivity of liquid steel (0.28 for steel at 2023 K);
   $\sigma$ = Stefan–Boltzman's constant (5.67 × $10^{-8}$ W/m$^2$K);
   $A$ = Cross-sectional area of furnace, m$^2$;
   $T_1$ = Room temperature, K;
   $T_2$ = Temperature of the molten bath, K.

   (b)

   Other heat losses from the furnace = @4% of total heat input (average)
$$= \{0.04(\text{Total } H_{in})\}$$

Hence,

$$\text{radiation and other heat losses } (H_{\text{loss}}) = \left[ q_r + \{0.04 \ (\text{Total } H_{\text{in}})\} \right] \qquad (6.129)$$

**Total heat output** = (i) Melting of charge material $\left( H_{\text{melting}} \right)$

+ (ii) Heat requires for increasing temperature of HM

$(H_{\text{HM to T3}})$ + (iii) Heat require to form slag $\left( H_{\text{slag}} \right)$

+ (iv) Radiation and other heat losses $(H_{\text{loss}})$.

Therefore,

$$\text{Total } H_{\text{out}} = H_{\text{melting}} + H_{\text{HM to T3}} + H_{\text{slag}} + H_{\text{loss}} \qquad (6.130)$$

**Therefore, Electric power require** = Heat output−Heat input

$$= \text{Total } H_{\text{out}} - \text{Total } H_{\text{in}} \qquad (6.131)$$

**Example 6.7** Calculate amount of hot metal (at 1350 °C) required along with 350 kg stainless steel scrap to produce one tonne SAF 2205 (AISI Grade) duplex stainless steel in CONARC process. Find out how much lime is required to get basicity of 3.0. How much oxygen required to produce one tonne steel. Find out amount of slag produce. Also find out how much electrical power required.

| Element/Compound% | Steel produced | SS Scrap | HM | Spar | Lime | Ni | Fe–Cr | Fe–Mn | Fe–Mo |
|---|---|---|---|---|---|---|---|---|---|
| C | 0.03 | 0.08 | 3.5 | | | | 7.0 | 7.0 | 0.15 |
| Si | 0.8 | 1.0 | 1.0 | | | | 4.0 | 1.5 | |
| Mn | 2.0 | 2.0 | 0.5 | | | | 0.18 | 72.5 | |
| P | 0.03 | 0.04 | 0.4 | | | | 0.05 | 0.4 | |
| S | 0.02 | 0.03 | 0.04 | 0.3 | 0.15 | | 0.05 | 0.05 | |
| Fe | 67.12 | 70.85 | 94.56 | | | | 26.22 | 18.55 | 44.85 |
| Cr | 22.0 | 18.0 | | | | | 62.5 | | |
| Ni | 5.0 | 8.0 | | | | 100 | | | |
| Mo | 3.0 | | | | | | | | 55.0 |
| CaO | | | | 0.5 | 95.85 | | | | |
| MgO | | | | | 2.5 | | | | |
| SiO$_2$ | | | | 1.5 | 1.5 | | | | |
| Al$_2$O$_3$ | | | | 0.7 | | | | | |

(continued)

(continued)

| Element/Compound% | Steel produced | SS Scrap | HM | Spar | Lime | Ni | Fe–Cr | Fe–Mn | Fe–Mo |
|---|---|---|---|---|---|---|---|---|---|
| FeO | | | | | | | | | |
| Fe$_{Tol}$ | | | | | | | | | |
| CaF$_2$ | | | | 97.0 | | | | | |
| Amount charged, Kg | 1000 | 350 | ? | 5.5 | ? | 22 | 261.2 | 20.0 | 55.0 |

Given: (i) 1.5% Fe loss in slag w.r.t. liquid steel, and 0.6% Fe in fumes, (ii) 90% C form CO and 10% C form $CO_2$, (iii) There is no loss of Ni in slag, (iv) Other heat losses (excluding radiation) from the furnace is 4% of total heat input.

**Solution**

**(A)   Material Balance**

**Fe Balance**:

$$Fe\ input = Fe\ output$$

(Fe from HM + Fe from SS scrap + Fe from HC Fe−Cr + Fe from HC Fe − Mn
   + Fe from Fe − Mo) = (Fe in Steel + Fe losses in slag + Fe losses in flumes)
$\rightarrow \{(0.9456 \times W_{HM}) + (0.7085 \times 350) + (0.2622 \times 261.2)$
   $+ (0.1855 \times 20) + (0.4485 \times 55)\}$
$= \{(0.6712 \times 1000) + (0.015 \times 1000) + (0.006 \times 1000)\}$
$\rightarrow$ **$W_{HM} = 367.34\ kg$**

$$Fe + 1/2\,O_2 = FeO$$
$$56 \quad 16 \quad\ 72$$

56 kg Fe form 72 kg FeO
So, 15 kg . . . . . . . . $\left(\frac{72}{56}\right) \times 15 =$ **19.29 kg of FeO** goes to slag.
Again, 56 kg Fe react with 16 kg oxygen to form 72 kg FeO
So, 15  kg . . . . . . . . . . . . . . $\left(\frac{16}{56}\right) \times 15 =$ **4.29 kg oxygen** require for FeO formation.

**Si Balance**: Si input = Si output

(Si from HM + Si from SS scrap + Si from HC Fe−Cr + Si from HC Fe − Mn
   + Si from lime + Si from Spar) = (Si in Steel + Si losses in slag)
$\rightarrow \{(0.01 \times 367.34) + (0.01 \times 350) + (0.04 \times 261.2) + (0.015 \times 20)$
   $+ \left(0.015 \times \left(\frac{28}{60}\right) \times W_{Lime}\right) + \left(0.015 \times \left(\frac{28}{60}\right) \times 5.5\right)\}$

$$= \{(0.008 \times 1000) + W_{Si\ in\ slag}\}$$
$$\rightarrow W_{Si\ in\ slag} = \{9.96 + 7 \times 10^{-3}W_{Lime}\}$$
$$\rightarrow W_{SiO_2\ in\ slag} = \left\{(9.96 + 7 \times 10^{-3}W_{Lime}) \times \left(\frac{60}{28}\right)\right\} \tag{1}$$

$$Si + O_2 = SiO_2$$
$$28 \quad 32 \quad 60$$

60 kg $SiO_2$ content 28 kg Si
So, X kg . . . . . . . . . . . . . . . $\left\{\left(\frac{28}{60}\right) \times X\right\}$ kg Si.
Again 28 kg Si form 60 kg $SiO_2$.
So, Y kg Si form $\left\{\left(\frac{60}{28}\right) \times Y\right\}$ kg $SiO_2$ goes to slag.

**CaO Balance**: CaO input = CaO output

$$\{(CaO\ from\ lime) + (CaO\ from\ spar)\} = (CaO\ in\ slag)$$
$$\rightarrow \{(0.9585 \times W_{Lime}) + (0.005 \times 5.5)\} = \left(W_{CaO\ in\ slag}\right)$$
$$\rightarrow (W_{CaO\ in\ slag}) = \{(0.9585 \times W_{Lime})) + 0.027\} \tag{2}$$

Since basicity $= 3.0 = (\frac{W_{CaO}}{W_{SiO_2}}) = (\frac{\{(0.9585 \times W_{Lime})+0.027\}}{\{(9.96+7\times10^{-3}W_{Lime})\times(\frac{60}{28})\}})$ (From Eqs. (1) and (2))

$$\rightarrow \left[3.0\left\{(9.96 + 7 \times 10^{-3}W_{Lime}) \times \left(\frac{60}{28}\right)\right\}\right] = \{(0.9585 \times W_{Lime}) + 0.027\}$$
$$\rightarrow W_{Lime} = (63.973/0.9135) = \mathbf{70.03\ kg}$$

From Eq. (1):

$$W_{SiO_2\ in\ slag} = \left\{(9.96 + 7 \times 10^{-3}W_{Lime}) \times \left(\frac{60}{28}\right)\right\}$$
$$= \left\{(9.96 + 7 \times 10^{-3} \times 70.03) \times \left(\frac{60}{28}\right)\right\}$$
$$= \mathbf{22.39\ kg}$$

$$\rightarrow W_{Si\ in\ slag} = \{(9.96 + 7 \times 10^{-3}\ W_{Lime}) = (9.96 + 7 \times 10^{-3} \times 70.03)$$
$$= 10.45\ kg$$

Again, 28 kg Si react with 32 kg oxygen to form 60 kg $SiO_2$

$$\rightarrow So\ \ 10.45\ kg\ Si . . . . . . . . . \left\{\left(\frac{32}{28}\right) \times 10.45\right\} = \mathbf{11.94\ kg\ oxygen}\ require$$

for $SiO_2$ formation

**Mn Balance**: Mn input = Mn output

(Mn from HM + Mn from SS scrap + Mn from HC Fe−Cr
+ Mn from HC Fe − Mn)
= (Mn in Steel + Mn losses in slag)
→ {(0.005 × 367.34) + (0.02 × 350) + (0.0018 × 261.2) + (0.725 × 20)}
= {(0.02 × 1000) + W_{Mn in slag}}
→ W_{Mn in slag} = **3.81 kg**

$$Mn + 1/2 \, O_2 = MnO$$
$$55 \quad 16 \quad 71$$

55 kg Mn form 71 kg MnO
So, 3.81 kg .........{($\frac{71}{55}$) × 3.81} = **4.92 kg MnO** goes to slag.
Again 55 kg Mn react with 16 kg oxygen to form MnO.
So, 3.81 kg ...........{($\frac{16}{55}$) × 3.81} = **1.11 kg oxygen** to form MnO.

**C Balance**: C input = C output
(C from HM + C from SS scrap + C from HC Fe–Cr + C from HC Fe–Mn + C from Fe-Mo).
= (C in Steel + C losses in gas)

→ {(0.035 × 367.34) + (0.0008 × 350) + (0.07 × 261.2) + (0.07 × 20)
+ (0.0015 × 55)} = {(0.0003 × 1000) + W_{C in gas}}
⇒ W_{C in gas} = **32.6 kg**

So, C consume for CO formation = 0.9 × 32.6 = 29.34 kg

$$C + 1/2O_2 = CO$$
$$12 \quad 16 \quad 28$$

12 kg C form 28 kg CO gas
So, 29.34 kg ........{($\frac{28}{12}$) × 29.34} = **68.46 kg of CO**.
Again 12 kg C react with 16 kg oxygen to form CO,
29.34 kg .......{($\frac{16}{12}$) × 29.34} = **39.12 kg oxygen** to form CO.
Again C consume for $CO_2$ formation = 0.1 × 32.6 = 3.26 kg.

$$C + O_2 = CO_2$$
$$12 \quad 32 \quad 44$$

12 kg C form 44 kg $CO_2$ gas
So, 3.26 kg .......{($\frac{44}{12}$) × 3.26} = **11.95 kg of $CO_2$**.

Again 12 kg C react with 32 kg oxygen

$$3.26 \, kg \ldots \ldots \left\{ \left( \frac{32}{12} \right) \times 3.26 \right\} = \textbf{8.69 kg oxygen} \text{ for } CO_2 \text{ formation}$$

**Oxygen consumption for gas formation** $=$ Oxy for CO formation

$$+ \text{ Oxy for } CO_2 \text{ formation}$$

$$= 39.12 + 8.69 = 47.81 \, kg$$

**P Balance**: P input $=$ P output

(P from HM $+$ P from SS scrap $+$ P from HC Fe $-$ Cr $+$ P from HC Fe $-$ Mn)

$=$ (P in Steel $+$ P goes in slag)

$\rightarrow \{(0.004 \times 367.34) + (0.0004 \times 350) + (0.0005 \times 261.2) + (0.004 \times 20)\}$

$= \left\{ (0.0003 \times 1000) + W_{P \text{ in slag}} \right\}$

$\rightarrow W_{P \text{ in slag}} = \textbf{1.52 kg}$

$$2P + 5/2 \, O_2 = P_2O_5$$
$$62 \quad 80 \qquad 142$$

62 kg P forms 142 kg $P_2O_5$

So, $1.52 \, kg \ldots \ldots \{ (\frac{142}{62}) \times 1.52 \} = \textbf{3.48 kg } \textbf{P}_2\textbf{O}_5$ goes to slag.

Again, 62 kg P reacts with 80 kg oxygen.

So, $1.52 \, kg \ldots \ldots \{ (\frac{80}{62}) \times 1.52 \} = \textbf{1.96 kg oxygen}$ require for $P_2O_5$.

**S Balance**: S input $=$ S output

(S from HM $+$ S from SS scrap $+$ S from HC Fe$-$Cr $+$ S from HC Fe $-$ Mn

$+$ S from lime $+$ S from spar) $=$ (S in Steel $+$ S goes in slag)

$\rightarrow \{(0.0004 \times 367.34) + (0.0003 \times 350) + (0.0005 \times 261.2) + (0.0005 \times 20)$

$+ (0.0015 \times 70.03) + (0.003 \times 5.5)\} = \{(0.0002 \times 1000) + W_S \text{ in slag}\}$

$\rightarrow W_{S \text{ in slag}} = \textbf{0.31 kg}$

$$CaO + S = CaS + O$$
$$56 \quad 32 \quad 72$$

32 kg S forms 72 kg CaS

$$0.31 \, kg \ldots \ldots \left\{ \left( \frac{72}{32} \right) \times 0.31 \right\} = \textbf{0.70 kg of CaS}$$

Again 72 kg CaS forms from 56 kg of CaO.

$$0.70 \text{ kg} \ldots\ldots\ldots\left\{\left(\frac{56}{72}\right) \times 0.70\right\} = \textbf{0.54 kg of CaO consumefor CaS.}$$

Therefore, $W_{\text{CaO in slag}} = 70.03 - 0.54 = \textbf{69.49 kg}$.

**Cr Balance**: Cr input = Cr output

$$\{(\text{Cr from HC Fe$-$Cr}) + (\text{Cr from SS scrap})\} = (\text{Cr in Steel} + \text{Cr goes in slag})$$

$$\rightarrow \{(0.625 \times 261.2) + (0.18 \times 350)\} = \{(0.22 \times 1000) + W_{\text{Cr in slag}}\}$$
$$\rightarrow W_{\text{Cr in slag}} = 6.25 \text{ kg}$$

$$2Cr + 3/2O_2 = Cr_2O_3$$
$$104 \quad 48 \quad 152$$

104 kg Cr forms 152 kg $Cr_2O_3$.
$6.25$ kg $\ldots\ldots\{(\frac{152}{104}) \times 6.25\} = \textbf{9.13 kg of } Cr_2O_3$
Again 104 kg Cr reacts with 48 kg oxygen to form.
$6.25$ kg $\ldots\ldots\{(\frac{48}{104}) \times 6.25\} = \textbf{2.88 kg oxygen}$ to form $Cr_2O_3$.

**Mo Balance**: Mo input = Mo output
(Mo from Fe–Mo) = {(Mo in Steel) + (Mo losses in slag)}.

$$\rightarrow (0.55 \times 55) = \{(0.03 \times 1000) + W_{\text{Mo in slag}}\}.$$
$$\rightarrow W_{\text{Mo in slag}} = \textbf{0.25 kg}.$$

**Ni Balance**: Ni input = Ni output

$$\rightarrow \{(\text{Ni from SS scrap}) + (\text{Ni from Ni metal})\} = (\text{Ni in Steel})$$
$$\rightarrow \{0.08 \text{ x } 350) + (W_{Ni}) = (0.05 \text{ x } 1000)$$
$$\rightarrow W_{Ni} = \textbf{22 kg}$$

**MgO Balance**: MgO input = MgO output
(MgO from lime) = (MgO in slag).

$$\rightarrow \{(0.025 \times 70.03) = W_{\text{MgO in slag}}$$
$$\rightarrow W_{\text{MgO in slag}} = \textbf{1.75 kg}$$

**Al$_2$O$_3$ Balance**: $Al_2O_3$ input = $Al_2O_3$ output

$$\rightarrow (Al_2O_3 \text{ from spar}) = (Al_2O_3 \text{ in slag})$$
$$\rightarrow (0.007 \times 5.5) = W_{Al_2O_3 \text{ in slag}}$$
$$\rightarrow W_{Al_2O_3 \text{ in slag}} = \textbf{0.04 Kg}$$

**CaF$_2$ Balance**: $CaF_2$ input = $CaF_2$ output

$$\rightarrow (CaF_2 \text{ from spar}) = (CaF_2 \text{ in slag})$$

→ $(0.97 \times 5.5) = W_{CaF_2 \text{ in slag}}$
→ $W_{CaF_2 \text{ in slag}} = \mathbf{5.33 \ kg}$

**Amount of slag** $= W_{FeO \text{ in slag}} + W_{SiO_2 \text{ in slag}} + W_{MnO \text{ in slag}} + W_{P_2O_5 \text{ in slag}}$
$\qquad + W_{CaS \text{ in slag}} + W_{CaO \text{ in slag}} + W_{Cr_2O_3 \text{ in slag}} + W_{Mo \text{ in slag}}$
$\qquad + W_{MgO \text{ in slag}} + W_{Al_2O_3 \text{ in slag}} + W_{CaF_2 \text{ in slag}}$
$\qquad = 19.29 + 22.39 + 4.92 + 3.48 + 0.70 + 69.49 + 9.13$
$\qquad + 0.25 + 1.75 + 0.04 + 5.33 = \mathbf{136.77 \ kg}$

**Total oxygen consumption** $=$ Oxy for FeO $+$ Oxy for SiO$_2$ $+$ Oxy for MnO
$\qquad + \text{Oxy for P}_2O_3 + \text{Oxy for Cr}_2O_3 + \text{Oxy for CO and CO}_2$
$\qquad = 4.29 + 11.94 + 1.11 + 1.96 + 2.88 + 47.81 = \mathbf{69.99 \ kg}$

## (B)  Heat Balance

**Heat input**:

(i)  Sensible heat carries by hot metal $(H_{HM}) = (0.22 \ T + 17)$ kWh/t $= \left( \frac{(0.22T+17)}{1000} \right)$ kWh/kg,

Sensible heat carries by hot metal at 1350 °C $= \left( \frac{\{0.22 \times 1350\}+17\}}{1000} \right) = 0.314$ kWh/kg.
So, Sensible heat carries by 367.34 kg hot metal at 1350 °C $(H_{HM}) = 0.314 \times 367.34 = \mathbf{115.34 \ kWh}$.

(ii)  Heat of Reactions (from Table 6.2):

Heat formation for 19.29 kg FeO $= 2038.54 \times 19.29 = 39{,}323.44$ kJ.
Heat formation for 22.39 kg SiO$_2$ $= 13{,}236.08 \times 22.39 = 296{,}355.83$ kJ.
Heat formation for 4.92 kg MnO $= 4077.93 \times 4.92 = 20{,}063.42$ kJ.
Heat formation for 68.46 kg CO $= 1909.64 \times 68.46 = 130{,}733.95$ kJ.
Heat formation for 11.95 kg CO$_2$ $= 6857.05 \times 11.95 = 81{,}941.75$ kJ.
Heat formation for 3.48 kg P$_2$O$_5$ $= 7340.27 \times 3.48 = 25{,}544.14$ kJ.
Heat formation for 9.13 kg Cr$_2$O$_3$ $= 5985.6 \times 9.13 = 54{,}648.53$ kJ

Total heat of reactions $\left( \Delta H^0_{f,T} \right) = 39323.44 + 296355.83 + 20{,}063.42$
$\qquad + 130{,}733.95 + 81{,}941.75$
$\qquad + 25{,}544.14 + 54{,}648.53$
$\qquad = 648{,}611.06 \text{ kJ} = \mathbf{180.17 \ kWh}$

(Since 1 kwh $= 3600$ kJ)

**Total heat input, without electric power** $(H_{in}) = \left[ H_{HM} + \left( \Delta H^0_{f,T} \right) \right]$
$\qquad = \mathbf{115.34 + 180.17}$

$$= 295.51 \, \text{kWh}$$

Total heat input $(H_{in}) = [295.51 + E]$.

where E is electric power.

**Heat Output**:

(i)

Melting of scrap $= \left\{ W_{Sc} \times q_{sc} \times (T_2 - T_1) \right\} + (W_{Sc} \times L_f)$
$$= 350 \times 0.68197 \times (2023 - 298) + (350 \times 271.95)$$
$$= 411739.39 + 95182.5 = 506921.89 \, \text{kJ} = \textbf{140.81 kWh}$$

Melting of HC Fe$-$Cr $= \{261.2 \times 0.67 \times (2023 - 298)\} + (261.2 \times 324.52)$
$$= 301,881.9 + 84,764.62 = 386,646.52 \, \text{kJ}$$
$$= \textbf{107.4 kWh}$$

Melting of HC Fe $-$ Mn $= \{20 \times 0.7 \times (2023 - 298)\} + (20 \times 534.65)$
$$= 24,150.0 + 10,693.0 = 34,843.0 \, \text{kJ} = \textbf{9.68 kWh}$$

Melting of Fe $-$ Mo $= \{55 \times 1.005 \times (2023 - 298)\} + (55 \times 486.78)$
$$= 95,349.38 + 26,772.9 = 122,122.28 \, \text{kJ} = \textbf{33.92 kWh}$$

For melting of 1 kg Ni at 2023 K require 0.295 kWh.

Therefore, for melting 22 kg Ni $= (22 \times 0.295) = \textbf{6.49 kWh}$.

Total heat requires for melting $(H_{melting}) = \textbf{140.81} + \textbf{107.4} + \textbf{9.68} + \textbf{33.92} + \textbf{6.49} = \textbf{298.3 kWh}$.

(ii)  Heat requires for increasing temperature from 1350 to 1750 °C of HM

$$= \{367.34 \times 0.51 \times (2023 - 1623)\} \times \left( \frac{1}{3600} \right) = \textbf{20.82 kWh}$$

(iii)  Heat requires for 1 kg slag formation at 2023 K $= 0.444 \, \text{kWh}$

Therefore, heat requires for 136.77 kg slag formation $= 136.77 \times 0.444 = \textbf{60.73 kWh}$.

(iv)  Heat losses: (a) It is found by calculation for EAF (7.6 m diameter furnace) of capacity 180 tonne, the radiation loss per tonne of liquid steel is 67 kWh (as shown in Example 6.5).

(b)

Other heat losses from the furnace $= \{0.04(\text{Total } H_{in})\} = 0.04 \times [295.51 + E]$
$$= [11.82 + 0.04E] \, \text{kWh}.$$

Hence,

radiation and other heat losses $(H_{loss}) = \left[ q_r + \{0.04 \, (\text{Total } H_{in})\} \right.$

$$= 67.0 + [11.82 + 0.04E]$$
$$= [78.82 + 0.04E] \textbf{ kWh}.$$

Total Heat output $(H_{out}) = \left[H_{melting} + H_{HM \, to \, T3} + H_{slag} + H_{loss}\right]$
$$= 298.3 + 20.82 + 60.73 + [78.82 + 0.04E]$$
$$= [458.67 + 0.04E] \textbf{ kWh}.$$

**Since heat input** $((H_{in}) = $ **heat output** $(H_{out})$.
**Therefore,** $[295.51 + E] = [458.67 + 0.04\,E]$.
**So,** $(1 - 0.04)\,E = 458.67 - 295.51 = 163.16$.
**Hence, E** $= (163.16/0.96) = 169.96$ **kWh.**
**Electric power requirement** $= 169.96$ **kWh.**

| Input (kWh) | | Output (kWh) | |
|---|---|---|---|
| Sensible heat carries by HM | 115.34 | Heat requires for melting | 298.30 |
| Heat of formations | 180.17 | Heat requires for rising temperature of HM | 20.82 |
| Power require | 169.96 | Heat require to form slag | 60.73 |
| | | Radiation and other heat losses | 85.62 |
| **Total** | **465.47** | **Total** | **465.47** |

**Example 6.8** Calculate amount of DRI to be charged along with 300 kg hot metal (at 1350 °C) and 114 kg MS scrap to produce one tonne duplex stainless steel (SAF 2205) in CONARC process.

Other materials are charged as follows: HC Fe–Cr → 363 kg/t, HC Fe–Mn→ 30 kg/t, Fe-Mo → 55 kg/t, Ni → 50 kg/t, Lime → 82.2 kg/t, and Spar → 5.5 kg/t.

How much oxygen required to produce 1 tonne steel and find out amount of slag produce. Also find out how much electrical power required to produce one tonne duplex stainless steel.

Assume: (i) 1.5 and 0.6% Fe loss in slag and flumes, respectively, (ii) 90% C form CO and 10% C form $CO_2$, (iii) total heat losses are 17% of total heat input.

Composition of charge materials and stainless steel produce are shown as following:

| Mat | Composition % | | | | | | | | | | | |
|---|---|---|---|---|---|---|---|---|---|---|---|---|
| | C | Mn | Si | S | P | Fe | Cr | Ni | CaO | MgO | SiO$_2$ | Al$_2$O$_3$ |
| HM | 3.5 | 0.5 | 1.0 | 0.04 | 0.4 | 94.56 | | | | | | |
| MS scrap | 0.26 | 0.25 | 0.25 | 0.04 | 0.03 | 99.17 | | | | | | |
| DRI | 1.25 | – | – | 0.01 | 0.05 | 85.0 | | FeO 9.64 | 1.0 | 1.0 | 1.5 | 0.5 |

(continued)

(continued)

| Mat | Composition % | | | | | | | | | | | |
|---|---|---|---|---|---|---|---|---|---|---|---|---|
| | C | Mn | Si | S | P | Fe | Cr | Ni | CaO | MgO | SiO$_2$ | Al$_2$O$_3$ |
| HC Fe–Cr | 7.0 | 0.18 | 4.0 | 0.05 | 0.05 | 26.22 | 62.5 | | | | | |
| HC FeMn | 7.0 | 72.5 | 1.5 | 0.05 | 0.4 | 18.55 | | | | | | |
| FeMo | 0.15 | – | – | – | – | 44.85 | | Mo 55 | | | | |
| Lime | – | – | – | 0.15 | – | – | – | – | 95.85 | 2.5 | 1.5 | |
| Spar | – | – | – | 0.3 | – | – | – | – | 0.5 | CaF$_2$ 97.0 | 1.5 | 0.7 |

| AISI Grade | %C | %Cr | %Ni | %Mo | %Mn | %Si | %S | %P | %Fe |
|---|---|---|---|---|---|---|---|---|---|
| SAF 2205 | 0.03 | 22.0 | 5.0 | 3.0 | 2.0 | 0.8 | 0.02 | 0.03 | 67.12 |

## Solution

### (A)   Material Balance

**Fe Balance**: Fe input = Fe output

$$\text{(Fe from HM + Fe from MS scrap + Fe from DRI + Fe from HC Fe–Cr}$$
$$\text{+ Fe from HC Fe – Mn + Fe from Fe – Mo)}$$
$$= \text{(Fe in Steel + Fe losses in slag + Fe losses in flumes)}$$
$$\rightarrow \{(0.9456 \times 300) + (0.9917 \times 114) + [(0.85 \times W_{DRI}) +$$
$$\left(0.0964 \times \left(\frac{56}{72}\right) \times W_{DRI}\right)]$$
$$+ (0.2622 \times 363) + (0.1855 \times 30) + (0.4485 \times 55)\}$$
$$= \{(0.6712 \times 1000) + (0.015 \times 1000) + (0.006 \times 1000)\}$$
$$\rightarrow \mathbf{W_{DRI} = 183.85 \ kg}$$

$$Fe + 1/2 \, O_2 = FeO$$
$$56 \quad 16 \quad 72$$

72 kg FeO content 56 kg of Fe

$$X \ kg \ldots\ldots\ldots\ldots \left\{\left(\frac{56}{72}\right) \times X\right\} kg \ Fe$$

Again, 56 kg Fe form 72 kg FeO

So, 15 kg . . . . . . . . $\left(\frac{72}{56}\right) \times 15 = \mathbf{19.29\ kg\ of\ FeO}$ goes to slag.

Again, 56 kg Fe react with 16 kg oxygen to form 72 kg FeO

So, 15 kg . . . . . . . . . . . . . . . $\left(\frac{16}{56}\right) \times 15 = \mathbf{4.29\ kg\ oxygen}$ require for FeO formation.

**Si Balance**: Si input = Si output

$$(\text{Si from HM} + \text{Si from MS scrap} + \text{Si from DRI} + \text{Si from HC Fe}-\text{Cr} + \text{Si from HC Fe}-\text{Mn}$$
$$+ \text{Si from lime} + \text{Si from Spar}) = (\text{Si in Steel} + \text{Si losses in slag})$$

$$\rightarrow \left\{ (0.01 \times 300) + (0.0025 \times 114) + \left[ 0.015 \times \left(\frac{28}{60}\right) \times 183.85 \right] + (0.04 \times 363) + (0.015 \times 30) \right.$$
$$\left. + \left[ 0.015 \times \left(\frac{28}{60}\right) \times 82.2 \right] + \left[ 0.015 \times \left(\frac{28}{60}\right) \times 5.5 \right] \right\} = \left\{ (0.008 \times 1000) + W_{\text{Si in slag}} \right\}$$

$$\rightarrow W_{\text{Si in slag}} = \mathbf{12.16\ kg}$$

$$\text{Si} + O_2 = SiO_2$$
$$28 \quad 32 \quad 60$$

60 kg $SiO_2$ content 28 kg Si

So, X kg . . . . . . . . . . . . . $\{(\frac{28}{60}) \times X\}$ kg Si.

Again, 28 kg Si form 60 kg $SiO_2$.

So, 12.16 kg Si form $\{(\frac{60}{28}) \times 12.16\} = \mathbf{26.05\ kg\ SiO_2}$ goes to slag.

Again, 28 kg Si react with 32 kg oxygen to form 60 kg $SiO_2$.

So, 12.16 kg Si . . . . . . . . . $\{(\frac{32}{28}) \times 12.16\} = \mathbf{13.9\ kg\ oxygen}$ require for $SiO_2$ formation.

**Mn Balance**: Mn input = Mn output

$$(\text{Mn from HM} + \text{Mn from MS scrap} + \text{Mn from HC Fe}-\text{Cr} + \text{Mn from HC Fe} - \text{Mn})$$
$$= (\text{Mn in Steel} + \text{Mn losses in slag})$$
$$\rightarrow \{ (0.005 \times 300) + (0.0025 \times 114) + (0.0018 \times 363) + (0.725 \times 30) \}$$
$$= \{ (0.02 \times 1000) + W_{\text{Mn in slag}} \}$$
$$\rightarrow W_{\text{Mn in slag}} = \mathbf{4.19\ kg}$$

$$\text{Mn} + 1/2\ O_2 = MnO$$
$$55 \quad 16 \quad 71$$

55 kg Mn form 71 kg MnO

So, 4.19 kg . . . . . . . . . $\{(\frac{71}{55}) \times 4.19\} = \mathbf{5.41\ kg\ MnO}$ goes to slag.

Again, 55 kg Mn react with 16 kg oxygen to form MnO.

So, 4.19 kg . . . . . . . . . . $\{(\frac{16}{55}) \times 4.19\} = \mathbf{1.22\ kg\ oxygen}$ to form MnO.

**C Balance**: C input = C output

$$(\text{C from HM} + \text{C from MS scrap} + \text{C from DRI} + \text{C from HC Fe} - \text{Cr}$$
$$+ \text{ C from HC Fe} - \text{Mn} + \text{C from Fe} - \text{Mo})$$
$$= (\text{C in Steel} + \text{C losses in gases})$$
$$\rightarrow \{(0.035 \times 300) + (0.0026 \times 114) + (0.0125 \times 183.85)$$
$$+ (0.07 \times 363) + (0.07 \times 30) + (0.0015 \times 55)\}$$
$$= \{(0.0003 \times 1000) + W_{\text{C in gas}}\}$$
$$\rightarrow W_{\text{C in gas}} = 40.39 \text{ kg}$$

$$\text{FeO} + \text{C} = \text{Fe} + \text{CO}$$
$$72 \quad 12 \quad 56 \quad 28$$

72 kg FeO reduced by 12 kg C
$(0.0964 \times 183.85)$ kg FeO (in DRI) reduced by $\{(\frac{12}{72}) \times (0.0964 \times 183.85)\}$ = 2.95 kg **of** C.
72 kg FeO reduced by C to form 28 kg CO

$$17.72 \text{ kg} \dots\dots\dots\dots\dots\dots\dots\dots\dots\dots \left\{\left(\frac{28}{72}\right) \times 17.72\right\} = 6.89 \text{ kg CO}$$

Hence, C available for CO and $CO_2$ formation = 40.39 – 2.95 = 37.44 kg.
So, C consume for CO formation = $0.9 \times 37.44 = 33.7$ kg.

$$\text{C} + 1/2 \text{ O}_2 = \text{CO}$$
$$12 \quad 16 \quad 28$$

12 kg C form 28 kg CO gas
So, 33.7 kg ........$\{(\frac{28}{12}) \times 33.7\} = \textbf{78.62 kg of CO}$.
Again, 12 kg C react with 16 kg oxygen to form CO.

$$33.7 \text{ kg} \dots\dots\dots \left\{\left(\frac{16}{12}\right) \times 33.7\right\} = \textbf{44.93 kg oxygen to form CO.}$$

Total CO formation = 6.89 + 78.62 = **85.51 kg**.
Again, C consume for $CO_2$ formation = $0.1 \times 37.44 = 3.74$ kg.

$$\text{C} + \text{ O}_2 = \text{CO}_2$$
$$12 \quad 32 \quad 44$$

12 kg C form 44 kg $CO_2$ gas
So, 3.74 kg .......$\{(\frac{44}{12}) \times 3.74\} = \textbf{13.73 kg of CO}_2.$

Again, 12 kg C react with 32 kg oxygen.

$$3.74 \text{ kg} \ldots \ldots \left\{ \left( \frac{32}{12} \right) \times 3.74 \right\} = \textbf{9.97 kg oxygen} \text{ for } CO_2 \text{ formation.}$$

$$\textbf{Oxygen consumption for gas formation} = \text{Oxy for CO formation}$$
$$+ \text{ Oxy for } CO_2 \text{ formation}$$
$$= 44.93 + 9.97 = \textbf{54.9 kg}$$

**Gas formation** $= CO$ formation $+ CO_2$ formation $= \textbf{85.51} + \textbf{13.73} = \textbf{99.24 kg}$

**P Balance**: P input $=$ P output

$$(\text{P from HM} + \text{P from MS scrap} + \text{P from DRI} + \text{P from HC Fe} - \text{Cr}$$
$$+ \text{ P from HC Fe} - \text{Mn})$$
$$= (\text{P in Steel} + \text{P goes in slag})$$
$$\rightarrow \{(0.004 \times 300) + (0.0003 \times 114) + (0.0005 \times 183.85)$$
$$+ (0.0005 \times 363) + (0.004 \times 30)\}$$
$$= \left\{ (0.0003 \times 1000) + W_{\text{P in slag}} \right\}$$
$$\rightarrow W_{\text{P in slag}} = \textbf{1.33 kg}$$

$$2P + 5/2 \, O_2 = P_2O_5$$
$$62 \quad 80 \quad 142$$

62 kg P forms 142 kg $P_2O_5$
So, 1.33 kg $\ldots \ldots \{ (\frac{142}{62}) \times 1.33 \} = \textbf{3.04 kg } \mathbf{P_2O_5}$ goes to slag.
Again, 62 kg P reacts with 80 kg oxygen.
So, 1.33 kg $\ldots \ldots \{ (\frac{80}{62}) \times 1.33 \} = \textbf{1.72 kg oxygen}$ require for $P_2O_5$.

**S Balance**: S input $=$ S output

$$(\text{S from HM} + \text{S from MS scrap} + \text{S from DRI} + \text{S from HC Fe} - \text{Cr}$$
$$+ \text{ S from HC Fe} - \text{Mn} + \text{S from lime}$$
$$+ \text{ S from spar}) = (\text{S in Steel} + \text{S goes in slag})$$
$$\rightarrow \{(0.0004 \times 300) + (0.0004 \times 114) + (0.0001 \times 183.85) + (0.0005 \times 363)$$
$$+ (0.0005 \times 30) + (0.0015 \times 82.2) + (0.003 \times 5.5)\}$$
$$= \left\{ (0.0002 \times 1000) + W_{\text{S in slag}} \right\}$$
$$\rightarrow W_{\text{S in slag}} = \textbf{0.32 kg}$$

$$CaO + S = CaS + O$$
$$56 \quad 32 \quad 72$$

32 kg S forms 72 kg CaS

$$0.32 \text{ kg} \ldots \ldots \left\{ \left( \frac{72}{32} \right) \times 0.32 \right\} = \textbf{0.72 kg of CaS}$$

Again, 72 kg CaS forms from 56 kg of CaO.

$$0.72 \text{ kg} \ldots \ldots \ldots \left\{ \left( \frac{56}{72} \right) \times 0.72 \right\} = \textbf{0.56 kg of CaO consume for CaS.}$$

**Cr Balance**: Cr input = Cr output

$$(\text{Cr from HC Fe}-\text{Cr}) = (\text{Cr in Steel} + \text{Cr goes in slag})$$
$$\rightarrow (0.625 \times 363) = \left\{ (0.22 \times 1000) + W_{\text{Cr in slag}} \right\}$$
$$\rightarrow W_{\text{Cr in slag}} = \textbf{6.875 kg}$$

$$2Cr + 3/2 \, O_2 = Cr_2O_3$$
$$104 \quad 48 \quad 152$$

104 kg Cr forms 152 kg $Cr_2O_3$

$$6.875 \text{ kg} \ldots \ldots \left\{ \left( \frac{152}{104} \right) \times 6.875 \right\} = \textbf{10.05 kg} \text{ of } Cr_2O_3 \text{ in slag}$$

Again, 104 kg Cr reacts with 48 kg oxygen to form

$$6.875 \text{ kg} \ldots \ldots \left\{ \left( \frac{48}{104} \right) \times 6.875 \right\} = \textbf{3.17 kg oxygen} \text{ to form } Cr_2O_3.$$

**Mo Balance**: Mo input = Mo output

$$(\text{Mo from Fe} - \text{Mo}) = \{(\text{Mo in Steel}) + (\text{Mo losses in slag})\}$$

$$\rightarrow (0.55 \times 55) = \{(0.03 \times 1000) + W_{\text{Mo in slag}})$$
$$\rightarrow W_{\text{Mo in slag}} = \textbf{0.25 kg}$$

**CaO Balance**: CaO input = CaO output

$$\{(\text{CaO from lime} + \text{CaO from spar} + \text{CaO from DRI})\}$$

$$= \{(CaO \text{ in slag}) + (CaO \text{ form } CaS)\}$$

$$\rightarrow \{(0.9585 \times 82.2) + (0.0005 \times 5.5) + (0.01 \times 183.85\} = \{(W_{CaO \text{ in slag}}) + 0.56\}$$
$$\rightarrow W_{CaO \text{ in slag}} = \textbf{80.07 kg}$$

**MgO Balance**: MgO input = MgO output

$$\{(MgO \text{ from lime} + MgO \text{ from DRI})\} = (MgO \text{ in slag})$$

$$\rightarrow \{(0.025 \times 82.2) + (0.01 \times 183.85)\} = W_{MgO \text{ in slag}}$$
$$\rightarrow W_{MgO \text{ in slag}} = \textbf{3.89 kg}$$

**Al₂O₃ Balance**: $Al_2O_3$ input = $Al_2O_3$ output
$$\rightarrow \{(Al_2O_3 \text{ from spar} + Al_2O_3 \text{ from DRI})\} = (Al_2O_3 \text{ in slag})$$
$$\rightarrow \{(0.007 \times 5.5) + (0.005 \times 183.85)\} = W_{Al_2O_3 \text{ in slag}}$$
$$\rightarrow W_{Al_2O_3 \text{ in slag}} = \textbf{0.96 kg}$$

**CaF₂ Balance**: $CaF_2$ input = $CaF_2$ output
$$\rightarrow (CaF_2 \text{ from spar}) = (CaF_2 \text{ in slag})$$
$$\rightarrow (0.97 \times 5.5) = W_{CaF_2 \text{ in slag}}$$
$$\rightarrow W_{CaF_2 \text{ in slag}} = \textbf{5.34 kg}$$

$$\begin{aligned}
\textbf{Amount of slag} &= W_{FeO \text{ in slag}} + W_{SiO2 \text{ in slag}} + W_{MnO \text{ in slag}} + W_{P_2O_5 \text{ in slag}} \\
&\quad + W_{CaS \text{ in slag}} \\
&\quad + W_{Cr_2O_3 \text{ in slag}} + W_{Mo \text{ in slag}} + W_{CaO \text{ in slag}} + W_{MgO \text{ in slag}} \\
&\quad + W_{Al_2O_3 \text{ in slag}} \\
&\quad + W_{CaF_2 \text{ in slag}} \\
&= 19.29 + 26.05 + 5.41 + 3.04 + 0.72 + 10.05 + 0.25 \\
&\quad + 80.07 + 3.89 + 0.96 + 5.34 = \textbf{155.07 kg}
\end{aligned}$$

$$\begin{aligned}
\textbf{Amount of oxygen require} &= \text{Oxy for FeO} + \text{Oxy for } SiO_2 + \text{Oxy for MnO} \\
&\quad + \text{Oxy for } P_2O_5 + \text{Oxy for } Cr_2O_3 \\
&\quad + \text{Oxy for gas } CO_2 \text{ formation} \\
&= 4.29 + 13.9 + 1.22 + 1.72 + 3.17 + 54.9 \\
&= \textbf{79.2 kg}
\end{aligned}$$

(B)    **Heat Balance**

**Heat input**:

(i)   Sensible heat carries by hot metal at 1350 °C = **0.314 kWh/kg**

Sensible heat carries by 300 kg hot metal at 1350 °C ($H_{HM}$) = 0.314 × 300 = **94.20 kWh**.

(ii)   Heat of Reactions (from Table 6.2):

Heat formation for 19.29 kg FeO = 2038.54 × 19.29 = 39,323.44 kJ.
Heat formation for 26.05 kg $SiO_2$ = 13,236.08 × 26.05 = 344,799.88 kJ.
Heat formation for 5.41 kg MnO = 4077.93 × 5.41 = 22,061.60 kJ.
Heat formation for 78.62 kg CO = 1909.64 × 78.62 = 150,135.90 kJ.
Heat formation for 13.73 kg $CO_2$ = 6857.05 × 13.73 = 94,147.30 kJ.
Heat formation for 3.04 kg $P_2O_5$ = 7340.27 × 3.04 = 22,314.42 kJ.
Heat formation for 10.05 kg $Cr_2O_3$ = 5985.60 × 10.05 = 60,155.28 kJ

**Heat absorbs** for FeO + C = Fe + CO is 1295.87 kJ/kg FeO.

$$(0.0964 \times 183.85) = 17.72 \text{ kg FeO (in DRI) reduced by C, heat absorbs}$$
$$= 1295.87 \times 17.72$$
$$= \mathbf{22,966.89 \ kJ}.$$

$$\text{Total heat of reactions} = 39,323.44 + 344,799.88 + 22,061.60 + 150,135.90$$
$$+ \ 94,147.30$$
$$+ \ 22,314.42 + 60,155.28$$
$$= 732,937.82 \text{ kJ}$$

$$\text{Total heat of formations} \left( \Delta H^0_{f,T} \right) = \text{Total heat of reactions}$$
$$- \text{ Heat absorbs for FeO reduction}$$
$$= 732,937.82 - 22,966.89$$
$$= 709,970.93 \text{kJ} = \mathbf{197.21 \ kWh}$$

**Total heat input, without electric power** ($H_{in}$) = $\left[ H_{HM} + \left( \Delta H^0_{f,T} \right) \right]$
$$= 94.20 + 197.21 = \mathbf{291.41 \ kWh}$$

$$\text{Total heat input } (H_{in}) = [\mathbf{291.41 + E}] \ \mathbf{kWh}$$

where E is electric power.

**Heat Output:**

$$\text{Melting of scrap} = \left[ \{ W_{Sc} \times q_{sc} \times (T_2 - T_1) \} + (W_{Sc} \times L_f) \right] \times \left( \frac{1}{3600} \right)$$

$$= \left[ \{ 114 \times 0.68197 \times (2023 - 298) \} \right.$$

$$\left. + (114 \times 271.95) \right] \times \left( \frac{1}{3600} \right)$$

(i)

$$= [134, 109.4 + 31, 002.3] \times \left( \frac{1}{3600} \right)$$

$$= 165, 111.7 \times \left( \frac{1}{3600} \right)$$

$$= \mathbf{45.86 \ kWh}$$

$$\text{Melting of DRI} = [\{ 183.85 \times 0.837 \times (2023 - 298) \}$$

$$+ (183.85 \times 271.95)] \times \left( \frac{1}{3600} \right)$$

$$= [265, 447.23 + 49, 998.01] \times \left( \frac{1}{3600} \right)$$

$$= 315, 445.24 \times \left( \frac{1}{3600} \right)$$

$$= \mathbf{87.62 \ kWh}$$

$$\text{Melting of HC Fe}-\text{Cr} = [\{ 363 \times 0.67 \times (2023 - 298) \}$$

$$+ (363 \times 324.52)] \times \left( \frac{1}{3600} \right)$$

$$= [419,537.25 + 117,800.76] \times \left( \frac{1}{3600} \right)$$

$$= 537,338.01 \times \left( \frac{1}{3600} \right)$$

$$= \mathbf{149.26 \ kWh}$$

$$\text{Melting of HC Fe} - \text{Mn} = [\{ 30 \times 0.7 \times (2023 - 298) \}$$

$$+ (30 \times 534.65)] \times \left( \frac{1}{3600} \right)$$

$$= \{ 36, 225.0 + 16, 039.5 \} \times \left( \frac{1}{3600} \right)$$

$$= 52, 264.5 \times \left( \frac{1}{3600} \right)$$

$$= \mathbf{14.52 \ kWh}$$

$$\text{Melting of Fe} - \text{Mo} = [\{55 \times 1.005 \times (2023 - 298)\}$$

$$+ (55 \times 486.78)] \times \left(\frac{1}{3600}\right)$$

$$= [95,349.38 + 26,772.9] \times \left(\frac{1}{3600}\right)$$

$$= 122,122.28 \times \left(\frac{1}{3600}\right)$$

$$= \mathbf{33.92\ kWh}$$

Heat require to melt of 1 kg Ni at 2023 K is 1062 kJ $= 0.295$ kWh. Therefore, for melting 50 kg Ni $= (50 \times 0.295) = \mathbf{14.75\ kWh,}$

$$\textbf{Total heat requires for melting } (H_{\text{melting}}) = \mathbf{45.86 + 87.62 + 149.26}$$

$$+ \ \mathbf{14.52 + 33.92 + 14.75}$$

$$= \mathbf{345.93\ kWh}$$

(ii)   The heat requires for increasing temperature of hot metal (kWh)

$$= \{W_{\text{HM}} \times q_{\text{HM}} \times (T_2 - T_3)\} \times \left(\frac{1}{3600}\right)$$

where $T_3$ is temperature of the hot metal (K).

Heat requires for increasing temperature of HM from 1350 to 1750 °C $(H_{\text{HM to T3}})$

$$= [\{300 \times 0.51 \times (2023 - 1623)\}] \times \left(\frac{1}{3600}\right)$$

$$= 61,200 \times \left(\frac{1}{3600}\right) = \mathbf{17.0\ kWh.}$$

(iii)   Heat require to form slag at 2023 K is 444.44 kWh/t, i.e. **0.444 kWh/kg.**

Therefore, heat requires for 155.07 kg slag formation $(H_{\text{slag}}) = 155.07 \times 0.444 = \mathbf{68.92\ kWh.}$

(iv)   Since total heat losses are 17% of total heat input $(H_{\text{loss}}) = \{0.17 \times (\text{Total } H_{\text{in}})\}$

$$= [0.17 \times (\mathbf{291.41 + E})]\ kWh = (\mathbf{49.54 + 0.17\ E})\ kWh$$

$$\text{Total heat output } (H_{\text{out}}) = \left[ H_{\text{melting}} + H_{\text{HM to T3}} + H_{\text{slag}} + H_{\text{loss}} \right]$$

$$= 345.93 + 17.0 + 68.92 + (\mathbf{49.54 + 0.17\ E})$$

$$= (\mathbf{481.39 + 0.17\ E})\ \mathbf{kWh}$$

**Since heat input** ($H_{in}$) = **heat output** ($H_{out}$)

So, $[291.41 + E] = (481.39 + 0.17 E)$ or $(1 - 0.17)E = 481.39 - 291.41 = 189.98$.

**Therefore, Electrical power requirement, E** $= (189.98/0.83) = 228.89$ kWh.

| Input (kWh) | | Output (kWh) | |
|---|---|---|---|
| Sensible heat carries by HM | 94.20 | Heat requires for melting | 345.93 |
| Heat of formations | 197.21 | Heat requires for rising temperature of HM | 17.00 |
| Power requirement | 228.89 | Heat require to form slag | 68.92 |
| | | Radiation and other heat losses | 88.45 |
| Total | **520.30** | Total | **520.30** |

## 6.10  Kinetic Model

Almost all metallurgical reactions are heterogeneous, i.e. the reactants and products are not in the same phase. So, the reaction takes place only on a surface where the reactants and products meet. Figure 6.7 shows two cases, (i) burning of carbon in oxygen and (ii) oxidation of metal. So, the transport of reactants or products by diffusion play an important role in heterogeneous reactions [8].

When billets are heated in the furnace, an oxide layer forms on the surface and below this layer a decarburized layer is found. The experimental study of kineties of oxidation shows that the thickness of oxide layer follows a parabolic growth law:

$$X^2 = kt \tag{6.132}$$

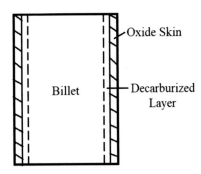

**Fig. 6.7**  Oxidizing and decarburized billet

where X is the thickness of the oxidized layer at time t and k is the rate constant which depends on the concentration of oxygen in the furnace, temperature, and composition of steel.

The carbon at the oxide/metal interface gets oxidized and its concentration there becomes zero. So, the carbon present in the billet diffuses out. The carbon profile in the billet is given by the equation:

$$\frac{\delta C}{\delta t} = D \cdot \frac{\delta^2 C}{\delta x^2} \tag{6.133}$$

where C is the concentration of carbon and D is the diffusivity of carbon.

Since the thickness of the layer, which is affected by carbon diffusion, is insignificant compared to the billet thickness, the problem can be considered as semi-infinity. So, the initial and boundary conditions are

$$\text{at } t = 0, \text{ for } x > 0, C = C_0 \tag{6.134}$$

$$\text{and } t > 0, \text{ at } x = X, C = 0 \tag{6.135}$$

Equation (6.135) states that the concentration of carbon at the oxide/metal interface is zero.

Assuming that there is no growth of oxide layer, the solution of Eqs. (6.133)–(6.135):

$$C = C_0 \, \text{erf}\left[x/2(Dt)^{1/2}\right] \tag{6.136}$$

Figure 6.8 shows the plot of $C/C_0$ as a function of $[x/2\,(Dt)^{1/2}]$.

The assumption made in deriving Eq. (6.136) are as follows:

1.  Diffusivity is constant,
2.  Oxide scale formed on the surface of billet does not grow during the diffusion of carbon.

**Fig. 6.8**  Concentration profile of carbon

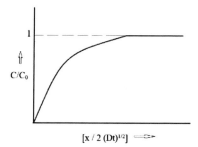

$[x / 2\,(Dt)^{1/2}]$

Diffusivity of carbon in steel depends on the percentage of carbon, steel composition, and temperature. So, Eq. (6.136) is not applicable for most of the actual cases, but it gives an idea of the affected thickness. The second assumption also is not valid. The growth of oxide layer will continue so long as furnace atmosphere is oxidizing.

## 6.11   Exercises

**Problem 6.1.** The following data are obtained from an experiment of reduction of iron ore fine by the hydrogen gas at 973 K. Find out (a) initial rate of reduction from fraction of reduction (R) w.r.t time (t) graph, (b) whether these data fit to the Mckewan equation?

| Time (s)      | 36   | 60    | 120   | 180   | 240   | 300   | 420   | 600   |
|---------------|------|-------|-------|-------|-------|-------|-------|-------|
| Wt. loss (mg) | 4.86 | 10.06 | 24.52 | 32.40 | 34.75 | 36.32 | 38.74 | 41.30 |

Given: Total removable oxygen present in iron ore is 56.18 mg.
[Ans: $1.54 \times 10^{-3}$ s$^{-1}$].

**Problem 6.2.** From an experiment on sulfur removal from molten iron to slag phase at 1500 °C, the following data are obtained:

| Time (min) | 0    | 4    | 15   | 30   | 60   | 90   | 150  |
|------------|------|------|------|------|------|------|------|
| [S] %      | 0.15 | 0.13 | 0.12 | 0.09 | 0.07 | 0.03 | 0.01 |

Calculate the first-order rate constant.

**Problem 6.3.** For reduction of iron oxide powder by CO gas, the following data are obtained. Find out activation energy for the reduction.

| Temp. (°C)                              | 800  | 900  | 1000 | 1100 |
|-----------------------------------------|------|------|------|------|
| Rate constant, k (s$^{-1}$) $\times 10^3$ | 1.02 | 1.18 | 1.70 | 2.50 |

**Problem 6.4.** In a rotary kiln for DRI production, how much ore (t/hr) is charged for 5 t/hr production?
Given: Fe in ore is 62.5%, $Fe_2O_3$ in fly ash is 5%, amount of fly ash is 0.35 t/hr, total Fe contain in DRI is 92%.
[Ans: 7.38 t/hr].

**Problem 6.5.** Calculate (i) the dissolved over-pressure needed for continued nucleation and growth of bubbles on the bottom of a 80 cm deep body of liquid: in (a) liquid Fe, (b) liquid Al; (ii) the size of bubble that detaches
Given: Diameter of crevice is 1.2 mm.

| | σ (N/m) | Contact angle, θ | ρ (kg/m³) |
|---|---|---|---|
| Fe | 1.76 | 90 | 7000 |
| Al | 0.90 | 90 | 2300 |

[Ans: (a) Fe: $0.61 \times 10^5$ Pa & 8.77 mm: (b) Al: $0.21 \times 10^5$ Pa and 10.94 mm].

**Problem 6.6.** Calculate radiation loss per m² area and per tonne of liquid steel at 1700 °C in an electric arc furnace. Given: (i) diameter of furnace is 7 m, capacity is 400 t, (ii) emissivity of liquid steel is 0.28, Stefan–Boltzmann constant is $5.669 \times 10^{-8}$ W/m² K⁴.

[Ans: 240.53 kW/m², 23.13 kW/t].

**Problem 6.7.** A ladle (3.5 m internal diameter and 4 m height) is filled with liquid steel leaving a freeboard of 0.2 m. Calculate the pressure developed on its bottom surface (in atm) and resultant force (in N). Given: Density of liquid steel 7000 kg/m³, 1 atm pressure $= 1.01 \times 10^5$ Pa and g = 9.81 m/s².

[Ans: 3.58 atm, $25.09 \times 10^5$ N].

## 6.12 Questions

Q1. Derive McKewan's model for gas–solid reaction.

Q2. Develop equation for gasification of carbon.

Q3. Develop Jander's model for solid–solid reaction.

Q4. Derive equation for growth and detachment of gas bubbles nucleating in liquid bath.

Q5. Discuss heat transfer in continuous casting.

Q6. Develop an expression to show how much slag powder should be added to a continuous caster to maintain a film of thickness δ, between mold and the slab.

Q7. What are the basic principles of material and heat balances?

## References

1. Edstrom JO (1953) J. Iron Steel Inst. 175, p 289
2. Mckewan WM (1965) Chipman Conference on *Steelmaking*, Ed. Elliott JF. MIT Press, Mass, p 141
3. Reif AE (1952) J Phys Chem 56, p 773
4. Ergun S (1956) J Phys Chem 60, p 480
5. Patel AB, Dutta SK (Guide): BE Thesis: Metallurgical Engineering Department., M. S. University of Baroda
6. Dutta SK, Chokshi Y (2020) Basic concept of iron and steel making. Springer Nature, Singapore, March 2020
7. Chakrabarti AK (2007) Steel making. Printice-Hall of India Pvt Ltd, New Delhi
8. Lahari AK (1999) Workshop on "Process Modeling Application for Steel Industry. IIS, Bangalore, Sept 1999

9.  Guthrie RIL (1992) Engineering in process metallurgy. Oxford University Press, USA
10. Dutta SK (2018) JPC bulletin on iron & steel, vol XVIII, no 10, October 2018, p 20
11. Dutta SK (2018) JPC bulletin on iron & steel, vol XVIII, no 11, November 2018, p 21
12. Dutta SK, Lele AB (2001) IE(I) J-MM, vol 82, April 2001, p 1
13. Patel RB, Dutta SK (2005) (Guide): ME Thesis, Metallurgical Engineering, M. S. University of Baroda
14. Sarnagi BM, Mishra PC (2001) Conference Proceedings on direct reduction and smelting, October 2001, Jamshedpur, p 151

# Appendix I

## (A) Physical Quantities and Their Dimensions

| Quantity | Symbol | Dimension |
|---|---|---|
| Mass | m | M |
| Length | L | L |
| Time | t | t |
| Temperature | T | T |
| Velocity | u | $Lt^{-1}$ |
| Acceleration | a | $Lt^{-2}$ |
| Area | A | $L^2$ |
| Force | F | $MLt^{-2}$ |
| Work | W | $ML^2t^{-2}$ |
| Energy heat | E, q | $ML^2t^{-2}$ |
| Density | $\rho$ | $ML^{-3}$ |
| Viscosity | $\mu$ | $ML^{-1}t^{-1}$ |
| Diffusion coefficient | D | $L^2t^{-1}$ |
| Specific heat | $C_p$ | $L^2t^{-2}\,T^{-1}$ |
| Thermal conductivity | k | $ML\,t^{-3}\,T^{-1}$ |
| Thermal diffusivity | $\alpha$ | $L^2t^{-1}$ |
| Heat transfer coefficient | h | $Mt^{-3}\,T^{-1}$ |
| Coefficient of thermal expansion | $\beta$ | $T^{-1}$ |

S. K. Dutta, *Fundamental of Transport Phenomena and Metallurgical Process Modeling*,
https://doi.org/10.1007/978-981-19-2156-8

## (B) Some Common Mathematical Functions

$$\frac{df}{dt} = \lim_{\Delta x \to 0} \left[ \frac{f(x + \Delta x) - f(x)}{\Delta x} \right]$$

$$\frac{d(x)}{dx} = 1, \ \frac{d(x^2)}{dx} = 2x, \ \frac{d(x^n)}{dx} = n \cdot x^{n-1}, \ \frac{d(x^{-1})}{dx} = -\left(\frac{1}{x^2}\right)$$

$$\int dx = x, \ \int x dx = \frac{x^2}{2}, \ \int x^2 dx = \frac{x^3}{3}, \ \int x^n dx = \frac{x^{n+1}}{n+1}, \ \int \frac{dx}{x} = \ln x$$

$$\ln(xy) = \ln x + \ln y, \ \ln\left(\frac{x}{y}\right) = \ln x - \ln y$$

If $\ln n = m$, then $n = e^m$ and $\log x = y$, then $x = 10^y$.

# Appendix
# II

## Equation of Continuity

$$\left[\left(\frac{\delta\rho}{\delta t}\right) + (\nabla\rho u) = 0\right]$$

Cartesian coordinates (x, y, z)

$$\frac{\delta\rho}{\delta t} + \frac{\delta(\rho u_x)}{\delta x} + \frac{\delta(\rho u_y)}{\delta y} + \frac{\delta(\rho u_z)}{\delta z} = 0 \qquad \text{(II.1)}$$

Cylindrical coordinates (r, θ, z)

$$\frac{\delta\rho}{\delta t} + \frac{1}{r}\frac{\delta(\rho r u_r)}{\delta r} + \frac{1}{r}\frac{\delta(\rho u_\theta)}{\delta\theta} + \frac{\delta(\rho u_z)}{\delta z} = 0 \qquad \text{(II.2)}$$

Spherical coordinates (r, θ, φ)

$$\frac{\delta\rho}{\delta t} + \frac{1}{r^2}\frac{\delta(\rho r^2 u_r)}{\delta r} + \frac{1}{r\sin\theta}\frac{\delta(\rho u_\theta \sin\theta)}{\delta\theta} + \frac{1}{r\sin\theta}\frac{\delta(\rho u_\varphi)}{\delta\varphi} = 0 \qquad \text{(II.3)}$$

© The Editor(s) (if applicable) and The Author(s), under exclusive license
to Springer Nature Singapore Pte Ltd. 2023
S. K. Dutta, *Fundamental of Transport Phenomena and Metallurgical Process Modeling*,
https://doi.org/10.1007/978-981-19-2156-8

# Appendix III

## (A) Equation of Motion (in Terms of τ)

$$\left[ \rho \left\{ \frac{Du}{Dt} \right\} = -\nabla p - (\nabla \tau) + \rho g \right]$$

Cartesian coordinates (x, y, z)

$$\rho \left( \frac{\delta u_x}{\delta t} + u_x \frac{\delta u_x}{\delta x} + u_y \frac{\delta u_x}{\delta y} + u_z \frac{\delta u_x}{\delta z} \right) = -\frac{\delta p}{\delta x} - \left[ \frac{\delta \tau_{xx}}{\delta x} + \frac{\delta \tau_{yx}}{\delta y} + \frac{\delta \tau_{zx}}{\delta z} \right] + \rho g_x$$

(III.1)

$$\rho \left( \frac{\delta u_y}{\delta t} + u_x \frac{\delta u_y}{\delta x} + u_y \frac{\delta u_y}{\delta y} + u_z \frac{\delta u_y}{\delta z} \right) = -\frac{\delta p}{\delta y} - \left[ \frac{\delta \tau_{xy}}{\delta x} + \frac{\delta \tau_{yy}}{\delta y} + \frac{\delta \tau_{zy}}{\delta z} \right] + \rho g_y$$

(III.2)

$$\rho \left( \frac{\delta u_z}{\delta t} + u_x \frac{\delta u_z}{\delta x} + u_y \frac{\delta u_z}{\delta y} + u_z \frac{\delta u_z}{\delta z} \right) = -\frac{\delta p}{\delta z} - \left[ \frac{\delta \tau_{xz}}{\delta x} + \frac{\delta \tau_{yz}}{\delta y} + \frac{\delta \tau_{zz}}{\delta z} \right] + \rho g_z$$

(III.3)

Cylindrical coordinates (r, θ, z)

$$\rho \left( \frac{\delta u_r}{\delta t} + u_r \frac{\delta u_r}{\delta r} + \frac{u_\theta}{r} \frac{\delta u_r}{\delta \theta} + u_z \frac{\delta u_r}{\delta z} - \frac{u_\theta^2}{r} \right)$$
$$= -\frac{\delta p}{\delta r} - \left[ \frac{1}{r} \frac{\delta(r\tau_{rr})}{\delta r} + \frac{1}{r} \frac{\delta \tau_{\theta x}}{\delta \theta} + \frac{\delta \tau_{zr}}{\delta z} - \frac{\tau_{\theta\theta}}{r} \right] + \rho g_r$$

(III.4)

$$\rho\left(\frac{\delta u_\theta}{\delta t} + u_r\frac{\delta u_\theta}{\delta r} + \frac{u_\theta}{r}\frac{\delta u_\theta}{\delta\theta} + u_z\frac{\delta u_\theta}{\delta z} + \frac{u_r u_\theta}{r}\right)$$

$$= -\frac{1}{r}\frac{\delta p}{\delta\theta} - \left[\frac{1}{r^2}\frac{\delta(r^2\tau_{r\theta})}{\delta r} + \frac{1}{r}\frac{\delta\tau_{\theta\theta}}{\delta\theta} + \frac{\delta\tau_{z\theta}}{\delta z} + \frac{\tau_{\theta r} - \tau_{r\theta}}{r}\right] + \rho g_\theta \qquad (III.5)$$

$$\rho\left(\frac{\delta u_z}{\delta t} + u_r\frac{\delta u_z}{\delta r} + \frac{u_\theta}{r}\frac{\delta u_z}{\delta\theta} + u_z\frac{\delta u_z}{\delta z}\right) = -\frac{\delta p}{\delta z} - \left[\frac{1}{r}\frac{\delta(r\tau_{rz})}{\delta r} + \frac{1}{r}\frac{\delta\tau_{\theta z}}{\delta\theta} + \frac{\delta\tau_{zz}}{\delta z}\right]$$

$$+ \rho g_z \qquad (III.6)$$

Spherical coordinates (r, θ, φ)

$$\rho\left(\frac{\delta u_r}{\delta t} + u_r\frac{\delta u_r}{\delta r} + \frac{u_\theta}{r}\frac{\delta u_r}{\delta\theta} + \frac{u_\varphi}{r\sin\theta}\frac{\delta u_r}{\delta\varphi} - \frac{u_\theta^2 + u_\varphi^2}{r}\right)$$

$$= -\frac{\delta p}{\delta r} - \left[\frac{1}{r^2}\frac{\delta(r^2\tau_{rr})}{\delta r} + \frac{1}{r\sin\theta}\frac{\delta(\tau_{\theta r}\sin\theta)}{\delta\theta} + \frac{1}{r\sin\theta}\frac{\delta\tau_{\varphi r}}{\delta\varphi} - \frac{\tau_{\theta\theta} + \tau_{\varphi\varphi}}{r}\right] + \rho g_r$$

$$(III.7)$$

$$\rho\left(\frac{\delta u_\theta}{\delta t} + u_r\frac{\delta u_\theta}{\delta r} + \frac{u_\theta}{r}\frac{\delta u_\theta}{\delta\theta} + \frac{u_\varphi}{r\sin\theta}\frac{\delta u_\theta}{\delta\varphi} + \frac{u_r u_\theta - u_\varphi^2\cot\theta}{r}\right)$$

$$= -\frac{1}{r}\frac{\delta p}{\delta\theta} - \left[\frac{1}{r^3}\frac{\delta(r^3\tau_{r\theta})}{\delta r} + \frac{1}{r\sin\theta}\frac{\delta(\tau_{\theta\theta}\sin\theta)}{\delta\theta} + \frac{1}{r\sin\theta}\frac{\delta\tau_{\varphi\theta}}{\delta\varphi} + \frac{(\tau_{\theta r} - \tau_{r\theta}) - \tau_{\varphi\varphi}\cot\theta}{r}\right]$$

$$+ \rho g_\theta \qquad (III.8)$$

$$\rho\left(\frac{\delta u_\varphi}{\delta t} + u_r\frac{\delta u_\varphi}{\delta r} + \frac{u_\theta}{r}\frac{\delta u_\varphi}{\delta\theta} + \frac{u_\varphi}{r\sin\theta}\frac{\delta u_\varphi}{\delta\varphi} + \frac{u_r u_\varphi - u_\theta u_\varphi\cot\theta}{r}\right)$$

$$= -\frac{1}{r\sin\theta}\frac{\delta p}{\delta\varphi} - \left[\frac{1}{r^3}\frac{\delta(r^3\tau_{r\varphi})}{\delta r} + \frac{1}{r\sin\theta}\frac{\delta(\tau_{\theta\varphi}\sin\theta)}{\delta\theta}\right.$$

$$\left. + \frac{1}{r\sin\theta}\frac{\delta\tau_{\varphi\varphi}}{\delta\varphi}\frac{(\tau_{\varphi r} - \tau_{r\varphi}) - \tau_{\varphi\theta}\cot\theta}{r}\right] + \rho g_\varphi \qquad (III.9)$$

# (B) Equation of Motion (*Navier–Stokes Equation* at Constant $\rho$ and $\mu$) $\left[\rho\left\{\frac{Du}{Dt}\right\} = -\nabla p + \mu\nabla^2 u + \rho g\right]$

Cartesian coordinates (x, y, z)

$$\rho\left(\frac{\delta u_x}{\delta t} + u_x\frac{\delta u_x}{\delta x} + u_y\frac{\delta u_x}{\delta y} + u_z\frac{\delta u_x}{\delta z}\right) = -\frac{\delta p}{\delta x} + \mu\left[\frac{\delta^2 u_x}{\delta x^2} + \frac{\delta^2 u_x}{\delta y^2} + \frac{\delta^2 u_x}{\delta z^2}\right] + \rho g_x$$

(III.10)

$$\rho\left(\frac{\delta u_y}{\delta t} + u_x\frac{\delta u_y}{\delta x} + u_y\frac{\delta u_y}{\delta y} + u_z\frac{\delta u_y}{\delta z}\right) = -\frac{\delta p}{\delta y} + \mu\left[\frac{\delta^2 u_y}{\delta x^2} + \frac{\delta^2 u_y}{\delta y^2} + \frac{\delta^2 u_y}{\delta z^2}\right] + \rho g_y$$

(III.11)

$$\rho\left(\frac{\delta u_z}{\delta t} + u_x\frac{\delta u_z}{\delta x} + u_y\frac{\delta u_z}{\delta y} + u_z\frac{\delta u_z}{\delta z}\right) = -\frac{\delta p}{\delta z} + \mu\left[\frac{\delta^2 u_z}{\delta x^2} + \frac{\delta^2 u_z}{\delta y^2} + \frac{\delta^2 u_z}{\delta z^2}\right] + \rho g_z$$

(III.12)

Cylindrical coordinates (r, θ, z)

$$\rho\left(\frac{\delta u_r}{\delta t} + u_r\frac{\delta u_r}{\delta r} + \frac{u_\theta}{r}\frac{\delta u_r}{\delta\theta} + u_z\frac{\delta u_r}{\delta z} - \frac{u_\theta^2}{r}\right)$$
$$= -\frac{\delta p}{\delta r} + \mu\left[\frac{\delta}{\delta r}\left\{\frac{1}{r}\frac{\delta(ru_r)}{\delta r}\right\} + \frac{1}{r^2}\frac{\delta^2 u_r}{\delta\theta^2} + \frac{\delta^2 u_r}{\delta z^2} - \frac{2}{r^2}\frac{\delta u_\theta}{\delta\theta}\right] + \rho g_r$$

(III.13)

$$\rho\left(\frac{\delta u_\theta}{\delta t} + u_r\frac{\delta u_\theta}{\delta r} + \frac{u_\theta}{r}\frac{\delta u_\theta}{\delta\theta} + u_z\frac{\delta u_\theta}{\delta z} + \frac{u_r u_\theta}{r}\right)$$
$$= -\frac{1}{r}\frac{\delta p}{\delta r} + \mu\left[\frac{\delta}{\delta r}\left\{\frac{1}{r}\frac{\delta(ru_\theta)}{\delta r}\right\} + \frac{1}{r^2}\frac{\delta^2 u_\theta}{\delta\theta^2} + \frac{\delta^2 u_\theta}{\delta z^2} + \frac{2}{r^2}\frac{\delta u_r}{\delta\theta}\right] + \rho g_\theta$$

(III.14)

$$\rho\left(\frac{\delta u_z}{\delta t} + u_r\frac{\delta u_z}{\delta r} + \frac{u_\theta}{r}\frac{\delta u_z}{\delta\theta} + u_z\frac{\delta u_z}{\delta z}\right)$$
$$= -\frac{\delta p}{\delta z} + \mu\left[\frac{1}{r}\frac{\delta}{\delta r}\left(r\frac{\delta u_z}{\delta r}\right) + \frac{1}{r^2}\frac{\delta^2 u_z}{\delta\theta^2} + \frac{\delta^2 u_z}{\delta z^2}\right] + \rho g_z$$

(III.15)

Spherical coordinates (r, θ, φ)

$$\rho\left(\frac{\delta u_r}{\delta t} + u_r\frac{\delta u_r}{\delta r} + \frac{u_\theta}{r}\frac{\delta u_r}{\delta\theta} + \frac{u_\varphi}{r\sin\theta}\frac{\delta u_r}{\delta\varphi} - \frac{u_\theta^2 + u_\varphi^2}{r}\right)$$
$$= -\frac{\delta p 8op}{\delta r} + \mu\left[\frac{1}{r^2}\frac{\delta^2}{\delta r^2}\left(r^2 u_r\right) + \frac{1}{r^2\sin\theta}\frac{\delta}{\delta\theta}\left(\sin\theta\frac{\delta u_r}{\delta\theta}\right) + \frac{1}{r^2\sin^2\theta}\frac{\delta^2 u_r}{\delta\varphi^2}\right] + \rho g_r$$

(III.16)

$$\rho\left(\frac{\delta u_\theta}{\delta t} + u_r\frac{\delta u_\theta}{\delta r} + \frac{u_\theta}{r}\frac{\delta u_\theta}{\delta \theta} + \frac{u_\varphi}{r\sin\theta}\frac{\delta u_\theta}{\delta \varphi} + \frac{u_r u_\theta - u_\varphi^2 \cot\theta}{r}\right)$$

$$= -\frac{1}{r}\frac{\delta p}{\delta \theta} + \mu\left[\frac{1}{r^2}\frac{\delta}{\delta r}\left(r^2\frac{\delta u_\theta}{\delta r}\right) + \frac{1}{r^2}\frac{\delta}{\delta \theta}\left\{\frac{1}{\sin\theta}\frac{\delta}{\delta \theta}(u_\theta\sin\theta)\right\}\right.$$

$$\left. + \frac{1}{r^2\sin^2\theta}\frac{\delta^2 u_\theta}{\delta \varphi^2} + \frac{2}{r^2}\frac{\delta u_\theta}{\delta \theta} - \frac{2}{r^2}\frac{\cot\theta}{\sin\theta}\frac{\delta u_\varphi}{\delta \varphi}\right] + \rho g_\theta \qquad (III.17)$$

$$\rho\left(\frac{\delta u_\varphi}{\delta t} + u_r\frac{\delta u_\varphi}{\delta r} + \frac{u_\theta}{r}\frac{\delta u_\varphi}{\delta \theta} + \frac{u_\varphi}{r\sin\theta}\frac{\delta u_\varphi}{\delta \varphi} + \frac{u_r u_\varphi - u_\theta u_\varphi \cot\theta}{r}\right)$$

$$= -\frac{1}{r\sin\theta}\frac{\delta p}{\delta \varphi} + \mu\left[\frac{1}{r^2}\frac{\delta}{\delta r}\left(r^2\frac{\delta u_\varphi}{\delta r}\right)\right.$$

$$+ \frac{1}{r^2}\frac{\delta}{\delta \theta}\left\{\frac{1}{\sin\theta}\frac{\delta}{\delta \theta}(u_\varphi\sin\theta)\right\} + \frac{1}{r^2\sin^2\theta}\frac{\delta^2 u_\varphi}{\delta \varphi^2}$$

$$\left. + \frac{1}{r^2}\frac{2}{\sin\theta}\frac{\delta u_r}{\delta \varphi} + \frac{2}{r^2}\frac{\cot\theta}{\sin\theta}\frac{\delta u_\theta}{\delta \varphi}\right] + \rho g_\varphi \qquad (III.18)$$

# Appendix
# IV

## Equations for Diffusion of Heat

Cartesian coordinates (x, y, z)

$$\left[\frac{\delta}{\delta x}\left\{k\left(\frac{\delta T}{\delta x}\right)\right\} + \frac{\delta}{\delta y}\left\{k\left(\frac{\delta T}{\delta y}\right)\right\} + \frac{\delta}{\delta z}\left\{k\left(\frac{\delta T}{\delta z}\right)\right\}\right] + q^{\cdot} = \rho\, C_p\left(\frac{\delta T}{\delta t}\right) \quad \text{(IV.1)}$$

Cylindrical coordinates (r, θ, z)

$$\frac{1}{r}\frac{\delta}{\delta r}\left(kr\frac{\delta T}{\delta r}\right) + \frac{1}{r^2}\frac{\delta}{\delta\theta}\left(k\frac{\delta T}{\delta\theta}\right) + \frac{\delta}{\delta z}\left(k\frac{\delta T}{\delta z}\right) + q^{\cdot} = \rho\, C_p\left(\frac{\delta T}{\delta t}\right) \quad \text{(IV.2)}$$

Spherical coordinates (r, θ, φ)

$$\frac{1}{r^2}\frac{\delta}{\delta r}\left(kr^2\frac{\delta T}{\delta r}\right) + \frac{1}{r^2 sin^2\theta}\frac{\delta}{\delta\varphi}\left(k\frac{\delta T}{\delta\varphi}\right) + \frac{1}{r^2 sin\theta}\frac{\delta}{\delta\theta}\left(k\sin\theta\frac{\delta T}{\delta\theta}\right) + q^{\cdot} = \rho\, C_p\left(\frac{\delta T}{\delta t}\right)$$
$$\text{(IV.3)}$$

© The Editor(s) (if applicable) and The Author(s), under exclusive license
to Springer Nature Singapore Pte Ltd. 2023
S. K. Dutta, *Fundamental of Transport Phenomena and Metallurgical Process Modeling*,
https://doi.org/10.1007/978-981-19-2156-8

# Appendix V

## (A) Equations for Mass Diffusion (Fick's Law for Binary Mixture)

Cartesian coordinates (x, y, z)

$$J'_{Ax} = -\rho D_{AB}\left(\frac{\delta m_A}{\delta x}\right) \tag{V.1}$$

$$J'_{Ay} = -\rho D_{AB}\left(\frac{\delta m_A}{\delta y}\right) \tag{V.2}$$

$$J'_{Az} = -\rho D_{AB}\left(\frac{\delta m_A}{\delta z}\right) \tag{V.3}$$

Cylindrical coordinates (r, θ, z)

$$J'_{Ar} = -\rho D_{AB}\left(\frac{\delta m_A}{\delta r}\right) \tag{V.4}$$

$$J'_{A\theta} = -\rho D_{AB}\frac{1}{r}\left(\frac{\delta m_A}{\delta \theta}\right) \tag{V.5}$$

$$J'_{Az} = -\rho D_{AB}\left(\frac{\delta m_A}{\delta z}\right) \tag{V.6}$$

Spherical coordinates (r, θ, φ)

$$J'_{Ar} = -\rho D_{AB}\left(\frac{\delta m_A}{\delta r}\right) \tag{V.7}$$

© The Editor(s) (if applicable) and The Author(s), under exclusive license
to Springer Nature Singapore Pte Ltd. 2023
S. K. Dutta, *Fundamental of Transport Phenomena and Metallurgical Process Modeling*,
https://doi.org/10.1007/978-981-19-2156-8

$$J'_{A\theta} = -\rho D_{AB} \frac{1}{r}\left(\frac{\delta m_A}{\delta \theta}\right) \qquad \text{(V.8)}$$

$$J'_{A\varphi} = -\rho D_{AB} \frac{1}{r\,\sin\theta}\left(\frac{\delta m_A}{\delta \varphi}\right) \qquad \text{(V.9)}$$

## (B) General Equations of Mass Diffusion

Cartesian coordinates (x, y, z)

$$\frac{\delta \rho_A}{\delta t} = \left[\frac{\delta}{\delta x}\left\{\rho\,D_{AB}\left(\frac{\delta m_A}{\delta x}\right)\right\} + \frac{\delta}{\delta y}\left\{\rho\,D_{AB}\left(\frac{\delta m_A}{\delta y}\right)\right\} + \frac{\delta}{\delta z}\left\{\rho\,D_{AB}\left(\frac{\delta m_A}{\delta z}\right)\right\}\right]$$
$$+ n'_A \qquad \text{(V.10)}$$

In terms of molar concentration:

$$\left[\frac{\delta}{\delta x}\left\{CD_{AB}\left(\frac{\delta X_A}{\delta x}\right)\right\} + \frac{\delta}{\delta y}\left\{CD_{AB}\left(\frac{\delta X_A}{\delta y}\right)\right\} + \frac{\delta}{\delta z}\left\{CD_{AB}\left(\frac{\delta X_A}{\delta z}\right)\right\}\right] + n'_A = \frac{\delta C_A}{\delta t}$$
$$\text{(V.11)}$$

If $\rho$ and $D_{AB}$ are constants:

$$\left[\frac{\delta^2 \rho_A}{\delta x^2} + \frac{\delta^2 \rho_A}{\delta y^2} + \frac{\delta^2 \rho_A}{\delta z^2}\right] + \frac{n'_A}{D_{AB}} = \frac{1}{D_{AB}}\cdot\frac{\delta \rho_A}{\delta t} \qquad \text{(V.12)}$$

If C and $D_{AB}$ are constants:

$$\left[\frac{\delta^2 C_A}{\delta x^2} + \frac{\delta^2 C_A}{\delta y^2} + \frac{\delta^2 C_A}{\delta z^2}\right] + \frac{n'_A}{D_{AB}} = \frac{1}{D_{AB}}\cdot\frac{\delta C_A}{\delta t} \qquad \text{(V.13)}$$

Cylindrical coordinates (r, θ, z)

$$\left[\frac{1}{r}\frac{\delta}{\delta r}\left\{C_{AB}\left(\frac{\delta X_A}{\delta r}\right)\right\} + \frac{1}{r^2}\frac{\delta}{\delta \theta}\left\{C_{AB}\left(\frac{\delta X_A}{\delta \theta}\right)\right\} + \frac{\delta}{\delta z}\left\{C_{AB}\left(\frac{\delta X_A}{\delta z}\right)\right\}\right] + n'_A = \frac{\delta C_A}{\delta t}$$
$$\text{(V.14)}$$

Spherical coordinates (r, θ, φ)

$$\left[\frac{1}{r^2}\frac{\delta}{\delta r}\left\{CD_{AB}r^2\left(\frac{\delta X_A}{\delta r}\right)\right\} + \frac{1}{r^2 sin^2\theta}\frac{\delta}{\delta\varphi}\left\{CD_{AB}\left(\frac{\delta X_A}{\delta\varphi}\right)\right\}\right.$$
$$\left.+ \frac{1}{r^2 sin\theta}\frac{\delta}{\delta\theta}\left\{CD_{AB}sin\theta\left(\frac{\delta X_A}{\delta\theta}\right)\right\}\right] + n'_A = \frac{\delta C_A}{\delta t} \qquad (V.15)$$

# Appendix VI

## Some Dimensionless Numbers

| Dimensionless numbers | Symbol | Formula |
|---|---|---|
| Froude number | Fr | $\frac{u^2}{gL}$ |
| *Galileo number* | Ga | $\left\{ \frac{\rho_f (\rho_s - \rho_f) g\, d_p^3}{\mu^2} \right\}$ |
| Grashof number | Gr | $\left\{ \frac{(\rho^2 L^3 g \beta \Delta T)}{\mu^2} \right\}$ |
| Nusselt number | Nu | $\frac{hL}{k}$ |
| Prandtl number | Pr | $\frac{\mu C_p}{k}$ |
| Schmidt number | Sc | $\frac{\mu}{D\rho}$ |
| Sherwood number | Sh | $\frac{K_m L}{D_{A-B}}$ |
| *Reynolds number* | Re | $\frac{\rho u L}{\mu}$ |
| Weber number | We | $\frac{L u^2 \rho}{\sigma}$ |

Printed in the United States
by Baker & Taylor Publisher Services